South Sefton
6th Form College

D0232084

A-Z

HANDBOOK 4TH EDITION

Geography

Ma
Da
Ge

PHILIP ALLAN
UPDATES

Philip Allan Updates, an imprint of Hodder Education, an Hachette UK company, Market Place, Deddington, Oxfordshire OX15 0SE

Orders

Bookpoint Ltd, 130 Milton Park, Abingdon, Oxfordshire OX14 4SB
tel: 01235 827720
fax: 01235 400454
e-mail: uk.orders@bookpoint.co.uk

Lines are open 9.00 a.m.–5.00 p.m., Monday to Saturday, with a 24-hour message answering service. You can also order through the Philip Allan Updates website: www.philipallan.co.uk

ISBN 978-0-340-99104-6

First published 1996
Second edition 2000
Third edition 2003
Fourth edition 2009

Impression number 5 4 3 2 1
Year 2014 2013 2012 2011 2010 2009

Typeset by Macmillan, India.

Printed by CPI Antony Rowe, Chippenham, Wiltshire.

Environmental information
Hachette UK's policy is to use papers that are natural, renewable and recyclable products and made from wood grown in sustainable forests. The logging and manufacturing processes are expected to conform to the environmental regulations of the country of origin.

Contents

How to use this book

The *A–Z Geography Handbook* is an alphabetical textbook designed for ease of use. Each entry begins with a one-sentence definition helping the user to add precision to the completion of assignments, whether coursework- or examination-based. Entries are then developed in line with the relative importance of the concept covered to provide sufficient material for topic building and essay writing. Entries are further developed through the use of worked examples and illustrations.

All entries are carefully cross-referenced through the use of bold italics. Your essay writing will benefit from following the topic pathways offered by these italicised entries.

To aid the revision process, carefully selected lists are provided at the back of this handbook for the units produced by the AQA, Edexcel and OCR examination boards. Information for examinations produced by CCEA, WJEC, CIE, Scottish Higher and the IB diploma can be found on the website that accompanies this book. These lists are broken down into various subject areas and some indication is given regarding the specification and modular paper to which they apply.

Students facing examinations can use the lists to make best use of the handbook during their revision time. Other guidance is given for geography examinations, including explanations of examiners' terms, developing essay-writing skills, techniques for learning and revision and synoptic assessment.

A–Z Online

This new digital edition of the *A–Z Geography Handbook* includes free access to a supporting website and a free desktop widget to make searching for terms even quicker. Log on to **www.philipallan.co.uk/a-zonline** and create an account using the unique code provided on the inside front cover of this book.

Once you are logged on, you will be able to:
- search the entire database of terms in this handbook
- print revision lists specific to your exam board
- get expert advice from examiners on how to get an A* grade
- create a personal library of your favourite terms
- expand your vocabulary with our word of the week

You can also add the other *A–Z Handbooks (digital editions)* that you own to your personal library on A–Z Online.

We hope that the *A–Z Geography Handbook* proves an invaluable resource to students, from the beginning of their course to its successful completion.

Acknowledgements

Geography at A-level has again moved on. The three main examination providers are still with us (AQA, Edexcel and OCR), but they have been reduced to one specification each. Other providers now feature in the A-level geography field and we have acknowledged this by offering revision lists for all of their specifications on the website that accompanies this book. A number of new terms have been added in light of these new specifications, particularly in the field of development, globalisation and in economic geography in general. Others have been rewritten or simply removed, in light of the authors' knowledge and involvement in the new A-level specifications.

Once again, we should like to thank our long-suffering families for their support and patience. Our thanks must also go to the team at Philip Allan Updates, especially Katie Blainey, who has been the driving force and inspiration behind this new generation of the *A–Z Handbooks*. We have been grateful for her suggestions and advice, but above all we should like to thank her for the enthusiasm she has brought to this project.

Researching, writing and editing a book of this size requires teamwork and above all hard work. A considerable amount of time went into checking the entries, but if any mistakes have slipped through, the authors accept full responsibility. Geography covers a huge range of topics, many of which change through time and space. The concepts of the subject are constantly being updated and revised, therefore it is inevitable that some of the ideas contained herein may already be outdated, or subject to review. This does not, however, alter the relevance or usefulness of geographical knowledge.

Malcolm Skinner, David Redfern, Geoff Farmer

ablation is the collective loss of water from an *ice sheet* or *glacier*. This loss can take a variety of forms:
- melting on the surface, melting internally, melting at the base or melting from the ice front
- *calving* of blocks of ice from the front of the ice mass into water to create *icebergs*
- *evaporation* of surface snow
- the process of **sublimation**
- the blowing away of snow by strong winds.

Ablation is dominant in the lower parts of a glacier or ice sheet and takes place more readily during warmer times.

abrasion (corrosion) is the scraping, scouring, rubbing and grinding action of materials being carried along by moving natural features such as rivers, *glaciers*, waves and strong winds. Rivers carry rock fragments either in the flow of the water, or they drag them along the river bed, and in doing so wear away at the banks and bed of the river channel. Similarly, glaciers use rock fragments embedded in their base; waves hurl pebbles and sand grains at a cliff face; and strong winds use rock fragments in a sand-blasting manner. In each case, the rate at which abrasion takes place depends on the amount of material being carried and on the relative resistance of the substances being eroded. Abrasion is very effective on less resistant surfaces, but is less so on harder surfaces.

abyssal plains are the relatively level areas of the ocean floor which can be found at approximately 5000 m below sea level. The North Atlantic abyssal plains, for example, lie at a depth of about 5500 m.

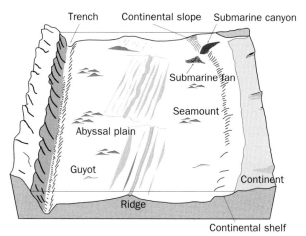

Major submarine features

Occasionally volcanic peaks (seamounts) rise steeply from the plain, sometimes reaching the surface as islands. Detailed surveying of the Pacific Ocean has revealed the presence of numerous flat-topped mountains (*guyots*) rising to within a kilometre of the surface.

accretion is the growth of a natural feature by enlargement. For example, hailstones grow by accretion. Droplets of water are carried upwards in a cumulonimbus *cloud* by strong vertical air currents and they freeze. These ice crystals then fall from the higher parts of the cloud and as they fall they enlarge by colliding with more supercooled water droplets (water droplets existing at temperatures below freezing), which freeze on contact. Subsequent updraughts of air carry the enlarged hailstone upwards again for the cycle to be repeated. In this way a hailstone grows by a series of concentric layers.

accumulation is the net gain in an ice mass. This gain can take place in a number of ways:
- by the *precipitation* of snow on to the surface of the ice mass
- by the refreezing of *meltwater*
- by snow avalanching from the surrounding slopes on to the *ice sheet* or *glacier*
- by strong winds drifting snow from other areas.

Accumulation is dominant in the upper parts of a glacier or ice sheet and takes place more readily during colder times.

acidification is the consequent effect of *acid rain* falling on an area. Rain is naturally acidic (a weak carbonic acid), but its acidity is increased by atmospheric pollution. By draining through soils acidified rainwater leaches out the soil bases, such as calcium, magnesium and potassium, and replaces them with hydrogen. Aluminium is mobilised, and is carried by *throughflow* to accumulate and become concentrated in lakes. This poisons fish stocks in such lakes. High concentrations of aluminium in a soil also poison tree roots. Underground water supplies also become more acidic as a result of passing through such soils. Acidification of soils may also be caused naturally in areas based on non-lime bedrocks such as granite.

acid lava contains a high proportion of silica. These lavas have a high melting point, are very viscous, solidify quickly and so do not flow very far. Acid lavas build high, steep-sided volcanic cones. These lavas may also solidify in the vent of the cone and cause recurrent and explosive eruptions.

acid rain (acid deposition) consists of the dry deposition of sulphur dioxide, nitrogen oxides and nitric acid, and the wet deposition of sulphuric acid, nitric acid and compounds of ammonium from *precipitation*, *mist* and clouds. Acid rain leads to direct damage to trees, particularly *coniferous* trees. It produces a yellowing of the needles and strange branching patterns. It also leads to the *leaching* of toxic metals (aluminium) from soils, and to their accumulation in rivers and lakes. This in turn leads to the death of fish. Acid rain is also blamed for damage to buildings, particularly those built of *limestone*, and to health problems in people such as bronchitis and other respiratory complaints. The major causes of acid rain are the burning of *fossil fuels* in power stations, the smelting of metals in older industrial plants and exhaust fumes from motor vehicles.

Various solutions to acid rain have been suggested:
- the use of catalytic converters on cars to reduce the amount of nitrogen oxides
- burning fossil fuels with a lower sulphur content
- replacing coal-fired power stations with nuclear power stations

- the use of flue gas desulphurisation schemes and other methods of removing sulphur either before coal is burned or after
- reducing the overall demand for electricity and car travel.

active layer is the name given to the upper layer of soil which undergoes seasonal thawing in *periglacial* areas which experience *permafrost* conditions. As summer temperatures rise thawing takes place from the surface downwards leaving a permanently frozen zone at depth in the soil. The thickness of the layer depends on local conditions but may extend to 5 or 6 m. As the ice in the active layer melts, large volumes of water are released; this water is unable to drain through the lower layers of the soil because the permafrost creates an *impermeable* lower boundary. The top layer may become saturated, which reduces the friction between soil particles and, as a result, even on slopes as low as 2°, the active layer can move downslope. Such movements, which are given the general name *solifluction*, can produce features such as solifluction lobes and terraces. Other *periglacial processes* such as *frost heave*, *ice wedges* and *patterned ground* take place in this active layer.

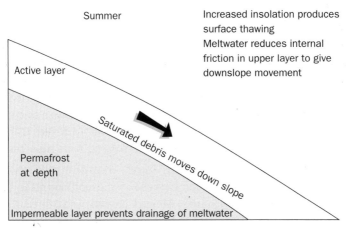

The active layer

adaptive capacity is the extent to which a *system* can cope with change. In human systems it depends on available human, physical and financial resources.

adiabatic temperature changes take place in rising (or sinking) air and these *lapse rates* are independent of the conditions in the surrounding environmental air. Adiabatic lapse rates refer to the temperature change for a parcel of air which has no exchange of heat or moisture with the surrounding environmental air; as it rises it will cool at a rate which is predictable and independent of environmental temperature. Conversely, sinking air will warm up according to adiabatic laws. The *dry adiabatic lapse rate*, which applies to air that is not saturated (i.e. *relative humidity* is less than 100%), is 1°C per 100 m. Saturated air has a lower lapse rate which varies according to the temperature of the air. As the air reaches 100% humidity *condensation* of water releases *latent heat* which offsets the normal decrease of temperature with height. The *saturated adiabatic lapse rate* is normally taken to be 0.5°C per 100 m.

adret is the slope of a hill or valley which faces towards the sun. In the northern hemisphere they face south or southwest and therefore they are warmer because they receive more *insolation*.

adsorption is the physical or chemical bonding of liquids or gases to solid particles. In the clay-*humus* complex of a soil nutrients can be held in combination with humus substances and attached to the small clay particles. The clays have a negative electrical charge at their surface and to balance this a layer of positive ions (cations) coats the clay particles. These ions are adsorbed (stuck) to the clay.

advance the line is a coastal defence option where new defences are built seaward of the existing line.

advection is the horizontal transfer of heat. When a warm, moist *air mass* moves over a cooler land or sea surface it is cooled from below. If the temperature of the air mass cools below *dew point* an *advection fog* is formed.

aeolian processes involve the erosion, transportation and *deposition* of particles by wind action and are particularly important in desert areas where there is insufficient vegetation and ground moisture to bind the soil together. There are two main processes of *wind erosion*: *deflation* and *abrasion*. Wind can move small particles in suspension and larger grained material by *saltation* or surface *creep*. Aeolian features include *deflation* hollows, rock pedestals, *yardangs*, *zeugens* and sand *dunes*.

aerial photographs are taken from above an area, looking down. Some aerial photographs have been taken with the camera pointing vertically down. Such photographs are very similar to large-scale maps, with a constant scale over the whole area shown. Their main weakness is that they present a level surface, so relief features are difficult to see. An oblique aerial photograph is where the camera has been pointing at an angle to the ground. The scale on such photographs is more variable. The foreground shows a smaller distance, whereas the areas further away at the top of the photograph have a greater 'real' distance between them. Heights and shapes of features and buildings are more easily seen.

afforestation is the deliberate planting of trees, usually where none have grown previously as undertaken by the *Forestry Commission* in heathland and moorland areas. Where planting takes place on land that was formerly wooded, but has been cleared, it is referred to as re-afforestation.

agglomeration is when several firms choose the same area as their location in order to minimise costs. This can be achieved by *linkages* between firms and their supporting services. Firms are also able to achieve *economies of scale* with such concentrations. Agglomeration was one of the factors that Weber suggested could affect the location of industry.

aggradation is the building up of deposits by rivers within channels when they are forced to drop the material (*load*) which they are carrying. This results in bed elevation.

agribusiness is the name for farming *systems* which are increasingly organised around scientific and business principles. Such businesses are linked to agricultural supply industries upstream from the farm which provide *inputs* such as chemicals, feedstuffs and machinery; secondly to the *food processing* industries downstream of the farm. These systems therefore exhibit *vertical integration* and are involved with most stages on the *agricultural chain*. The companies concerned are often *transnationals* such as Unilever and Nestlé, who in many cases seek out enterprising farmers with whom they can place contracts.

agricultural chain: at its simplest level this is a sequence of events linking the cultivation of crops or the tending of *livestock* to the food that is consumed on the table. In *subsistence* economies this can be relatively simple as the farmer's family eats all that s/he produces. On the other hand this can be extremely complex, especially where *agribusinesses* are concerned. Such chains include growing, processing, collection, packaging and distribution.

agricultural revolution (second): the name given to the period of widespread changes to the agriculture of Britain in the post-war years. There have been several major features:

- agriculture has become more intensive, aided by *mechanisation*
- greater use of *fertiliser* and *pesticides*
- new and more disease-resistant varieties of crops together with selective animal breeding
- farms in general have become larger, but there are fewer of them
- the number of farm labourers has declined
- a greater involvement of the government in agriculture followed later by the introduction of the *Common Agricultural Policy (CAP)*
- a growth in *agribusinesses*.

Some of the results of these changes have been:

- a different agricultural landscape
- war-time shortages of food being replaced by surplus production in basic foodstuffs
- a move from a dependence on imports to self-sufficiency in products such as cereals, poultry, milk and meat.

agro-chemicals is the term that describes the chemical *inputs* into farming which are used in order to increase productivity through:

- fertilisation of the soil (see *fertiliser*)
- the control of pests and diseases by *pesticides*, *herbicides* and *fungicides*.

Several problems have occurred with the use of chemicals in farming:

- contamination of water courses and water supplies (*eutrophication*)
- the growth of *algal blooms*
- damage caused to wildlife, some of which can be beneficial to the farmer
- poisoning of people through indiscriminate or accidental spraying
- the concentration of *pesticides* etc. in organic tissues in the *food chain* endangering aquatic life and human health
- possible links to diseases such as cancer
- the rise of resistant wildlife or 'super-pests'.

aid is the giving of resources by one (donor) country, or organisation, to another poorer (recipient) country with the main aim of improving the economy and/or living standards in the latter country. The resources given may be in the form of money, food, goods, technology, education or skilled people. There are three main types of aid:

- bilateral aid – aid from one country to another country. Often such aid is tied. The donor country imposes conditions on the recipient country such as contracts for buildings and projects, and preferential trade links

- multilateral aid – richer countries give money to an organisation (e.g. **World Bank**, **IMF**, **European Union**) which then redistributes the money to poorer countries. These organisations may withhold aid if they disagree with the recipient country's economic and/or political system
- voluntary aid – organisations such as Oxfam, CAFOD and Comic Relief collect money from the general public and then spend it on specific, usually small-scale projects in poorer countries.

AIDS is the abbreviation for Acquired Immune Deficiency Syndrome. It is the result of a viral infection called HIV (Human Immuno-deficiency Virus) which causes a breakdown in the immune system of the human body. It is transmitted through the exchange of body fluids, notably blood, semen and vaginal fluid. It is associated with sexual activity and drug abuse, but has also been transmitted via blood transfusions.

The world distribution of AIDS is:
- those countries where the disease is mainly associated with drug abusers, homosexuals and bisexuals – North America, Western Europe, Australia
- those countries where the disease is widespread among the heterosexual population and where vertical transmission (mother to child) is common – African countries south of the Sahara
- those countries where the disease has been introduced by travellers from the other two types of country.

Geographers are concerned with the study of AIDS and HIV as the virus has not spread evenly throughout the world and is having a disproportionate effect socially, culturally and economically in certain countries. By 2006, it was estimated that over 39 million people in the world were living with HIV or full-blown AIDS. In every global region the number of people living with HIV has risen. The situation is most serious in sub-Saharan Africa as over 62% of people living with HIV/AIDS in the world are to be found there. Some estimates say that over 6% of the population of sub-Saharan Africa is HIV-positive. The **United Nations** has estimated that, by 2020, 70 million people will have died from the disease.

air mass: an extensive body of air in which there is only a gradual horizontal change in temperature and **humidity** at a given height. **Lapse rates** are also almost uniform. These properties result from the air stagnating over a particular part of the Earth's surface for several days or even weeks. They are usually associated with large stationary **anticyclones**, often over ocean surfaces, deserts, ice-covered areas and large plains.

Several air masses may approach the British Isles and bring a range of weather conditions:
- Arctic – from the north bringing cold temperatures and snow
- **Polar** Maritime – from the northwest Atlantic bringing cold, moist weather
- Polar Continental – from the east bringing very cold temperatures in winter and possibly snow to eastern England
- Tropical Maritime – from the southwest bringing mild, cloudy wet weather in winter and cool moist conditions in summer
- Tropical Continental (in summer only) – from the southeast bringing very hot and dry (heatwave) conditions.

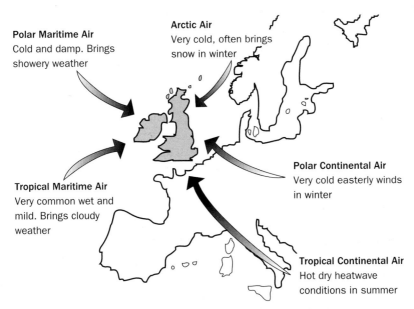

Air streams affecting the British Isles

alas: a steep-sided, flat-floored depression, sometimes containing a lake, found in areas where local melting of **permafrost** has taken place. It is one expression of **thermokarst**.

albedo is the amount of incoming solar **radiation** (**insolation**) that is reflected by the Earth's surface and the atmosphere. The planetary average for the albedo is 32% of insolation, but this amount varies considerably from place to place. Dark-coloured areas of the world, such as coniferous forests, reflect small amounts of insolation (10%), whereas light-coloured areas, such as deserts, reflect larger amounts (35%). Fresh snow and ice have very high albedos, up to 90%. Water surfaces also have variable rates, but here the angle of incidence of the sun is also important. Still water surfaces, with a vertical sun, reflect low amounts (3%), yet a low angle of incidence (15°) results in a high albedo (50%). Disturbed water surfaces have low albedos.

alcohol fuels are created by the fermentation of crops such as sugar cane, cassava and maize to produce ethanol. Ethanol has been used as an alternative to petrol in motor vehicles in Brazil, such that 75% of Brazilian cars now run on ethanol. The main advantages that have arisen are its cheapness relative to imported oil, employment creation in sugar **plantations** and reduced levels of air pollution. (See also **biofuel**.)

algal blooms are the excessive growth of **phytoplankton** in water bodies, many of them producing toxins which are harmful to wildlife. Such toxins are known to affect humans and in some cases fatalities have been reported. The increased occurrence of these blooms is often the result of **pollution** through sewage and agricultural **runoff** containing high levels of nitrogenous **fertiliser**. Several lakes in Western Europe, including the British Isles, are now affected by blue-green algae which is largely attributed to high levels of nutrients entering the water from domestic wastes, industry and farms.

alien species are species that are introduced and become established in a natural, semi-natural or modified *ecosystem* or *habitat* that is outside the species' native distributional range. Alien species may damage the *system* into which they have been introduced.

alluvial fans are cones of *sediment* deposited by rivers emerging from steep-sided valleys in an upland area on to a lowland plain. At such points the rivers spread out, decreasing their velocity, dissipating energy and therefore depositing sediment. Fans derived from less resistant rocks are larger than those derived from more resistant rocks. Consequently, their size is variable, ranging from 100 m to several kilometres across. In *arid* areas alluvial fans tend to be deposited by *mudflows*. Because of the variability of these flows, and the lack of vegetation cover, they are subject to frequent changes in character, and may be deeply dissected and unstable.

alluvium is the *sediment* deposited by a river. It can be deposited in two ways:
- when a river overflows its banks during times of flood, clays, silts and fine sands are deposited across the *floodplain*
- through the migration of a river *meander* which leaves behind the remains of the *point bar* created on the inside bank of the meander. Such deposits consist of sands and gravels.

alternative energy sources are those which do not rely on the burning of *fossil fuels*. There are a number of such sources: *tidal energy*, *solar energy*, *wind power*, *geothermal energy*, wave power and various forms of *biofuel*. Some of these energy sources are *renewable*, whereas others are not.

alternative technology is a feature of *economically less developed countries* where high-tech industries are both expensive and inappropriate for the needs of the people and the environment in which they live. Projects tend to be:
- *labour-intensive* because of the high rates of unemployment
- low-cost, using technologies which are based on local resources and skills
- in harmony with the local environment creating a more sustainable way of life.

(See also *intermediate technology*.)

anabatic (valley wind) is the name given to the motion of air up slopes resulting from *convection*. In upland areas by day, heated air will move towards the head of the valley and also up the valley sides.

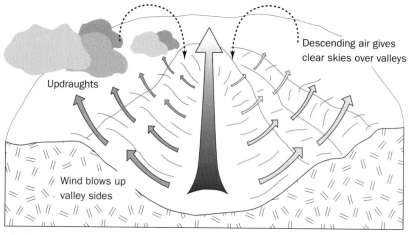

Updraughts

Descending air gives clear skies over valleys

Wind blows up valley sides

Anabatic flow

anaerobic refers to conditions within the soil where free oxygen is not present. Such conditions are usually the result of waterlogging where the pore spaces are filled with stagnant water which becomes de-oxygenised. The reddish-coloured ferric iron compounds of the soil are then chemically reduced to the grey-blue ferrous compounds. Under such conditions the rate of **decomposition** of organic matter will be slowed down which leads ultimately to the formation of **peat**.

antecedent drainage is a form of discordant drainage which occurs when an ancient river is able to maintain its course across more recently uplifted fold structures. The rate of downcutting by the river must be equal to, or greater than, the rate of uplift. If the upfolding is more rapid, but not fast enough to divert the river, a convexity may develop in the **long profile** of the river where it crosses the fold system.

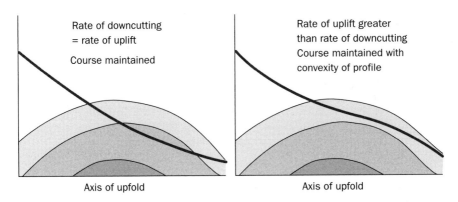

antecedent rainfall is the rainfall that fell in a previous period. If heavy rain falls on land that is saturated from a previous period of wet weather, it will produce a steep rising limb on a storm **hydrograph**.

anthropogenic heat is caused by human activities such as industrial processes, heating buildings and burning fuel in motor vehicles, and is a major factor in the development of **urban heat islands**. People can also contribute to this source.

anticyclone is an area of high atmospheric pressure that is usually slow moving or stationary. Anticyclones are generally larger than **depressions** (up to 3000 km across) and are dominated by subsiding air which produces warming and a decrease in **relative humidity**. In summer they bring hot, sunny conditions with little **cloud** or rain although clear skies at night can lead to **inversions** that produce **dew**, **mist** and coastal **fogs**. In winter these stable conditions favour the development of fog and frost, and pollution may be trapped in the lower layers of the atmosphere by the inversion.

anvil clouds are associated with **thunderstorms**. The rapidly rising air in a thunderstorm is forced to spread out sideways when reaching the **tropopause** and consequently the clouds produced by the rising air create a characteristic anvil shape.

AONBs see *Areas of Outstanding Natural Beauty*.

apartheid was the policy of separate development that operated in South Africa after 1948. It involved planned racial **segregation** and banned the mixing of races through marriage. The population was divided into three groups: the whites who were regarded as 'first class' and had full political, social and economic rights; the coloureds and Indians who enjoyed some rights but were seen as 'second class' and blacks who had virtually no rights outside the 'homelands' where they were forced to live. Apartheid, which means 'living apart', established segregation in housing, the blacks being forced to live in townships such as Soweto (Johannesburg) and Crossroads (Cape Town), which were well away from white residential areas. These settlements lacked services and *infrastructure* and were usually surrounded by large *squatter settlements*. Schools and universities were segregated by law and whites had their own buses and sections on beaches, in restaurants and in places of entertainment. To achieve residential segregation, African homelands (Bantustans) were established in rural areas. In these areas the blacks were allowed some of the political rights that they were not able to exercise in the rest of the country. The system finally came to an end in 1990 when the South African government repealed discriminatory laws and lifted the ban on certain political parties, particularly the ANC (African National Congress). This paved the way for the election of 1994, the first in South Africa with universal suffrage and the result of which, for the first time, gave a black majority in parliament.

appropriate technology is designed with special consideration to the environmental, ethical, cultural, social and economic aspects of the community for which it is intended. Ideally, appropriate technology involves fewer resources, is easier to maintain, has a low overall cost and has less of an impact on the environment compared to more sophisticated technology. In the developing world, it usually describes simple technologies suitable for use in those nations – the use of *biogas* obtained from dung fermentation systems across much of rural India is a good example. (See also *intermediate technology*.)

aquaculture is the management of water environments, other than the sea, for the purpose of increasing production and controlled harvesting of plant and animal food sources, for example *fish farming* in manmade tanks, pools, rivers and lakes.

aquifer is the name given to a *permeable* rock that can store and transmit water. The accumulation of water is most effective when rocks such as *limestone*, sandstone or *chalk* are underlain by an *impermeable* rock such as clay. The water can percolate through the permeable layer but its downward movement is prevented by the impermeable boundary. When tapped by wells, aquifers are an important source of water supply, although in the London Basin in recent years the increases in demand have been greater than the rate of replacement and this has led to a lowering of the *water table*.

arable is a type of agriculture that concentrates on the cultivation of plant crops such as grasses, cereals, vegetables, root crops and animal foodstuffs.

Areas of Outstanding Natural Beauty (AONBs) cover about 10% of the total land area in England and Wales and are given special protection under the *National Parks* and Access to the Countryside Act of 1949. They are controlled by local planning authorities who consider developments carefully so that the natural beauty of these areas may be preserved. They are generally smaller than National Parks and include areas such as the North and South Downs, the Cotswolds, the Chilterns, and areas of coastal scenery such as north Norfolk, Cornwall and the Northumberland coast north of Alnmouth. In 1995 the Tamar Valley was designated, bringing the total number of AONBs to 40.

arête: the name given to a narrow, knife-edged ridge with steep sides found in upland glaciated regions. Arêtes result from the formation of *cirques* (corries), the arête representing the area remaining between two of these features after they have been enlarged through glacial erosion and *freeze–thaw* action.

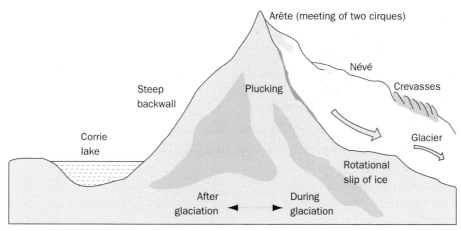

Formation of an arête

arid/aridity refers to areas where the climate is extremely dry. Such climates were defined as being areas with less than 250 mm of *precipitation* per year but, in reality, this is far too simple. Several climatologists have attempted to devise a quantitative index expressing the relationship between precipitation and *evapotranspiration* which determines aridity. In those areas of the world where there is little precipitation annually or where there is seasonal *drought* the calculation of *potential evapotranspiration* is used. The best known aridity index was put forward in 1931 by C. W. Thornthwaite.

Thornthwaite's Aridity Index:

$$\text{Moisture index (Im)} = \frac{100 \times \text{water surplus} - 60 \times \text{water deficit}}{\text{PE (potential evapotranspiration)}}$$

Definition of climates

	Im	
	0 to −20	dry subhumid
	−20 to −40	semi-arid
	under −40	arid

aridosols: the *zonal soil* type in *arid* and semi-arid regions. They range from yellow-red to grey-brown in colour and are generally very thin. Mineral ions accumulate, causing them to have a high salinity or alkalinity.

arithmetic average (mean): the arithmetic mean or average of a distribution is a *measure of central tendency* which is calculated by dividing the total value by the number of occurrences.

$$\text{FORMULA: } \frac{\text{sum of values of the variables}}{\text{sum of numbers in the set}}$$

The mean is a useful measure and can be subject to further calculations, but it is distorted by extreme values and may not be a whole number or the same as one of the items in the distribution.

arroyo: also known as a *wadi*, this is a trench with a roughly rectangular cross-section, excavated in valley-bottom *alluvium* with a through-flowing stream channel on the floor of the trench. It is generally similar to a gully, which has a V-shaped cross-section. Arroyos are typically associated with semi-*arid* areas where various changes could have resulted in increased *runoff*. They were first described after geomorphological research in western USA in the nineteenth century.

Asian Tigers: the name given to four countries of the *Pacific Rim* which have experienced phenomenal *economic growth* in the last 25 years. The countries are Hong Kong, Singapore, South Korea and Taiwan. Much of the growth has been based on the development of *manufacturing industry*.

aspect is the direction towards which a slope faces. In the northern hemisphere, north-facing slopes (*ubac*) receive less *insolation* than south-facing slopes (*adret*) and are therefore generally cooler.

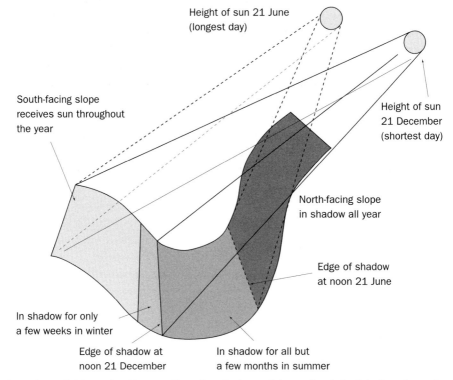

Aspect: sunlight received in a northern hemisphere alpine valley (orientated east–west)

asthenosphere is a layer in the Earth's *mantle* in which the rocks are soft and easily deformed. It is several hundred kilometres thick and lies below the *lithosphere*, which has a higher viscosity and is much more resistant to deformation. Temperatures increase with depth in the mantle until, at about 80 km, they reach 1400°C. Although different minerals have slightly different melting points, this temperature is very close to the melting point of the mantle rocks and therefore these rocks will soften. The temperatures in the asthenosphere are high enough to cause deformation. There is a sharp boundary between the two zones at about 80 km depth, but as this can vary it is more accurate to describe the boundary as occurring at the 1400°C isotherm.

asylum seeker is a person seeking to be classed as a refugee. The *United Nations* defines a refugee as someone who has a well-founded fear of being persecuted for reasons of race, religion, nationality, membership of a particular social group or political opinion, is outside their country of nationality and is unable to or, owing to such fear, is unwilling to avail him/herself of the protection of that country.

atlantic-type coastline occurs when tectonic processes have created folding of rock strata that is at right angles to the trend of the coastline. A series of bays and *headlands* develops as in the coastline of southwest Ireland.

atmosphere is a mixture of transparent gases held to the Earth by gravitational force. It consists of mainly nitrogen (78.09%) and oxygen (20.95%) by volume. Other gases include argon, *carbon dioxide* and traces of hydrogen, neon, helium, krypton, xenon, ozone, *methane* and radon. By international convention the upper limit of the atmosphere is assumed to be 1000 km, but, due to gravity and compression, most of the atmosphere is concentrated near to the Earth's surface. About 50% lies within 5.6 km of the surface and 99% within 40 km. Most of our climate and weather processes operate within 16–17 km of the surface in the zone of the lower atmosphere known as the *troposphere*. In this layer temperatures generally decrease with height (averaging 6.5°C per km). The top of this layer is marked by the *tropopause* where temperature remains fairly constant. This layer, which can act as a *temperature inversion*, forms an effective ceiling to any *convection* in the troposphere and an upper limit to weather *systems*. Carbon dioxide absorbs long-wave *radiation* from the Earth and is important in plant photosynthesis. Ozone absorbs and filters out ultraviolet radiation leading to a warming of the upper layer of the atmosphere, the stratosphere, the zone which lies above the tropopause, extending to about 50 km above the Earth's surface.

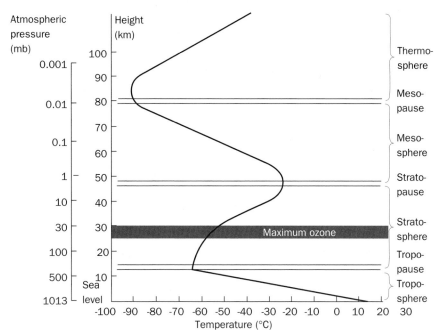

The vertical structure of the atmosphere

atmospheric circulation models are theories put forward to explain the general circulation of the atmosphere. This circulation, in the form of global wind patterns, results from the differential distribution of solar *radiation* over the Earth's surface. Because year to year changes in global temperature are only very small, the incoming solar radiation must be balanced by out-going radiation from the Earth. However, this balance varies with latitude. There is a net surplus of radiation at lower latitudes between 35°S and 40°N and a net deficit on the poleward sides of these latitudes. If the tropical regions are not getting progressively warmer, and the higher latitudes are not getting colder, there must be a continuous transfer of energy from the tropics to *polar* latitudes. Global wind patterns and *ocean currents* transfer this heat.

The first model of part of the circulation was suggested by Halley (1686) with warm tropical air rising and spreading towards the poles at high altitude with a return flow towards the equator at low level. Hadley modified this in 1735 to include the effects of the Earth's rotation, deflecting winds towards the right in the northern hemisphere and to the left in the southern hemisphere to produce the northeast and southeast trade belt. This part of the *system* is referred to as the *Hadley cell*; it was a single-cell model that did not extend into the westerly wind belts of the middle latitudes. Further improvements were made by Ferrel in 1856 when he developed a three-cell model giving a reasonably complete system of global winds. This was further refined by Rossby in 1941. Although this tri-cellular model forms the basis of our understanding of global circulation, it does not allow for the influence of *depressions*/*anticyclones* or high-level *jet streams* in the redistribution of energy. More recent approaches, known as wave theory models, have been developed to explain the behaviour of the upper air westerly air streams (*Rossby waves*) and jet streams. (See also *Ferrel's law*.)

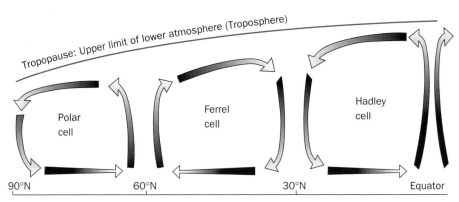

Tri-cellular model

atmospheric particulates: the name given to constituents of the *atmosphere* which are solid rather than gaseous. They are generated from a number of sources. In rural areas they are made up mainly of soil and sand (also volcanic dust and salt from sea spray). In urban areas they derive mainly from power stations and vehicle exhausts, particularly those burning diesel fuel. In such areas, particulates cause health problems such as asthma and eye irritation and are responsible for the increased occurrence of *fogs* and *smog*.

atolls: see *coral reef*.

attitudes are sets of beliefs that predispose a person, group or organisation to perceive and act towards people, environments and situations in a particular way.

autonomy is the right of self-government. Separatist groups, for example the Basques, Quebecois, Bretons and the Scottish Nationalists, seek autonomy from their respective national governments.

avalanche is the term applied to rapid downslope *mass movements* of rock, ice and snow. They occur most commonly in winter and spring when snowmelt helps to lubricate snow and rock *debris*. Snow avalanches form from recent falls of uncompacted snow on steep slopes, or from partially thawed layers of older snow.

azonal soil is an *immature soil* that lacks a well-developed profile because it has been little affected by soil forming processes. There are no well-developed soil horizons. These soils are strongly influenced by factors such as *parent material* or relief and in the American 7th Approximation Classification they are divided into lithosols, regosols and alluvial soils. Lithosols develop at high altitude or in exposed sites where a resistant parent material (*scree*) reduces the impact of *weathering* and where steep slopes promote downslope movement which prevents the formation of horizons. Regosols form on unconsolidated materials (sand *dunes* and volcanic ash) where profiles are disrupted by the accumulation of fresh material. Alluvial soils (river *floodplain* or *salt marsh*) are subjected to periodic *flooding*, the addition of superficial deposits and poor drainage, which restrict soil processes.

A–Z Online

Log on to A–Z Online to search the database of terms, print revision lists and much more. Go to **www.philipallan.co.uk/a-zonline** to get started.

baby boom is the brief increase in *birth rate* that takes place in a population after a period of economic or social uncertainty, such as an economic *recession* or war. The UK experienced a baby boom at the end of the Second World War when returning soldiers and their wives started families. There is often a cyclical effect, or 'echo' of such an event, a generation later when the original 'baby boomers' themselves start to have families.

backshore is the name given to that part of a beach which lies above the high water mark and is usually beyond the reach of wave action.

backward integration: see *vertical integration*.

backwash: the term has two meanings:
- the movement of water back down the beach towards the sea after the *swash* has reached its highest point
- the movement of resources from peripheral areas to a *core*, particularly to be seen in developing countries. Such resources include *raw materials*, food, people with skills and ideas and capital from savings. This was described in the *core-periphery* theories of *Friedmann* and the work of *Myrdal*. Backwash ultimately leads to *polarisation*, where there is uneven development between the core of a country and the peripheral areas.

bacteria are microscopic single-celled *organisms* which are key decomposers of the organic matter of the soil. There are many specialist types that are able to convert mineral compounds into forms that can be absorbed by plants. Bacteria are also involved in the process by which nitrogen in the air is transformed into *nitrate*, an essential nutrient for plant growth.

badlands are areas in semi-*arid* environments where soft and relatively *impermeable* rocks have been moulded by rapid *runoff* which results from heavy but irregular rainfall. General features of badlands landscapes include:
- extensive development of *wadis* of all sizes with steep sides and *debris*-covered bottoms
- gullies that erode headwards on hillsides, cutting into them and contributing to their collapse
- slope failure and *slumping*
- *alluvial fans* at the foot of steep slopes where wadis or smaller gullies emerge
- pipes, which are formed when water passes through surface cracks and carves out eroded passageways. Pipes may also form caves when surface runoff is directed towards them
- natural arches, which are created by the *erosion* of a cave over a period of time.

An example of a badland landscape occurs in South Dakota, USA.

bahada/bajada is a gently sloping plain found against a cliff in a desert area. It is made up of many *alluvial fans* occurring at the mouths of *wadis*, which have merged to form this larger feature.

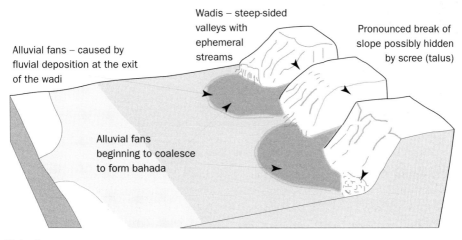

Bahada

balance of payments: the sum of the nation's income and expenditure on foreign trade. Within the overall balance of payments there is the balance of *invisible trade* and the *balance of trade*.

balance of trade is the term used to describe that part of the *balance of payments* account that registers exports and imports of *visible trade* goods. The UK has been importing more visible goods than exports since the last quarter of the nineteenth century and this has been paid for by the surplus on *invisible trade*.

bankfull is when the channel of a river contains all the water that it can carry, i.e. its *carrying capacity* is at a maximum. If a river exceeds its bankfull *discharge*, then *flooding* occurs.

barchan: a crescent-shaped sand *dune* that is concave in the direction towards which the wind is blowing and has a pair of horns which project downwind. Barchans are associated with *arid* areas that have winds blowing from a constant direction. The windward slope is quite gentle, whereas the leeward slope is steeper (34°). Barchans can be as large as 400 m wide and 30 m high, although many are smaller. The horns tend to move faster than the centre, but they move into the shelter of the main body of the dune. Thus, a shape is created that can only be altered by a shift in the direction of the wind. Sand is driven up the windward slope and is deposited near the top, steepening the leeward slope. When the angle of the leeward slope exceeds 34°, *slumping* takes place and so the barchan moves forwards. In Peru, barchans have been measured as moving at 25–30 m a year. Although barchans have a distinctive shape, their occurrence is relatively rare.

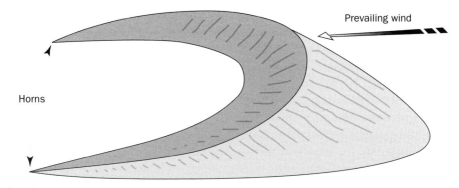

Barchan

bar chart: a diagram consisting of a series of vertical rectangles rising from the horizontal x-axis. All of the rectangles are of the same width. The height of the bar is proportional to the quantity represented. The x-axis is often used to show time intervals (e.g. years, months) and the vertical y-axis to show values such as amounts and frequencies. All bar charts should begin at zero on the vertical axis.

There are various forms of bar chart:
- a standard bar chart
- a multiple bar chart
- a divided bar chart (often using percentage data)
- a divergent bar chart

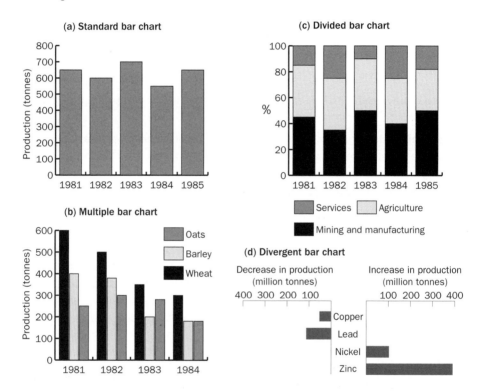

barrage: the term has two meanings:
- a dam constructed across an **estuary** with the intention of harnessing **tidal energy**. Barrages have been built across estuaries with a large tidal range. The tides pass through sluices in the barrage, each of which contains a turbine to generate electricity. The turbines can operate using each of the incoming and receding tides. Examples of barrages are the ones in the Bay of Fundy (eastern Canada) and the Rance Estuary (northern France). It is also proposed to build one across the Severn Estuary.
- a form of **hard engineering** of **coastal management** whereby the **flooding** of major estuaries and sea inlets is prevented. It acts as a dam across an estuary, stopping the incursion of sea water. The Thames Barrier is an example of such a barrage, as is the one in Cardiff Bay. The latter has created a large freshwater lake behind it, which is now the centre of **regeneration** with new housing and leisure facilities as well as being the site of the National Assembly for Wales.

barrier beach: a series of elongated low islands consisting of coral and/or sand lying parallel to a coastline. They therefore have ocean on one side and marsh and lagoons on the landward side. Sand is deposited by the sea under normal low energy conditions. Wind may then move the sand to build *dunes* further up the beach, which in turn become colonised by stabilising plants. Breaks in the islands are maintained by the scour of tidal currents and rising tides spilling into the lagoons behind. Storm waves may wash over the beach, but after a while the beach and lagoon reestablish themselves during low energy conditions. Some barrier beaches may stretch across a bay connecting one *headland* with another. Examples of barrier beaches are found near Porthleven (Cornwall) and at regular intervals along the Atlantic coast of the USA, for example at Miami Beach. The formation of the latter is attributed to a *post-glacial* rise in sea level that flooded the land behind a pre-existing stretch of sand dunes.

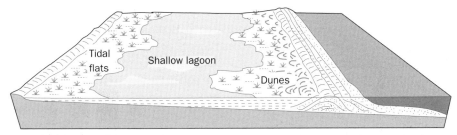

Barrier beach

barrier island: one element of a *barrier beach*.

bars: deposits of sand and shingle situated some distance from a coastline. They usually lie below the level of the sea, only appearing above the level of the water at low tide. There are two theories on where and how they form:
- in shallow seas where the waves break some distance from the shore
- where steep waves break on a beach, creating a strong *backwash* that carries material back down the beach to form a ridge.

When a bar begins to appear above the level of the sea for most of the time, it becomes a *barrier beach*.

basal sapping occurs when *erosion* is concentrated along the base of a slope, causing undercutting and retreat of the slope. Headwall retreat in a *cirque* (corrie) may involve basal sapping; weathered rock fragments become incorporated in the ice and are removed. It can also occur at the foot of a *scarp* where the accumulation of *groundwater* produces a concentration of moisture that leads to increased *chemical weathering*.

basal slipping/sliding is an important process of *glacier* flow, particularly in temperate glaciers. In areas with milder summers, which allow melting to occur, the temperature at the base of the glacier is close to the *pressure melting point* of the ice. As the ice begins to move, there will be an increase in pressure and friction with the underlying bedrock. This increases the temperature and causes the basal ice to release *meltwater* which acts as a lubricant between the base of the glacier and the bedrock enabling the glacier to move more rapidly.

base flow is that part of a river's *discharge* which is produced by *groundwater* seeping slowly into the bed of the river. It is the main contributor to a river's flow during dry conditions. It is relatively constant in amount, although may increase slightly during a wet spell of weather.

base level is the lowest limit to which *erosion* can take place. For rivers, the ultimate base level is regarded as sea level. Exceptions occur when rivers drain into inland basins which are themselves below sea level. When the course of a river crosses a band of more resistant rock, this can produce a local base level. This is a temporary feature; in time the river will erode the resistant layer and the local base level will be removed.

base year: where time series data are put into *index number* form, the year chosen to have a value of 100 in the index series is called the base year.

battery farming is a very intensive type of *commercial farming* in which poultry are reared in specially designed units that enable very high levels of output. The birds are kept in cages and are provided with controlled amounts of water and food delivered automatically. Although the production *system* generates lower priced food for consumers, animal rights groups oppose these methods.

beach: an accumulation of sand and shingle that often occurs in a sheltered position on a coastline, but may also be found in more exposed conditions where there is a plentiful supply of *sediment*. The upper part of a beach is often composed of coarser materials such as pebbles and has a steeper slope (10–20°). The lower part of the beach is composed of sand or mud, with a low gradient (2°). (See also *storm beach* and *berm*.)

beach depletion: the reduction in the amount of sand and/or shingle on a beach. It can result from the process of *longshore drift* or from the loss of natural replenishment methods by obstructions such as a *groyne*. Groynes are frequently built across a beach to prevent the movement of *sediment*. The sediment builds up on one side of the groyne until it is high enough to overtop it. However, this prevents sediment moving along the coastline and protecting it. Such areas then become liable to *erosion* by the sea, which can cause significant problems.

It is also thought that other natural replenishment methods are not as active as they once were:

- many of the world's beaches are composed of material deposited on the *continental shelf* at the end of the last glaciation, when sea level was 120 m, lower than today. This material has been transported shorewards by rising sea levels since that time. However, since sea levels have become much more stable, the movement of such material has now ceased
- the extraction of large quantities of this material from offshore zones for commercial purposes has led to a further reduction in its availability
- many rivers do not transport the same quantities of gravels as they once did during the latter stages of the last glaciation
- the colonisation of beaches by vegetation has also stabilised much sediment, thereby removing it from the *system*.

beach nourishment: the replacement of beach material that has been removed by *beach depletion*. There are three main sources of such material:

- land-based sand and gravel pits
- dredged deposits from offshore
- recycled material from other beaches.

In some extreme cases, sand and shingle may be regularly transported from one end of a beach to the other.

Beaufort scale: a numerical system for identifying and measuring the speed of a wind by examining the effects of the wind on natural and structural features. A version of the scale is:

Scale force	Wind name	Speed (km/hr)	Wind effects
0	Calm	0–1	Smoke rises vertically
1	Light air	1–3	Smoke deviates to show direction
2	Light breeze	4–11	Leaves rustle, wind vane moves
3	Gentle breeze	12–19	Leaves move, small flags extend
4	Moderate breeze	20–29	Raises dust, small branches move
5	Fresh breeze	30–39	Small trees sway, waves form
6	Strong breeze	40–50	Large branches move, whistling effect
7	Moderate gale	51–61	Whole trees move, difficult to walk into
8	Fresh gale	62–74	Small branches break off trees
9	Strong gale	75–86	Slight structural damage e.g. slates blown off roofs
10	Whole gale	87–101	Structural damage likely, trees blown down
11	Storm	102–115	Widespread damage
12	Hurricane	116+	Devastation

bedding plane: the line of division between each layer (strata) in a *sedimentary rock*. It usually indicates where one phase of *deposition* has ended and another begun. Bedding planes can provide openings or cracks along which *weathering* processes can operate.

bedload: larger particles (sands, gravels and pebbles) that are forced to roll, slide or bounce along the bed of a river by the force of the moving water in that river. Large particles only form part of the *load* of a river during and immediately after extreme events that lead to significant increases in stream *discharge*, such as prolonged heavy rainfall and snowmelt. In these circumstances, the *competence* of the river increases and allows larger particles to be carried.

beneficiation is the concentration of *raw materials*, usually metallic ores, close to where they are extracted. The waste content in the ore is reduced, as is its volume. Consequently, the costs of transporting the ore to the processing plant are lowered considerably.

Benioff Zone: the boundary between an oceanic plate that is undergoing *subduction* beneath an overriding continental plate. It is a sloping plane (usually between 30° and 60°). The sinking oceanic plate is much colder than the *crust* into which it is sinking and sudden stresses along the Benioff Zone may trigger *earthquakes*. It is also within the Benioff Zone that remelting of the subducting oceanic plate may take place. The andesitic lavas created in this way may rise to the surface to create a volcanic *island arc*.

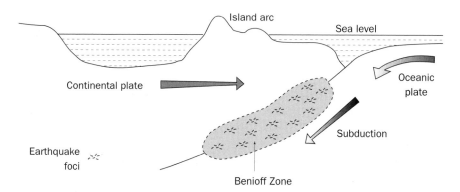

Bergeron–Findeisen theory is one of the ways in which it is suggested that raindrops can be formed. These meteorologists noted that in air temperatures of between –5°C and –25°C, supercooled water droplets and ice crystals can co-exist. When this happens, the air is oversaturated in terms of the ice crystals and vapour is then *sublimated* on to their surfaces. To compensate for the removal of this vapour, the water droplets evaporate into the air and this increased vapour is again sublimated on to the ice surface. The ice crystals therefore grow at the expense of the water droplets. Eventually, as the ice crystals lock together to form snowflakes, they become large enough to fall. If the flakes melt in the higher temperatures nearer the ground, they form drops of water. This process is the basis of the artificial creation of rain through *cloud seeding*.

bergschrund: a large crevasse near the upper limit of a *cirque* (corrie) *glacier* formed as the ice pulls away from the headwall. The bergschrund *hypothesis* suggests that *freeze-thaw* action at the base of the crevasse helps to break up the rock of the headwall, producing fragments that become incorporated into the glacier. These are transported away subjecting the headwall to a form of *plucking*, which helps to maintain the steepness of the backwall. Recent research on temperature variations within the bergschrund have indicated that the changes may be too small to produce freeze–thaw action and this has led to a reassessment of the ideas; for many geographers the hypothesis is no longer acceptable in its original form.

berm: this is a ridge of material that is found running across the back of a beach (*backshore*). It marks the highest line on a beach that the waves have reached at a previous high tide.

best-fit line: on a *scatter graph*, a line can be drawn as close to all the plotted points as is possible, indicating the *trend* in the pattern that is under investigation. Some points may well lie at some distance from this line and are therefore anomalous, these being known as *residuals*. The line is usually drawn by eye. If all the points fit exactly on to the best-fit line, the *correlation* between the two variables is perfect.

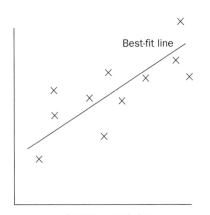

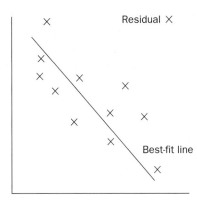

Examples of scatter graphs with best-fit lines

bias occurs in **sampling** when there is some distortion or error in the sampled data. The sample is not representative of the total (parent) **statistical population**, i.e. it does not have the same characteristics as the base population. Bias may result from poor choice of sampling method or when an insufficient number of samples have been selected.

bid-rent theory: in an attempt to explain **land use** within urban areas, this theory states that in a free market the highest bidder will obtain the use of the land. The highest bidder is likely to be the one who can obtain the maximum profit from that site and so can pay the highest rent. In an urban area, competition for land is seen as being greatest at the centre because of the accessibility of central areas. In more recent times though, sites on the edge of urban areas have become more sought after and have seen the growth of industrial areas and massive edge-of-town **shopping centres** such as the MetroCentre in Gateshead.

bifurcation ratio: this is the relationship between the number of **streams of one order** of **magnitude** in a **drainage basin** and those of the next highest order. It is obtained by dividing the number of streams in one order by the number in the next highest order. As the ratio is reduced so the risk of **flooding** within the basin increases.

Worked example: bifurcation ratio

$$\frac{n1 \text{ (number of first order streams)}}{n2 \text{ (number of second order streams)}} = \frac{28}{6} = 4.66$$

Having found all the bifurcation ratios within the basin:

$$\frac{4.66 + 3.00 + 2.00}{3} = 3.22 = \text{bifurcation ratio for the basin}$$

bilharzia (schistosomiasis) is a disease caused by a parasitic flatworm, the blood fluke. The worm originally develops in a snail host within fresh water and enters the bloodstream when people work, bathe or swim in the water that contains the snails. The disease is found all over the world, but is particularly prevalent in Africa. The **World Health Organization (WHO)** has estimated that around 200 million people are infected with the worm, making the disease humanity's most serious parasitic infection after **malaria**.

biodiversity is the range of wild and cultivated species within an *ecosystem*. Such diversity has recently become a major environmental issue because environments are being degraded at an accelerating rate and much diversity is being lost through the destruction of natural *habitats*. The number of species of *organism* on Earth is still imperfectly understood, as is the rate at which they are being lost, but particular fears have been expressed about loss of species caused by *rainforest* exploitation.

biofuel is that part of the *biomass* that can be converted into energy. At its simplest, this involves the burning of *fuelwood*, dung and crop residue for cooking and heating, but this is mostly inefficient. In developed countries, biofuels also include landfill gas and municipal waste. Brazil has developed a system that converts sugar cane to ethanol for motor vehicle use, which by 2007 had provided at least 25% of all fuels sold at filling stations.

biogas: this is the process by which *methane* gas is obtained from animal dung, human excreta and crop residues. The gas can then be used either directly as a cooking or engine fuel or, in a high-tech application, to fuel high-efficiency gas turbines in order to generate electricity.

biogeochemical cycles: this is the way in which various chemicals (e.g. carbon, nitrogen, phosphorus) are circulated around the *ecosystem* and recycled continually. At its simplest, each cycle consists of plants taking up chemical *nutrients*, which they pass on to grazing animals and carnivores. As animals at each *trophic level* die, they decompose and the nutrients are returned to the soil. These cycles can operate on land, in the sea or in the air.

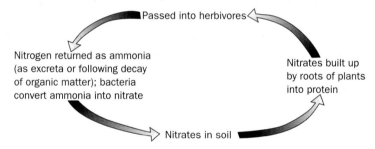

An example of biogeochemical cycle – nitrogen

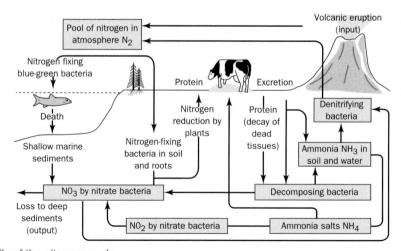

Details of the nitrogen cycle

biological controls are natural predators, parasites, *bacteria* and viruses that are used to control weeds and pests in agriculture. This approach reduces the need for chemical controls, which may accumulate in the *food chain*. However, effective control is rarely achieved solely through biological methods and some application of *pesticides/herbicides* is needed.

biological oxygen demand (BOD) is the amount of dissolved oxygen needed to enable the *decomposition* of organic material in polluted water. It is measured in milligrams of oxygen per litre of water at 25°C. Polluted streams have low values of dissolved oxygen content and therefore such streams have a high BOD. Heavily polluted water courses may have a BOD of 350 to 400.

biomass: the total amount of organic material both above and below the ground surface and in water bodies. Plants and vegetable matter, both living and decaying, comprise the greatest bulk of the Earth's total biomass. Animal biomass is small in comparison and mostly consists of micro-*organisms* that live in the soil.

biome: a naturally occurring organic community of plants and animals. Each biome derives its name from the dominant type of vegetation found within its physical environment or *habitat* (e.g. *savanna*, grassland, coniferous forest) and consists of the *ecosystem* of plants, soils and animals.

biosphere reserve: an international *conservation* designation given by UNESCO under its programme on Man and the Biosphere (MAB). According to the *Statutory Framework of the World Network of Biosphere Reserves*, biosphere reserves are created 'to promote and demonstrate a balanced relationship between humans and the biosphere'. Under article 4, biosphere reserves must 'encompass a mosaic of ecological systems' and thus consist of combinations of terrestrial, coastal or marine *ecosystems*. Through appropriate zoning and management, the conservation of these ecosystems and their *biodiversity* is sought to be maintained. The design of the reserve must include a legally protected *core* area, a buffer area where non-conservation activities are prohibited and a transition zone where approved practices are permitted.

biotechnology is the application of biological knowledge and research to technological development. It has emerged as a *high-tech industry* that uses the properties of living cells in specialised areas such as the discovery and production of vaccines and antibiotics.

biotic factors are those that result from the action of living *organisms* (plants and animals) influencing plant growth and distribution. Plants can modify the physical conditions of a *habitat* by providing shade and shelter; they alter light intensity, temperature and *humidity* conditions and they can reduce wind speeds near to ground level. In this way, they can create their own particular microclimate and, through their use of water, mineral nutrients and the return of organic matter, they also influence the condition of the soil. The development of a given species in a site depends on its ability to compete for space, light, water and soil nutrients. The term also includes factors related to the activity of animals within the habitat. Many plants depend on animals for seed dispersal and cross-pollination of their flowers. However, plants are also basic food producers and may be damaged by animals that feed on them removing leaves or bark.

bi-polar test: this is a method of determining *attitudes* towards a particular phenomenon. It is based on two extremes of attitude, which are said to be 'poles apart' or 'bi-polar' views. It is possible to ask people to judge a phenomenon by scoring somewhere between these two poles on a grid like the one below:

Bi-polar scoring chart for .... industrial estate

Location .... Time ....

Date ....

The layout is	ATTRACTIVE	5	④	3	2	1	UGLY
The amount of greenness is	POOR	1	2	3	4	⑤	EXCELLENT
The traffic flow is	CONGESTED	1	2	③	4	5	LIGHT
This place is	QUIET	5	4	③	2	1	NOISY
Smells are	PLEASANT	5	④	3	2	1	OFFENSIVE
Street and premises are (e.g. litter, graffiti)	DIRTY	1	2	3	④	5	CLEAN

Bi-polar scoring chart

Note: All the 'bad' words should not be placed on one side of the chart as this may bias responses.

birth control programmes are designed to limit births and therefore lower the *birth rate*. These programmes are typical of *Third World* countries, particularly in Asia, such as China, Sri Lanka and Singapore. The programme in Singapore was typical of many in having several strategies:
- to establish family planning clinics and to provide contraceptives at minimal cost
- to advertise in the media the need for, and the advantages of, smaller families
- to legislate to allow abortions and sterilisation
- to introduce social and economic incentives such as paid maternity leave, income tax benefits, housing priority, cheaper health treatment and free education, which would cease if family size grew.

birth rate is a measure of a country's fertility. It is expressed in the number of live births per 1000 people in 1 year.

blockfields are extensive sheets of large angular rock fragments that are formed by frost action in *periglacial* areas. When areas of exposed bedrock are subjected to fluctuations of temperature through freezing point, *freeze–thaw* processes are very effective in producing accumulations of shattered *debris*. These are also called felsenmeer. Where blocks have moved downslope to form lines of angular debris, they are called block streams.

blocking high: an *anticyclone* that prevents *depressions* from following their normal paths by diverting them around the edge of the high pressure. They occur when an anticyclone breaks away from the Azores High and establishes itself in more northerly areas (latitude

50°N–70°N) for several days. Once in place, the system may remain for several weeks in extreme cases and this alters the usual weather pattern. The effects depend on the season, the position of the blocking high and the length of time it persists. In summer, warm/hot conditions with above average sunshine and low rainfall may be experienced. In winter, blocking prevents milder influences reaching the British Isles and low temperatures, cold easterly winds and frequent snowfalls may be the pattern.

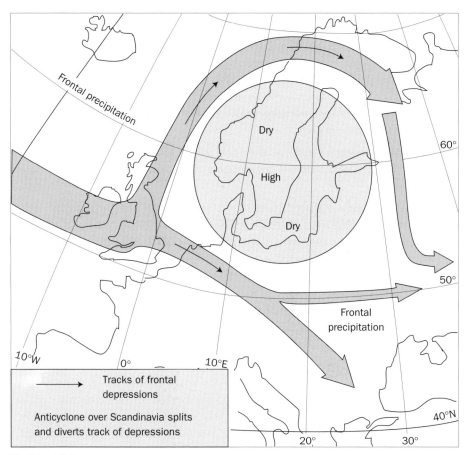

Blocking high

blow hole: a feature located on a coastal *headland* where *erosion* has extended a cave such that it emerges at the top of the cliff. On some occasions, at very high tides and in stormy conditions, water can be seen to spray out from the hole created.

blue collar: a term used to describe employment in manual work.

bluff: a steep slope resulting from lateral *erosion* by a river cutting back interlocking spurs. This may create a bluff line on the edge of a wide *floodplain*.

bog: blanket bogs develop under conditions of high rainfall in areas that are too wet for tree development. They are areas of waterlogged, spongy ground dominated by herbaceous plants such as bog moss (sphagnum) and cotton grass, bell heather, bog myrtle and rushes. These decay slowly to form highly acidic peat with *pH* values just above 4. Blanket bog is widespread throughout upland regions in the British Isles, but becomes more extensive and occupies lower levels towards the west where it extends on to undulating lowland areas. The acidity is due to the high rainfall maintaining a downward flow in the groundwater. This creates **anaerobic** conditions in which **bacteria** cannot survive and organic **decomposition** is extremely slow. Many of these peat deposits in Britain are relic features dating from the Atlantic stage of the **post-glacial** period when the climate was milder but much wetter than the present day.

Raised bogs are so called because their central regions are at a higher level than their margins. They are commonly found on silted **glacial lakes** or on estuarine muds which were originally colonised by **haloseres**. This is due to the fact that after the bog plants had invaded they continued to develop producing more and more peat, which accumulated more rapidly in the central area. These are found in many parts of central Ireland as well as in the lowlands of the western counties in Britain, where heavy rainfall and water retaining peat keep the **water table** near to the surface. This helps to promote the further accumulation of peat.

boom irrigation: water is introduced to the field using a sprinkler that rotates around a central pivot. This creates a pattern of circular higher yielding crop areas separated by lower yield or fallow areas.

boreal: this term means northern. It is an alternative name for the largely coniferous forests which extend across large areas of Europe and North America between latitudes 45°N and 75°N. These boreal forests experience severe climatic conditions with cold winters and short summers. The **growing season** is less than 6 months, and may be as low as 3 months on the northern limit. The trees, which are mainly evergreen species such as pine, fir and spruce, have adapted to the conditions with flexible branches to shed snow and needle-like leaves to reduce water loss through **transpiration**. This is particularly important in winter when freezing of the soil restricts availability of water; the trees experience a physiological **drought**.

The term is also used for the Boreal climatic period which extended from approximately 7000 years **BP** to 5500 years BP. This was a period which followed the cold and wet pre-Boreal when pine and birch trees dominated the vegetation. The Boreal was cold but drier. As a result, the species that favoured 'wet' conditions, such as birch and pine, declined, and trees, like elm and hazel, reached their peak. Oak trees appeared during this stage and became increasingly important as the climate continued to get warmer. They reached their dominance in the warmer Atlantic period that followed the Boreal.

Boserup, E.: a Danish economist who in 1965 suggested a theory on the relationship between population and resources which opposed the views put forward by **Malthus**. Whereas Malthus argued that food supply acted as a ceiling to population growth, she suggested that in a pre-industrial society an increase in population stimulates a change in agricultural methods, which results in greater food production. If groups do not produce more food, then starvation follows; this is often stated in the saying 'necessity is the mother of invention'. From her studies on different agricultural **systems**, Boserup suggested that farming became more intensive as population pressure increased; fallow periods became shorter, harvesting one crop a year with a fallow period of only a few months. With further population growth, multi-cropping is stimulated as the most intensive system. (See also **Simon, J.**)

bottomset beds consist of the finest clays deposited in a *delta*. They are deposited in near horizontal layers at some distance from the mouth of a river. The settling rates of such clays is very slow, but in salt water the particles are subject to *flocculation* causing them to sink to the sea bed.

bottom-up: development initiatives that begin at a local level and then spread across a nation as a whole. Such schemes often concern *self-help* programmes, which aim to both improve matters in the short term and also publicise/encourage further development in the future.

boulder clay is the unsorted and unstratified *debris* stranded or deposited over a landscape by the action of ice. It is composed of fragments of rock of all shapes and sizes, ranging from large boulders to fine clay particles. The stones in boulder clay tend to be angular or sub-angular. The composition of boulder clay in an area reflects the character of the rock over which the ice has moved to that area. It is also known as *till*.

BP: an abbreviation for 'before present'. It is used for stating the age of relatively recent deposits from the glacial or *post-glacial* periods where dates given as BC would not present such an accurate impression as to how long the deposit had been formed. It is applied particularly to the results of *carbon dating*, where absolute dates plus or minus margins of error are given.

braiding is when a river is forced to divide into several channels with islands separating them. It is a feature of rivers that are supplied with large *loads* of sand and gravels. The banks formed from these materials are unstable and consequently the channel becomes very wide in relation to its depth. The river becomes choked, with several sandbars and channels which are constantly changing their location. Braiding occurs in a number of environments where there are rapidly fluctuating *discharges*:
- semi-*arid* areas of low relief which receive rivers from mountainous areas
- glacial *outwash* plains
- *periglacial* areas underlain by *permafrost*.

branch plant is a factory built by a *transnational corporation (TNC)* which has its headquarters overseas. These are common features in the motor vehicle industry where branch plants can occur in a wide number of countries.

Brandt Report: published in 1980 to highlight the growing gap in social and economic development between the developed countries of the world, the *North*, and the less developed countries, the *South*. It was compiled by an independent group of statesmen headed by Herr Willy Brandt, the former West German Chancellor. The Report discussed a range of issues: disarmament, political corruption, violation of human rights, *overpopulation*, world health, industrialisation, world trade, environmental pollution, urbanisation and forms of communication between countries. It concluded that the North and the South were mutually dependent on each other and it warned against the North establishing economic barriers against the growing industrialisation of the South.

break of bulk: a point where cargo is unloaded from some form of bulk carrier and transferred to smaller units of transport for further movement. It applies where the mode of transport changes, e.g. sea to land. These are attractive points in terms of economic location because they offer potential savings in transport costs. If *raw materials* are processed at the break of bulk point, there is no need to transfer materials from sea-going to land-based transport – they can be unloaded directly into the processing plant without further transhipment

29

costs. In this way, expensive unloading and reloading costs can be reduced. Heavy industries that process imported raw materials find tidewater break of bulk locations advantageous.

BRIC: the collective term for Brazil, Russia, India and China – the economic powerhouses of the twenty-first century. Some authors also refer to these as *recently industrialised countries (RICs)*.

brown earth: a type of soil associated with the northwest European-type climate and *deciduous woodland*. The considerable amount of leaf litter that accumulates in the autumn is decomposed relatively quickly by a range of soil *organisms* to create less acidic *mull humus*. This is incorporated into the upper horizons of the soil by earthworms, giving it a dark brown colour. There is a downward movement of soil water due to *precipitation* exceeding *evapotranspiration*, but the degree of *leaching* is limited. However, some clay particles are washed down through the *soil profile* (*lessivage*), and so the lower horizons are enriched with clay. The different horizons in the soil merge gradually due to the active mixing by soil fauna. The soil tends to become lighter in colour downwards through the profile. Brown earth soils are potentially fertile, though they benefit from liming. Brown earth soils on *parent material* such as granite and sandstones tend to be more acidic, whereas those on *calcareous soils* (*limestone*, *chalk*) are less acidic.

brownfield sites: land and buildings within urban areas that could be redeveloped, particularly for new housing. Some organisations have urged the UK government to put greater emphasis on such land as an alternative to *greenfield sites*. Examples of the types of brownfield sites being developed within the UK include:
- old industrial sites such as brickworks and gasworks
- old factories being converted into housing units
- upper floors of shops being converted into flats
- warehouse conversion into flats, together with the redevelopment of derelict dock areas for other housing
- derelict hospital sites.

Since 2000 the UK government has modified the term 'brownfield site' to 'previously developed land', most of which is in urban areas. Farmland, parks, allotments and playing fields are excluded. The government has set a target stating that, by 2008, 60% of new houses built each year will be on such land. Some say this is too low. However, there is still a mismatch between where land is available (the industrial areas of northern England) and where the pressure for housing is greatest (southeast England). A significant feature of this trend is the purchase and subsequent demolishing of large houses and their gardens in established suburban areas by private developers, to be followed by the building of several new houses (with much smaller gardens) on the same land.

budget airlines are low-cost airlines operating scheduled services. They reduce the cost of flying by:
- operating many services to and from smaller regional airports (where they have often received subsidies)
- promoting online booking and ticketing procedures
- operating tight schedules with rapid turn-arounds
- providing few (if any) in-flight services.

Prominent operators in the UK include Ryanair, easyJet and Jet2.

business cycle: the regular pattern of upturns and downturns in demand and output within the economy that tend to repeat themselves every 5 years or so. The causes of this cyclical pattern to economic activity are not fully known, but are partly explained by:
- the bunching of investment spending which, by definition, does not need to be repeated for several years
- government policies that aim for rapid growth just prior to election dates, which lead to inflation and therefore the need to constrain the economy post-election
- confidence in the future, which means that firms will invest and expand contributing to the economic upswing. Alternatively, if firms foresee an economic slowdown in the economy, they cease to expand and postpone investment plans.

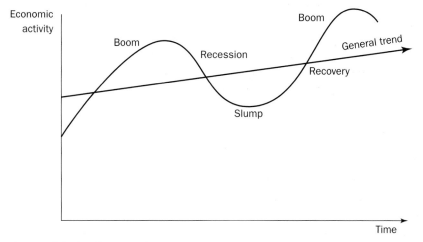

The phases of the business cycle

business park: a term adopted by property developers in order to attract firms needing office accommodation rather than industrial units, as well as *high-tech industries*. Some business parks may include leisure activities such as golf courses and riding centres.

Butler's model suggests that tourist areas develop through a number of stages:
- the initial development is slow and often caters for a social and economic elite
- development accelerates as the area becomes more widely known and increasingly accessible
- a peak point is reached where the resort is widely developed but stagnation sets in
- the response to stagnation determines the resort's future. Some enter terminal decline where there is no way back; some rejuvenate with new facilities and attractions; some go into a slow decline but still attract a reasonable number of tourists; some try to appeal to a different segment of the tourism market.

buttes are tall thin pillars of rock with a distinctive flat top found in *arid* or semi-arid areas. They are formed by water *erosion* in those areas where the strata is formed of *sedimentary rock* with horizontal *bedding planes*. Water erosion has removed most of the rock, leaving only a thin pillar. The lower slopes of buttes are covered in *scree* resulting from *physical weathering* and rockfall. The most spectacular examples are in the USA in Monument Valley National Park in Arizona.

Buys Ballot was a Dutch scientist who, in 1857, proposed a law that states that if an observer stands with his back to the wind in the northern hemisphere, low pressure lies to the left. In the southern hemisphere, the reverse holds true. This can be used to predict wind direction in a weather system if the position of high and low pressure is known. The law is based on the effects of the *coriolis force*.

Do you need revision help and advice?

Go to pages 325–53 for a range of revision appendices that include plenty of exam advice and tips.

calcareous soils are those that are derived from a *parent material* such as *limestone* and *chalk*. They are naturally alkaline.

calcification occurs in soils in areas of low *precipitation*, where rates of *evaporation* are high, and where there is a water deficit for a large proportion of the year. When rain falls it is sufficient to penetrate the upper horizons, dissolve some calcium and percolate downwards. However, there is insufficient rainfall to perform *leaching* of the soil effectively and soon the water is evaporated, leading to the *deposition* of calcium carbonate.

calorie intake: a kilocalorie is a measurement of the energy-producing capacity of food and a way of showing the *diet* of populations in different parts of the world. It has been calculated that the average adult in temperate latitudes requires 2600 kilocalories per day compared to 2300 for people in tropical areas. The average daily consumption is 3300 in the *First World*, but only 2200 in the *Third World*. Modern methods of measurement have replaced calories with megajoules.

calving is the process of *ablation* by which small masses of floating ice break away from an *ice sheet* or *glacier*. This can produce *icebergs* if the edge of the *ice cap* extends into the sea or smaller masses if a glacier terminates in a lake.

Cambridge phenomenon: the concentration of *high-tech industry* in the Cambridge area of England, especially those related to software, electronics and *biotechnology*. The region is sometimes known as Silicon Fen. In 2004, 24% of venture investment in the UK was received by this area. The growth of these small-scale industries has been due to a number of factors:
- the existence of a pool of highly educated and technologically qualified workers and scientists
- the high income of many of these people, allowing them to set up their own businesses
- the demand for research and development by the local university
- the high-quality *communication systems* in the area, notably the M11 motorway and the electrification of the rail link to London.

canopy: the highest layer of foliage in a woodland formed by the crowns of fully developed trees. A well-developed canopy can significantly reduce light intensity, which will restrict the growth of smaller trees and shrubs.

canyon: a narrow river valley with near-vertical sides cut into solid rock. Canyons are a common feature of mountains and plateaux in hot and semi-*arid* areas. (See also *gorge*.)

CAP: see *Common Agricultural Policy*.

capacity: the largest amount of *load* that a river can carry for a given velocity. Research has shown that a river's capacity increases according to the third power of the velocity of that river, i.e. if a river's velocity doubles, then its capacity increases by eight times (2^3).

capillary action is the movement of water upwards through a substance. It is caused by the adhesive attraction that water molecules have for the walls of the surrounding surfaces. The smaller this space, then the greater the degree of capillary action.

capital represents the finance invested in a company either to start up that business or for production and expansion. Capital is obtained either from shareholders (share capital) or from lenders (loan capital). Capital can also said to be fixed. This is the investment in buildings and equipment and is not mobile compared with money capital. Many mills in the former Lancashire textile working areas were converted to other industrial uses and these premises were then said to represent fixed capital. Some geographers argue that there is a third form of capital – social capital. This is represented by housing, hospitals, schools, shops and recreational amenities that may attract a firm, particularly its management, to an area.

capitalism is the social and economic system that relies on the market mechanism to distribute *factors of production* in the most efficient way. Most of the capital or wealth is owned and controlled by individual people or companies rather than by the state or government.

carbon credits are a key component of national and international emissions trading schemes that have been implemented to mitigate *global warming*. Credits can be exchanged between businesses or bought and sold in international markets at the prevailing market price. Credits can also be used to finance carbon reduction schemes between trading partners and around the world.

carbon dating (radio-carbon dating) is a means of determining the age of prehistoric organic remains (e.g. wood, bone) up to about 50,000 years *BP*. It is based on the fact that radioactive carbon or carbon-14 decreases at a known constant rate after the death of the *organism* (half-life of 5,730 ± 40 years – i.e. half of the carbon-14 present will decay during that period).

carbon dioxide (CO_2): one of the gases that occurs naturally in the Earth's *atmosphere* and usually takes up only 0.03% by volume. Carbon dioxide absorbs long-wave *radiation* from the Earth and is one of the factors that in the past has kept the temperature steady. In the twentieth century, however, carbon dioxide has substantially increased (on some estimates by at least 15% in the last hundred years and it could double by the middle of the twenty-first century), mainly through the burning of *fossil fuels*, although *deforestation* has played some part. The overall effect of this is that more long-wave radiation will be trapped and therefore lead to an increase in the atmospheric temperature, the *greenhouse effect*.

carbon footprint is a measure of the impact that human activities have on the environment in terms of the amount of *greenhouse gases* produced, measured in units of *carbon dioxide*.

Carboniferous limestone: the calcareous rock laid down in the geological period of that name (280–345 million years *BP*), which gives rise to its own particular type of scenery known as *karst*. The *limestone* was laid on the bed of a warm, clear sea and is part of the evidence that the British Isles was at one time to be found in warmer latitudes. Major features to be found associated with this limestone include underground drainage systems, limestone pavements, swallow holes and *dry valleys*. The rock has developed its own particular type of scenery because:

- it is found in well-defined, often thick, beds which are almost horizontal and are well jointed at right angles to the *bedding planes*

- calcium carbonate is soluble through carbonic acid in rainwater, which combines with other acids from upland vegetation
- it is pervious but not **porous**, which means that water can pass along the bedding planes and down the **joints** but not through the rock itself.

Carboniferous limestone scenery can be best seen in Britain in the Yorkshire Dales and Peak District **National Parks**. (See also **limestone**.)

carbon neutral: being carbon neutral is the equivalent of having a zero **carbon footprint** by achieving a net zero carbon emission.

carbon tax is a method of raising energy prices by increasing the revenue obtained by a government from petrol and heating oil sales. Such taxes are one of the ways in which it is hoped that the use of **fossil fuels** will be reduced.

carbon trading is where economically developed countries can buy **carbon credits** from developing countries, for example by helping them to modernise old, inefficient power stations, thereby reducing their greenhouse gas emissions.

carrying capacity is the largest population that the resources of a given environment can support. The term has its origins in **ecology** where the population is related to plants, but it is also used to describe the maximum number of **livestock** that can be supported per unit area. **Malthus** put forward the concept of a population ceiling where saturation level is reached when the population equals the carrying capacity of that environment.

cartel: a group of producers who make an agreement to limit output in order to keep prices high. In order to do this they must control a large proportion of the output and they must agree on levels of production. Probably the best-known cartel is the **Organization of Petroleum Exporting Countries (OPEC)**. The major problems with cartels are:

- that if they do force a high price, it encourages other producers to enter the market
- the members of the cartel may cheat by secretly producing more than laid down in the cartel agreement in order to gain more revenue.

In most countries cartels are illegal because of their potential to exploit customers.

cartography is the science and skill of map and chart production.

cash cropping is the growing of crops for sale as distinct from a crop grown for consumption by the farmer and his family. Cash cropping operates well when there are large domestic markets, opportunities for foreign trade and well-developed transport systems.

caste: a social system where a society is divided up into a number of groups based on hereditary rank, profession and wealth and where position at birth stays for the rest of one's life. It is typical of the traditional Hindu society of India, but the system is now illegal.

catastrophism is the belief that the Earth's features are produced by sudden catastrophic events rather than by slow, more continuous processes such as plate movement, weathering, **erosion** and **deposition**. Until the middle of the nineteenth century, the recognised explanation for geological deposits was related to the deluge theory linking these materials to the biblical record of Noah's flood. Although the term is outmoded, large events, which are very rare, may produce greater physical changes over a long period of time than common day-to-day processes.

catena is a sequence of soils down a slope where each soil type is different but linked to its adjacent types. Catenas illustrate the way in which soils can change on a slope where there are no real changes in climate or the *parent material*. Catenas develop over a long period of time and are therefore best established in areas with stable environments. Parts of Africa, particularly East Africa, show good catena development.

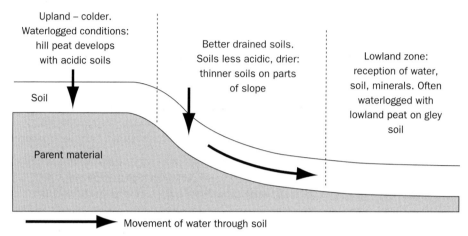

Upland – colder. Waterlogged conditions: hill peat develops with acidic soils

Better drained soils. Soils less acidic, drier: thinner soils on parts of slope

Lowland zone: reception of water, soil, minerals. Often waterlogged with lowland peat on gley soil

Soil

Parent material

Movement of water through soil

Typical soil catena in upland Britain

cation exchange capacity is the ability of a soil to retain cations (positively charged ions), particularly those of calcium, magnesium, potassium and sodium. Sandy soils have a much lower ability to do this than clay soils, which in turn have a lower capacity than *humus*. This effects the *fertility* of the soil, sandy soils being very much less fertile than ones that contain a great deal of humus.

cavitation is a process that takes place in streams flowing at high velocity or under great pressure, for example in *subglacial* channels. Air bubbles form and collapse within the water, causing shock waves against the channel bed and sides. This produces a sand-blasting effect where particles in the stream are thrown against the margins of the channel, polishing and smoothing the sides and creating pot holes in the bed.

CBD: see *central business district*.

census: when the UK first introduced a census in 1801, it was little more than a headcount, but the aim has changed considerably over time so that nowadays it constitutes the collection of information for a broad data bank. This information is of use to both the *public sector* and the *private sector*:

- government and local government use the census data as the basis on which to allocate resources. Government departments include Health, Employment, Education, Home Office
- non-government users could include retailers, advertisers, financial services, property developers, *utilities*.

The UK census now collects a wide variety of data relating to the characteristics of the population, covering:

- demographic information – such as *birth rate*, *death rate* and family size
- housing standards – such as access to bath/WC and overcrowding data, such as number of occupants per room

- social patterns – such as country of birth, mother tongue, religion and educational achievement
- economic data – such as car ownership, occupation and journey to work; it has also been suggested that for future UK censuses information on income should be collected
- religion – first introduced in the census of 2001.

Some research has been carried out to try to find other ways of collecting data. In Denmark, Holland and Sweden, for example, rolling registers of population are compiled, but there has to be a legal requirement that everyone is registered and that they record any changes such as moving house.

central business district (CBD) of an urban area contains the principal commercial streets and major public buildings and is the centre for business and commercial activities. The CBD role was based on its accessibility from all parts of the urban area and as a result contains the highest land values in the area. The CBD is not, however, static. It can be moving outwards in some directions (zones of assimilation) and retreating in other parts (zones of discard). In some CBDs, *retailing* is declining due to competition from *out-of-town locations*, giving a greater emphasis to offices and other services. In a sizeable urban centre there is often segregation of different types of businesses within the CBD to form distinct quarters. Retailing tends to separate from commercial and professional offices to form a distinct inner *core*. The outer core is made up of offices, entertainment centres with some smaller shops. Beyond this, the outer part of the CBD is known as the frame and contains, among other features, service industries, wholesaling and car parks.

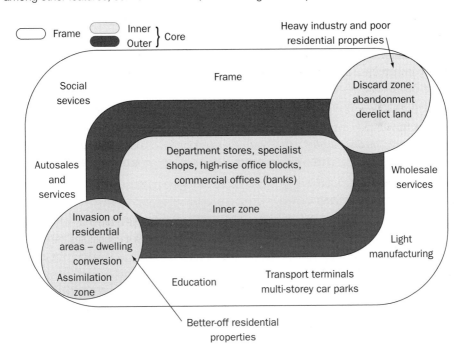

CBD

centrally planned economies: see *command economies*.

central place theory was an attempt to explain a settlement pattern of market centres in a regular order. Perhaps the best-known theory of this kind was developed by the German,

W. Christaller, who suggested that there was a pattern in the distribution and location of settlements of different sizes and also in the ways in which they provide services to the inhabitants living within their **spheres of influence**. Each settlement, no matter what its size or level of service provision, was to be regarded as a central place, i.e. a centre providing goods and services to its surrounding area.

central tendency: the tendency of values within a set of data to group around a particular value or values. Measures such as the *range*, *interquartile range* and *standard deviation* can be calculated to describe the extent to which data is clustered about the *mean*.

CFCs: see *chlorofluorocarbons*.

chalk is a relatively pure form of *limestone* containing a high proportion of calcium carbonate. It is a **sedimentary rock** that forms as a result of the compression of the shells of marine **organisms** which have accumulated on the floor of a shallow sea. Chalk is a porous rock allowing **groundwater** to pass through pore spaces, although **permeability** is greatly assisted by the presence of joints and **bedding planes**. Although chalk is a relatively soft rock, when folded or tilted and flanked by other rocks, it can form upland areas because its permeability allows water to pass through reducing the amount of surface **erosion**. Where the dip of the chalk is gentle **escarpments** are formed as in the North and South Downs. Chalk does contain some impurities, notably flints which are produced from the chemical precipitation of silica. These dry, upland areas have many remnants of occupation by early man, hard flints were used for tools and weapons.

channel efficiency is measured by the *hydraulic radius*, which is defined as the ratio of the cross-sectional area divided by the length of the *wetted perimeter*. The higher the ratio, the more efficient the channel and the smaller the loss of energy due to external friction. Under normal conditions, about 95% of a stream's energy is used in overcoming friction between the water and the margins of the channel and as a result of internal shearing between turbulent currents. Other measures that reflect the efficiency of the stream to move *sediment* are *capacity* and *competence*.

channel flow is the runoff of surface water within a well-defined channel, as in a river or stream channel. Flow is the *discharge* (Q) measured in terms of the volume of water (in cubic metres) passing a point in the river channel in a unit of time (one second) and is usually given in 'cumecs'.

channelisation is the modification of river channels for the purposes of *flood control*, land drainage, navigation and the reduction or prevention of *erosion*. River channels may be modified by engineering works including realignment or by maintenance measures such as clearing the channel. Channelisation can also influence the morphological and ecological characteristics of a river downstream from the site of modifications.

channel morphology is concerned with the shape and dimension of channel cross-sections, which can be influenced by a number of factors: the *discharge*; the quantity and calibre of *sediment load*; and the materials which form the bed and banks. Channel shape may be described by the *hydraulic radius*. Where both banks are being eroded, channels tend to be broadly rectangular or trapezoidal in shape. Where one bank undergoes more rapid *erosion*, as in a *meander*, the channel is likely to be more asymmetrical. The shape of the channel can be expressed in the form ratio: FR = d/w (where d is depth of channel and w is width). Generally, form ratio decreases downstream as width tends to increase more rapidly than depth. River channels also show great variation in plan: some are relatively straight, some are sinuous and some are fully meandering. The *sinuosity* can be measured to assess the degree to which a river is meandering. Many channels also have the flow subdivided by depositional material, which produces braided patterns. (See also *braiding*.)

channel processes operate within rivers to transport, erode and deposit *sediment*. Energy remaining after the river has overcome friction can be used to transport sediment. This *load* is transported by three main processes: suspension, *solution* and *bedload*. Bedload may be moved by *saltation*, when *debris* is temporarily lifted up by the current and is bounced along the bed, and traction, where larger pebbles or boulders are rolled along the bed. The critical values for the pick up and settling velocities of different sized particles are displayed in a *Hjulström curve*. Velocity influences the *competence*, which is the maximum size of material the river is capable of transporting, and the *capacity* which is the total load transported. *Erosion* of new material, or the wearing down of existing sediment, can be achieved through a variety of processes: *abrasion*, where the river uses the load to wear away the bed and banks; evorsion, where the force of the water prises away new sediment; *cavitation*, a pressure effect under high velocity flows; *solution* or corrosion, which is particularly effective when the river flows through areas of *limestone* or *chalk*, which are prone to solution by acidic water. The existing load may be worn down by attrition; boulders collide with other material and angular debris becomes progressively more rounded.

chaparral is a type of scrub vegetation found in the area of California that has a climate similar to the Mediterranean region. It is a *biome* characterised by short, woody and dense bushes that have sclerophyllous leaves – small, thick, leathery with thick cuticles. Some species have tomentose leaves, that is, they have a covering of fine hairs that restrict wind *evaporation* and help to reduce the rate of water loss by *transpiration*. The main species are evergreen oaks and a *drought*-resistant conifer, the pinon. Other *xerophytic* bushes and herbaceous plants are to be found and bare soil occurs between the plants. The vegetation is closely related to the *maquis* of southern Europe.

chelation is an important biochemical process in soil formation. Mineral ions, particularly those of aluminium, magnesium, iron and calcium can be held in the molecular structure of organic compounds. If the soil has a thick layer of *humus* on its surface, percolating water will contain humic acids and organic compounds which will take up the mineral ions from the soil solids. Elements can be moved from the upper layers and washed down the *soil profile*; this process is termed cheluviation.

chemical waste includes substances that are either toxic, ignitable, corrosive or irritant and are potentially dangerous to humans and animals. The leakage of such wastes has been linked to a variety of birth defects, cancers, brain damage and blood and nervous disorders. A number of methods have been used to dispose of chemical waste:
- sealing waste in drums and then storing them, but inappropriate labelling is a problem
- using *landfill sites*, but the chemicals may penetrate *groundwater* supplies
- incineration both on land and at sea, but this releases toxic gases into the atmosphere
- discharging into the sea or rivers, but this does not remove the problem.

(See also *toxic waste*.)

chemical weathering involves the decay or *decomposition* of rock in situ. It usually takes place in the presence of water, which acts as a dilute acid. The end products of chemical weathering are either soluble and are therefore removed in solution, or they have a different volume, usually bigger, than the mineral they replace. The rate of chemical weathering tends to increase with rising temperature and *humidity* levels, except in the action of carbonic acid (carbonation) where lower temperatures produce greater rates of weathering on **limestones**. Chemical weathering can also occur from the action of dilute acids resulting from both atmospheric pollution (sulphuric acid) and the decay of plants and animals (organic acids). There are a number of different types of chemical weathering including **hydrolysis**, carbonation and **oxidation**.

chernozem: a type of soil associated with the continental interior-type climate and **temperate grassland**. The thick grass cover provides plentiful **mull humus** which forms a black upper horizon with a crumb structure. There is an abundance of soil fauna which rapidly decay and incorporate the organic matter into the upper horizons of the soil during the warm summers. However, as the winters are much colder, the process of decay is greatly reduced. The snowmelt in late spring and the early summer rainstorms cause some **leaching** to take place, but in the later hot summer there is a **capillary** upward movement of water. This alternating pattern of soil water movement causes nodules of calcium carbonate to be deposited at depths of about 1 m. The lower horizons are paler due to the reduction of humus content. Chernozems are naturally very fertile.

Chi-squared test: this technique is used to analyse the distributions of dots or points. Normally it is used to compare an actual distribution of points with a **random** distribution of the same number of points. First of all a **null hypothesis** (H_0) is formulated to the effect that there is no significant difference between the observed pattern in the distribution and the expected pattern of distribution (which is usually regarded as being random). The alternative hypothesis to this will be that there is a difference between the observed and the expected patterns, and therefore there must be some factor responsible for this difference.

The method of calculating the value of chi-square is shown below. The letters A to D in the table refer to the areas A to D in the map alongside the table. In the column headed O are listed the numbers of points in each of the areas A to D on the map (the observed frequencies), the total number of points in this case being 20. Column E contains the list of expected frequencies in each of the areas A to D assuming that the points are randomly spaced. In the columns O–E each of the expected frequencies is subtracted from the observed frequencies, and in the last column the result is squared. The relevant values are then inserted into the expression for chi-square, and the resultant value is 2.0.

	O	E	O–E	$(O–E)^2$
A	4	5	−1	1
B	7	5	2	4
C	3	5	−2	4
D	6	5	1	1
Σ	20	20	0	10

$$X^2 = \frac{\Sigma(O - E)^2}{E} = \frac{10}{5} = 2.0$$

Chi-squared test

The aim of a chi-squared test, therefore, is to find out whether the observed pattern agrees with or differs from the theoretical (expected) pattern. This can be measured by comparing the calculated result of the test with its level of significance. There are two levels of significance: 95% and 99%. At 95% there is a 1 in 20 probability that the pattern being considered occurred by chance and at 99% there is only a 1 in 100 probability that the pattern is a chance one. The levels of significance can be found in a book of statistical tables. If the calculated value is the same or greater than the values given in the table, then the null hypothesis can be rejected and the alternative hypothesis accepted. In the case of our example, however, the value of chi-square is very low (2.0) showing that there is little difference between the observed and the expected pattern. The null hypothesis cannot therefore be rejected. (See also *significance testing*.)

chlorofluorocarbons (CFCs) are chemicals used in foams, refrigerators, aerosols and air-conditioning units. They are held responsible for the destruction of the **ozone layer** between 24 and 40 km above the Earth's surface. Following growing concern over **ozone depletion**, a **United Nations** conference in 1987 (the Montreal Protocol) agreed that there should be a 50% reduction in the production of CFCs by 1999. There is still considerable concern that this reduction will not be adequate because of the slow speed with which they disintegrate within the atmosphere. CFCs are also thought to be responsible for part of the so-called **greenhouse effect**.

cholera: an infectious gastroenteritis caused by the bacterium *Vibrio cholerae*. Transmission to humans occurs through eating food or drinking water contaminated with the bacterium. The disease can spread rapidly in areas with inadequate treatment of sewage and drinking water.

choropleth maps are shaded or coloured to show varying spatial distributions within administrative areas. They usually show groupings or classes of data, with a shading system or colour allocated to each group. Shading varies from dark to light for high to low values. Choropleth maps should avoid the use of black or white. Black implies completeness and is difficult to write or print over. White implies emptiness and is often used to represent areas where there is no data. The main weaknesses with choropleth mapping are that it is very dependent on administrative boundaries and shows only average values within that administrative area. It is also unlikely that abrupt changes will take place between such areas, as implied by changes in the types of shading at the administrative boundary – they are much more likely to be gradual.

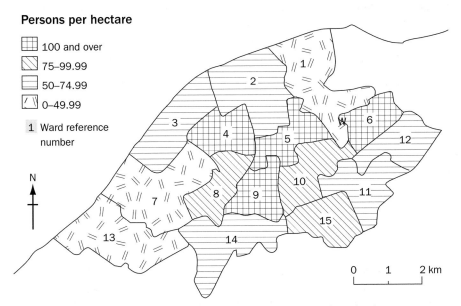

Persons per hectare

- 100 and over
- 75–99.99
- 50–74.99
- 0–49.99

1 Ward reference number

N

0 1 2 km

An example of a choropleth map showing the population density of wards of a town

cirque: also known as corrie or cwm, this is a semi-circular hollow high up at the head of a *glacial valley* on the flanks of a glacial mountain. It was formerly a massive collecting ground for ice which flowed out of it into a glacial valley below. A cirque has a steep headwall to the rear, a bowl-shaped rock basin in its centre (sometimes occupied by a small circular lake called a tarn) and a rock lip at its lower end. This lower end of the cirque often marks the point where a stream plunges steeply down into the main glacial valley below. Cirques develop from an initial hollow in which a snow bank has accumulated. The presence of the snow bank would increase the amount of diurnal and seasonal frost weathering caused by *meltwater* in the hollow, so that it would gradually be enlarged until it was big enough to hold a small *glacier*. This glacier would then start to cause *plucking* and quarrying at its head, probably as a result of the downward *percolation* of water and the *frost shattering* of the rocks in the headwall. The *rotational movement* of the ice in the hollow also helped to create the rock basin and rock lip.

In Europe and North America most cirques are orientated between northwest and southeast. This is because:

- north-facing slopes receive less insolation and this helps to preserve small glaciers
- the main snow-bearing winds are from the west and eddies would ensure that snow banks were preserved on the lee slopes, i.e. those facing east.

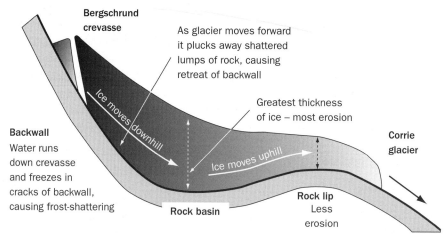

Formation of a cirque/corrie

clapotis occur where the sequence of advancing sea waves coincides exactly with the sequence of reflected waves from a cliff or sea wall. The result is the appearance of a series of 'standing' crests and troughs in the waves some distance from the cliff face or sea wall.

Clarke–Fisher model: as an economy develops through time, this shows in theory how the relative importance of the sectors of employment changes. In pre-industrial times an economy will be dominated by the *primary sector*, but over time people move from this sector as *manufacturing* develops. To support the growing industrial base and the demands of a more affluent population, there is a need for a whole range of services including transport and utilities, consumer and financial services, leading to an expansion of *tertiary* employment. In more advanced economies a *quaternary sector* develops, which encompasses services such as *research and development* and information processing.

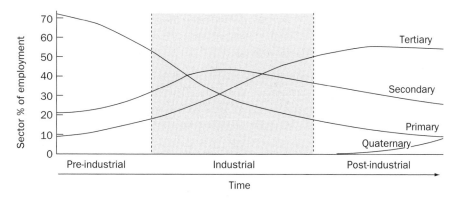

Clarke-Fisher model

class intervals are the dividing lines chosen in order to group data into categories for the purpose of analysis. (See also *dispersion measures*.)

Clean Air Act of 1956: the high incidence of *smog*, which often lasted for several days in the cities of the UK, particularly London, prompted the government to pass legislation to control the amount of smoke polluting the atmosphere. This Act first introduced smoke-free zones into the UK's urban areas, which reduced the health hazard from this phenomenon.

climate: the mean atmospheric conditions of an area measured over a substantial period of time. Different parts of the world have recognisable climatic characteristics with distinctive seasonal patterns.

climatic change: evidence shows that change has always been a feature of the Earth's climate. Apart from the evidence from the *Pleistocene ice ages*, recent research has revealed the existence of a whole series of climatic trends on a variety of timescales. In North Africa, for example, there is evidence to suggest that within the last 20,000 years there have been periods when the climate was much wetter than it is today. At the present time there is great concern that human activity has the potential to bring about climatic change through *global warming* or the *greenhouse effect*. Some of the major causes of climatic change that have been suggested include orbital fluctuations, sunspot activity, major volcanic activity, shifts in the broad pattern of *ocean currents* and variations in the level of atmospheric *carbon dioxide*.

climax vegetation: as the vegetation of an area develops and changes naturally through time, the characteristics and species of plants will alter until they reach a balance with the environmental conditions (soil and climate). This state is known as the climax community and will not change as long as the environmental conditions remain unchanged. (See also *succession*.)

clouds are visible masses of water droplets and/or ice crystals in the atmosphere. They are formed when air cools to *dew point* and vapour condenses. They generally develop when air is forced to rise either because of a relief barrier or where *air masses* are converging at a *front* or as a result of *convection*. Clouds are classified into four types, based on shape and height: cirrus (wispy, hair-like), *stratus* (layer), cumulus (heaped) and nimbus (rain-bearing). The prefix alto- is used to indicate middle-level clouds.

cloud seeding involves the introduction into *clouds* of *condensation* nucleii, salt particles or water droplets in order to induce greater *precipitation*. Solid *carbon dioxide* pellets, or dry ice, and silver iodide smoke can promote cloud growth and increase precipitation by triggering the freezing stage of the *Bergeron–Findeisen theory*, but the results of seeding are not predictable or reliable. Although modest increases have been reported, several experiments have resulted in a decrease in rain.

Club of Rome: an international team of economists, scientists, social scientists, civil servants and philosophers drawn from non-communist countries. It was formed in 1968 to consider possible solutions to major world problems such as the gap between the *North* and the *South*, the pollution of the environment, unemployment, inflation, rich and poor regions. Their first considerations were published in *The Limits to Growth* in 1972. They predicted that if present trends in population growth, industrialisation, food output, pollution and resource depletion were maintained, the limits to growth on the planet will be reached within the next 100 years, but they suggested that the trends could be altered to establish stability.

coastal landforms can be divided into two main groups: erosional and depositional. However, not all landforms can be explained in terms of present-day processes and therefore a third group reflects the effects of *sea level change*. Erosional landforms include cliffs, *wave-cut platforms*, coves and features produced by the retreat of *headlands* such as pinnacles, *stacks* and arches. The main depositional landforms are beaches, spits, and bars but this group would also include features such as sand dune systems and *salt marshes*. Sea level changes may be *eustatic*, i.e. general changes that affect all oceans equally, or they may be *isostatic*, i.e. localised changes resulting from *post-glacial* uplift of land. When sea level rises, lower parts of river valleys may be flooded to form *rias* or cliffs may be subjected to renewed attack. When sea level falls, raised *beaches* and raised platforms may result where features produced at sea level are abandoned. Former cliff lines may become degraded as they are now removed from the constant undercutting by marine processes and *sub-aerial weathering* reduces the angle of the cliff face.

coastal management involves the use of strategies to combat the effects of *erosion*, *flooding* and cliff falls. During the nineteenth and early twentieth centuries, large-scale engineering works were undertaken in many parts of the British coast with the construction of piers, *groynes*, breakwaters and sea walls. One group of responses can be described as the *hard engineering* approach, i.e. constructing defences. Sea walls are designed to absorb wave energy and to protect the cliff base from erosion; methods include bull nose concrete walls, stone and concrete blocks, and revetments. Groynes aim to trap and stabilise sand and shingle by interfering with *longshore drift*. This encourages the formation of a beach as a means of absorbing wave energy. 'Soft engineering approaches' or non-structural responses try to maximise the natural process of development. *Beach nourishment* involves the replacement of material lost by longshore drift. Beaches are replenished with sand brought in from other parts of the coast. Beaches may also be stabilised by revegetating and reducing the slope angle. There is, however, another view which suggests that we should not fight nature and that it might be cheaper to let nature take its course and provide compensation to those people affected instead. Sea defences need constant attention and this costs a lot of taxpayers' money. A more radical approach would be to prohibit development in coastal areas; such policies of 'coastal retreat' have been introduced in parts of the USA.

coastal processes include the direct action of *constructive* and *destructive waves* and their influence on *erosion*, transportation and *deposition* of *sediment*. The characteristics of waves, height, velocity, wave length and wave period, are affected by wind speed and *fetch*. Waves erode material through hydraulic pressure, when air is trapped and compressed in a joint in a cliff and the increase in pressure weakens the rock. *Abrasion* also occurs when previously eroded *debris* is thrown against the cliff. In some coasts corrosion or *solution* can take place when minerals are dissolved. Material is transported by longshore drift and deposition of this sediment occurs in sheltered areas where there is low wave energy, for example in a bay or where the coastline changes direction. Coasts may also be attacked by non-marine processes. *Sub-aerial weathering* by water, wind and frost can lead to forms of *mass movement* such as *slumping*, *landslides* and soil *creep* which can remove material from the cliff face.

cold front: the boundary between a warm and cold *air mass* where the cold air is undercutting the warm air causing it to rise. The gradient of the cold front is steeper than that of the *warm front* and this produces a more rapid uplift of air giving rapid cooling and *condensation*. This is called an ana-cold front which gives cumulonimbus *clouds* and heavy, but short-lived rain showers sometimes accompanied by *hail* and *sleet*. Due to friction with the ground, the cold front may slow down at ground level creating an overhang which produces turbulent and squally conditions. The passage of a cold front is marked by a change in wind direction from southwest to northwest and a drop in temperature. When the warm air in the warm sector of the *depression* is subsiding (kata-cold front) conditions are more stable producing stratocumulus cloud and light *precipitation*. The weather changes are more gradual than with ana-cold fronts.

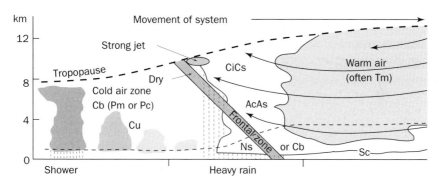

Cold front

collective farming was a type of agricultural system first developed in Russia after the revolution and practised in other former communist states. The land was owned by the state and the collective leased to a large group of workers who operated it as a single farm holding, sharing its profits. Collective farms were intended to be self-governing but government intervention often set *quotas* and production targets.

colloids are very fine grained substances consisting of molecules dispersed in a liquid or gas. Clay colloids are finely divided clays dispersed in water which when combined with humus form the clay *humus* complex in the soil.

colonialism is the establishment and maintenance of rule, for an extended period, by a sovereign power over a subordinate and alien people that is separate from the ruling power. Characteristic features of the colonial situation include political and legal domination over an alien society, relations of economic and political dependence and often a racial and cultural *inequality*. Many of the countries of Europe established colonial empires in Africa, Asia, Oceania and the Americas, beginning with Spain and Portugal in the fifteenth century and followed by Britain, France, the Netherlands, Belgium, Germany and Italy. In the nineteenth and twentieth centuries, both the USA and Japan also established colonial rule over others. In the twentieth century, many areas have ceased to be colonial, as independence movements arose, but there is a general feeling throughout much of the *Third world* that decolonisation has not resulted in meaningful economic or political independence. The persistence of primary export production and of dependent political elites linked to former colonial powers, suggests that colonialism has only been transformed into *neo-colonialism*.

coloniser plants are the first plants to move into an area as part of the *vegetation succession*. These species form part of the *pioneer* community which is the first stage (*sere*) in a sequence that ends with the climatic *climax vegetation*. They gradually modify the microclimate and soil enabling other species to migrate in as conditions change. In a *lithosere*, a primary succession formed on bare rock, the initial colonisers are *bacteria* and single cell plants, together with mosses and *lichens*.

command economies: economic systems controlled by decision-making at the central government level. The former USSR and the Eastern European bloc with countries such as East Germany and Hungary operated command economies before their collapse in the 1980s. In each system almost the whole economy is planned, with targets being set for every point of production. The systems sought to distribute resources fairly among their citizens, but became highly bureaucratic and inefficient. They also became less able to supply goods of the right quality and quantity to meet demand. They were characterised by lower living standards than Western countries, which operate a *market economy*.

commercial farming is where the production is intended for sale in markets. Commercial farmers seek to maximise yields so as to maximise their profits.

comminution is the reduction in size of particles caused by their movement in rivers, *glaciers*, seas and in the wind. They may be reduced in size by striking one another, or by striking other objects as they are moved.

Common Agricultural Policy (CAP) is the scheme by which agricultural production within the *European Union* is organised. It was established by the Treaty of Rome with a number of aims:
- to increase agricultural productivity within member states
- to ensure a fair standard of living for their farmers
- to stabilise agricultural markets within and between the member states
- to ensure reasonable consumer prices
- to maintain jobs in agricultural areas.

These aims have replaced existing national policies and have often caused conflict between member states. Farmers are given guaranteed prices for their produce and therefore have produced as much as is possible. This has created surpluses in a range of agricultural products, known as 'mountains and lakes'. Some 70% of the EU's budget was spent supporting farming, yet farming provided only 5% of the EU's total income. The net gainers from the CAP were those countries with inefficient farmers, such as France and the countries of southern Europe. The net losers were those with an efficient farming sector, such as the UK. Supporters point to the facts that the EU is now largely self-sufficient in food, that more marginal farmers (e.g. hill sheep farmers of north and west Britain) are still in business and that farming is much more productive. Those against the CAP state that its bureaucracy is highly inefficient, and open to corruption, and that it has caused food prices to be higher than they should be on a world scale.

Since 1992 the CAP has undergone reformation in order to answer its critics, particularly on the argument of overproduction and the resulting **food surpluses**. Measures taken include:

- a decrease in the support for cereals, beef and sheep
- the introduction of **quotas** on production, particularly of milk
- an increase in **set-aside** policies
- an increase in environmentally sensitive farming by encouraging the decreased use of ferti- liser and pesticides and by the correct management of land taken out of production
- early retirement plans for farmers aged 55 and above.

In 2002, proposals were put forward to gradually switch funds from intensive production to schemes that promote rural life, safer food, animal welfare and a greener environment. These new plans are geared towards consumers and taxpayers, but they give EU farmers the freedom to produce what the market wants.

communication systems are mechanisms by which information, goods and people are exchanged or moved from one area to another. Examples of communication systems include:

- telephone, facsimile and the internet for information
- road, rail, air and water transport systems for goods and people.

communism is a socioeconomic structure and political **ideology** that promotes the establishment of an egalitarian and classless society based on common ownership and control of the means of production and property. The former **USSR** was an example of a country run along such lines.

community forests were established in 1990 with the main aim of improving neglected parts of **urban fringes** through the encouragement of local landowners to plant tracts of woodland that can be used by the local community. There is no compulsory purchase of land. Help is given by the Countryside Commission and the **Forestry Commission**, and it is hoped that there will be opportunities for:

- recreation – walking, riding and other leisure activities
- education – as an outdoor resource
- **habitat** development for wildlife.

An example is the East London Forest (Thames Chase), stretching north to south from Brentwood to Stifford. Much of this land consists of derelict gravel workings and poor quality farmland. Much of the tree planting will shield the stretch of the M25 which runs through the area.

commuting is the daily movement of people from their place of residence to their place of work and back again. There are two types of commuting:

- rural to urban – where the person lives in a small town or village and travels to work in a larger town or city. The area surrounding a large town or city where the workforce lives is called the commuter hinterland
- intra-urban – where a person lives in one part of a town or city and travels to work in another part of the same town or city. This often involves movement between the outer suburbs and the centre, but also includes movement between inner-city housing areas and edge-of-town **industrial estates**.

comparative advantage: the principle that countries can benefit from specialising in the production of goods at which they are relatively more efficient or skilled. In this way, the consumers within each country gain the maximum benefit from international trade.

Worked example: comparative advantage

	Cost in days' work	
	Country A	Country B
To produce 1 unit of food	1	3
To produce 1 unit of clothing	2	1

If Country A specialises in food and Country B specialises in clothing, both can have higher living standards. If 1 unit of food exchanges for 1 unit of clothing, then trade allows Country A to obtain 1 unit of food and 1 unit of clothing in 2 days' work (instead of 3) by producing 1 unit of food for itself and another unit of food to exchange for 1 unit of clothing. Similarly, Country B saves 2 days' work (2 instead of 4).

competence: the diameter of the largest particle that a river can carry for a given velocity. Research has shown that a river's competence increases according to the sixth power of the velocity of that river, i.e. if a river's velocity doubles, then its competence increases by 64 times (2^6). This is because fast-flowing rivers have greater turbulence and are therefore better able to lift particles from the stream bed.

competition exists where more than one company or organisation has an opportunity to meet the demands for a service or good. In this situation there is not a *monopoly*. Where competition is high, consumers usually benefit in the short term due to falling prices. However, in the longer term, one of the companies may not survive and its subsequent closure will result in less choice for the consumer. In some industries and services, competition may lead to duplication which is wasteful. A recent example of this is the *deregulation* of buses in some areas. Here the duplication of buses by differing companies on profitable routes, and their absence on less profitable routes, has led to the overall decline of services.

composition of a vegetation type refers to the *species* which make it up. Where a vegetation is made up of only one species (for example, a pine plantation) it is called a plant society. However, vegetation is usually made up of a collection of species and is called a plant community. *Tropical rainforests*, where there are warm and moist conditions, have several thousand species of plants.

compression flow is one of the ways in which ice moves within a *glacier*. This occurs when there is a reduction in the gradient of the valley floor, leading to ice deceleration and a thickening of the ice mass. At such points *erosion* is at its maximum.

concentric urban model: this was devised by Burgess in 1923 in an attempt to explain the pattern of social areas within the city of Chicago, but it was later seen to have a wider application. The model is based on the ideas that both the growth of a city and the socioeconomic groupings of people spread outwards from its central area to form a series of concentric zones. At the centre of the city is the *central business district*, the focus of commercial, social and civic life. This is surrounded by the *transition zone* containing industrial premises, obsolete housing and slum property occupied by lower social groups of people and a high proportion of immigrants. This is surrounded in turn by a zone of working-class housing, occupied largely by people who have migrated out from the transition zone, but still need to live close to their place of work. Second generation immigrants also form a significant proportion of the population of this zone. The next zone moving outwards is the zone of middle-class housing consisting of single family dwellings interspersed with exclusive residences and luxury apartment buildings. Finally, at the fringe of the urban area is the commuter zone. This is separated from the continuously built-up area by a **green belt**, but includes villages which are changing their character and function to become **dormitory settlements** for commuters who travel to work in the city each day. Burgess's model has been widely criticised. For example, it makes insufficient reference to the siting of industry which rarely forms a concentric zone anyway. Also, it does not account for the effects of both topographical features and transport systems.

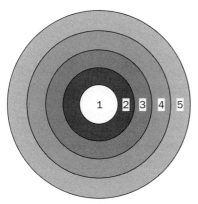

1 = CBD
2 = Zone of transition (inner city)
3 = Zone of working men's homes
4 = Residential zone
5 = Commuter zone

Burgess's model of urban land use

condensation is the process by which droplets of water or ice are formed when water vapour is cooled to **dew point**. During the process, **latent heat** of condensation is released causing a lowering of the **adiabatic** lapse rate in rising air. In the **atmosphere**, **hygroscopic** nucleii, such as salt particles, may attract water and cause condensation before the air is cooled to dew point. This produces **cloud**, **fog** and **mist**.

confidence levels indicate the degree of confidence that can be placed in the results of a statistical exercise. These are stated as the 68%, 95% and 99% confidence level. They indicate the level of probability that the result obtained is the correct one. For example, if after **sampling** the **land use** in an area, 40% of land is estimated to be woodland, and there is a sampling error of 2%, we can be 68% confident that the actual amount of woodland lies within ±1 sample error of the estimate, i.e. the actual amount of woodland lies between 38% and 42%. To reach the 95% confidence level we extend the range to ±2 sampling errors and for 99% confidence we extend to ±3 sampling errors. So, in this example, we can be 99% confident that the actual amount of woodland lies between 34% and 46%. In statistical tests, 95% is an acceptable level of confidence.

conflict: this may result from opposing views over the ways in which a **resource** might be developed. Different individuals or groups may have different attitudes towards the exploitation of a specific resource. For example, the intensification of agriculture to increase production could lead to conflicts; the use of **fertiliser** may increase nitrate pollution of **groundwater** and streams which will arouse opposition from the National Rivers Authority, the water supply industry and fishing interests. Removing hedgerows destroys **habitats** and may detract from the beauty of the area. This may generate opposition from wildlife groups, conservationists, ramblers and leisure interests.

coniferous woodland: in addition to the extensive areas of coniferous (*boreal*) forests of the high latitudes, large areas of woodland have been created by *afforestation*, especially in Europe, mainly for the purpose of commercial timber production. These manmade woodlands are predominantly coniferous softwoods which produce timber with more varied uses than deciduous hardwoods and, in similar environmental conditions, softwoods grow more quickly. The *Forestry Commission*, along with private landowners, has established many *plantations* using native species such as the Scots Pine, as well as Sitka Spruce, Douglas Fir and Lodgepole Pine which are not native.

conservation is the protection and possible enhancement of natural and manmade landscapes for future use. In urban areas individual buildings or areas of settlement may be protected because of their historic interest; in rural areas species, *habitats* and landscapes may be protected. Conservation also encourages sensible use of resources and a reduction in the rate of consumption of *non-renewable resources* as a means to achieving *sustainable development*. Efficient use of resources includes the adoption of less wasteful extraction methods, more efficient use of energy in processing and *recycling* waste material.

conservative plate margin: in *plate tectonics*, when two crustal plates slide past one another the movement of the plates is parallel to the plate margin. The San Andreas Fault in California forms a boundary between the Pacific Plate and the North American Plate. Both plates are moving towards the northwest but the faster rate of movement in the Pacific Plate creates the impression that they are moving in opposite directions. The stresses caused by the movement of sections of *crust* past each other can trigger *earthquakes*.

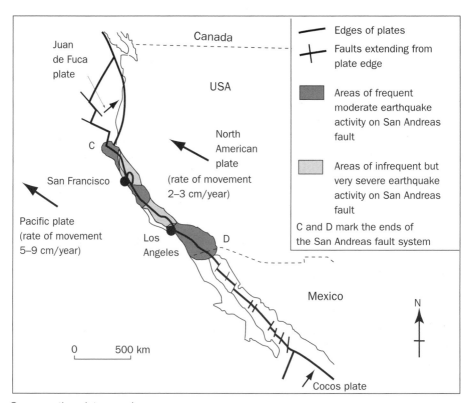

Conservative plate margin

constructive plate margin: in *plate tectonics*, when two plates are moving away from each other *magma* flows upwards and spreads creating new areas of crustal material. In ocean areas this produces *ocean ridges* such as the mid-Atlantic ridge and in continental areas it results in *rifts* such as the East African Rift Valley. They are also known as divergent margins.

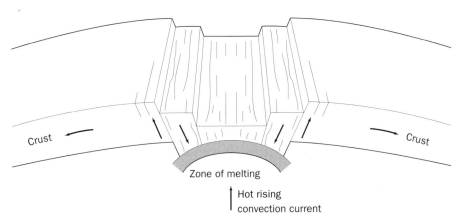

Crust

Crust

Zone of melting

Hot rising
convection current

Constructive plate margin

constructive waves build up material on the beach and contribute to the formation of beach ridges and *berms*. They are waves of relatively low frequency, usually between 6 and 8 per minute, and low height. When the wave breaks there is a strong *swash* carrying material up the beach. The *backwash*, moving material down the beach, is much weaker and therefore less capable of removing sand and shingle. These tend to occur on low angle beaches where energy is dissipated over the beach. Material is constantly moved up the beach increasing the gradient of the beach profile.

contagious: a disease that is capable of being passed on by direct contact with a diseased individual or by handling that person's clothing.

containerisation was an *innovation* in the handling and transportation of cargo by ship in which goods were packed into containers of specified sizes at factories, taken by train and lorry to a container port to be loaded on to specialised container ships. The international specified sizes for containers are 3, 6 or 12 m long, 2.5 m wide and 2, 2.5 or 3 m high. Their introduction resulted in a large reduction in the traditional dock *labour* force and was not achieved without some industrial unrest in the UK. The advantages of containerisation were:
- the turn-around time (to load and unload cargo) for ships was considerably reduced
- with standard containers, standard handling facilities could be developed at ports
- easy storage of freight on dockside
- greater security for cargo, once containers had been sealed
- reduction in labour force required at docks.

continental climate is a general term covering the climate of those areas protected from or unaffected by maritime influence. It is most marked in the temperate to high latitudes of the northern hemisphere. In these regions the climate is generally characterised by:

- warm/hot summers with cold/very cold winters
- large annual ranges of temperature
- low levels of *precipitation*
- summer maximum of precipitation.

continental drift: the *hypothesis* that the present distribution of the continents is the result of the break-up of larger land masses followed by their drifting apart over long periods of geological time. The theory was put forward by the German, *Wegener*, who suggested that the Earth's continents had once been joined together as one large land mass which he named *Pangaea*. This 'supercontinent' had two main components: *Laurasia* to the north and *Gondwanaland* to the south.

The evidence for such movements put forward by Wegener was:
- the general jigsaw fit of today's continents
- geological evidence of rocks of similar type, age and formation that occur in South America and Africa and also between the eastern part of North America and Europe
- fossil and *palaeoclimate* evidence that similar animals were once found in now widely separated areas along with similar climates.

Wegener's ideas were initially viewed with a great deal of scepticism as he was unable to suggest a mechanism for the movement. Later research, however, particularly in the field of *palaeomagnetism*, has led on to the now widely accepted theories of *plate tectonics*.

continental shelf is the relatively shallow belt of sea bottom bordering a continental mass, the outer edge of which (continental slope) sinks rapidly to the ocean floor. It generally extends to a depth of about 100 fathoms or 200 m. In recent years its definition and delimitation have assumed increasing importance in international law in connection with the ownership of minerals, particularly oil and gas, which lie beneath it.

contract farming is where large *agribusiness* companies, increasingly involved in the *food chain*, offer contracts to farmers to supply them with produce. It is estimated that 40–50% of British food is produced on contract, many producers dealing direct with supermarket companies. Contract farming is particularly found with products such as sugar beet, vegetables, fruit, and pig and poultry products. Large *transnational* firms are often involved, e.g. Unilever and Nestlé.

conurbation: a large and almost continuous *urban area* built up from separate centres which through urban growth and sprawl have joined.

convection occurs when the lower air heated from the ground expands and rises. As long as the rising air is warmer than its surrounds, it will continue to rise. This gives rise to *instability* within the atmosphere and cumulonimbus *cloud* formation. Convection can also occur within water bodies.

convection currents are movements within the *asthenosphere* which move the crustal *plates*.

convergence is where air flows into an area from different directions. When this happens at or near the surface, the air is forced to rise. In the tropics, the *inter-tropical convergence zone* is an area of surface convergence.

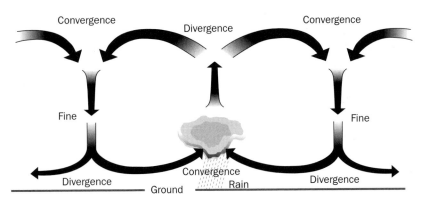

Convergence/divergence

cooperative agriculture: individual farmers who are not in a good position to meet market demand, particularly that of major *food retailers*, can improve their position by joining with other farmers in a producer cooperative. The advantages to the farmers are:
- they can negotiate favourable contract prices with retailers
- it provides a continuous supply of uniform quality produce
- processing, grading and packaging costs are shared
- farm *inputs* can be obtained in bulk.

coral reef is an accumulation of coral formed around the edges of landmasses such as the eastern coast of Australia and around the shores of islands. Corals are tiny animals (polyps) related to sea anemones, which exist in large colonies. As they die they leave behind a hard skeleton consisting of calcium carbonate which appears like rock. The accumulations of such deposits create coral reefs with living coral on the top. The growth of coral takes place in sea water with temperatures not less than 21°C, and only within 30–40 m of the sea surface. They need clear oxygenated water with plentiful supplies of microscopic zooplankton on which to feed. They cannot live in fresh or silt-laden water.

The formation of coral reefs and atolls is still very much a source of debate. Charles Darwin put forward a theory that still has some credence today. He imagined that a volcanic island would have been created in the middle of the ocean (a *hot spot*). Coral would have grown around the fringes of this island, but with time the island may have subsided or sea level may have risen, creating a rising reef of coral around it (so long as the rate of land subsidence/sea level rise was not greater than the rate of coral growth). The result of such movements would cause first a fringing reef, then a barrier reef and then possibly an atoll (see diagram overleaf).

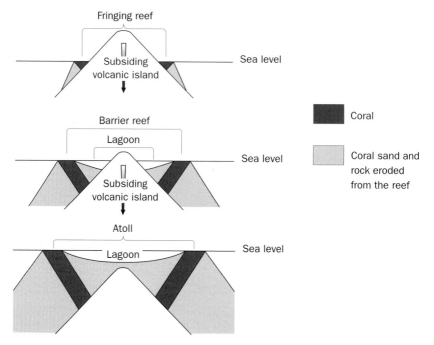

Formation of a coral reef and atoll

Coral is under threat in a number of areas of the world. As coastal **mangrove** swamps are cleared, their ability to trap **sediments** that cleanse coastal waters has reduced, and the off-shore polyps are being smothered by dirty water. Tourists are also creating problems through powered water craft increasing turbidity and toxicity, and by souvenir hunting. A natural predator, the Crown of Thorns jellyfish, is also on the increase, although the reasons for this are less clear.

core is the name given to the interior of the Earth. There are thought to be two parts to the core: the inner core, which has properties similar to that of a solid, and the outer core, which has properties similar to that of a liquid. It is thought that the core is very dense, and that it is probably composed of iron, with lesser amounts of elements such as nickel. The core also has very high temperatures (5500°C) and very high pressures.

Core also refers to an area of concentration of economic development, associated with the models of **Myrdal**, **Hirschman** and **Friedmann**.

coriolis force is the effect of the Earth's rotation on air flow. In the northern hemisphere, the coriolis force causes a deflection in the movement of air to the right, whereas in the southern hemisphere it is to the left. This helps to explain why the **prevailing winds** blowing from the tropical high pressure zone approach the British Isles from the southwest rather than the south.

corrasion: an alternative name for **abrasion**.

correlation is the degree of association between two sets of data. This involves the comparison of one set of variables with another. This can be done in two main ways:
- the calculation of a correlation coefficient which summarises the relationship between the two sets of variables. The **Spearman Rank correlation** coefficient technique is one of the most useful and is quite easy to use. It is based on the ranks of the individual values

of the two variables rather than the values themselves. All such calculations should also be tested for their statistical *significance*

● the construction of a *scatter graph* with the *independent variable* on the x-axis, and the *dependent variable* on the y-axis.

A positive correlation indicates that, as one variable increases, so does the other. A *negative correlation* indicates that, as the independent variable increases, the dependent variable decreases. It is also important to note that a high correlation between two sets of data does not necessarily prove that there is a causal relationship between the variables. It cannot be assumed that a change in variable A will cause a change in variable B. Further investigation may be required.

corrie: see *cirque*.

cost–benefit analysis means evaluating the financial and social costs of a course of action against the financial and social benefits. To obtain this, an estimate of the external costs (such as environmental damage) must be made along with the benefits (such as increases in employment). This is often difficult to calculate. A person in work can save the government a known average amount, but it is more difficult to quantify the damage to the environment. A cost–benefit analysis of a nuclear power plant, for example, would consider the possible damage to the countryside (both in construction and a possible nuclear accident) against the cost of the power produced. The analysis would also have to look at competitors in electricity generation and the cost–benefit, for example, of the closure of a coal-fired station and its effect on the coal *mining* industry.

cost of living: the amount of money spent by the average household over a period of time. This is difficult to evaluate precisely because people have differing spending priorities. Younger people with children, for example, spend a high proportion of their income on essentials such as food and housing, whereas older people may spend a lot more on things like entertainment and holidays. Changes in housing costs will affect the younger people's cost of living more than the older people. The government attempts to measure the cost of living using a 'basket of goods' which research has identified as the average household's expenditure. The best-known measure for the cost of living is the retail price index (RPI).

cottage industry is an industry in which employees work in their own homes, often using their own equipment.

Council of Ministers: the *European Union*'s main decision-making body. Its membership comprises one government minister from each of the member states. The minister chosen will depend on the issue under discussion. If the meeting is focusing on the European economy, the Chancellor of the Exchequer is likely to be the UK's representative.

counterurbanisation: the process of population *decentralisation* as people move from major urban areas such as *conurbations* to much smaller urban settlements and rural areas. This was first noted in the USA in the early 1970s, but has spread to most industrialised countries. The explanation for this movement is to be found in a combination of reasons but, above everything else, counterurbanisation seems to be a reaction against large cities. There is also the rise of the *new technology*, particularly in the area of communications, which seems to be reducing the need for people and activities to agglomerate in towns and cities.

craters are the holes at the top of a volcanic cone. They are usually circular depressions surrounded by low rims of ejected *debris* and they may become occupied by a lake. Craters can vary in size from hundreds of metres in width to several kilometres.

cratons (shield areas) are the oldest (over 550 million years old) and most stable interior parts of continents. They are located away from tectonic plate margins and form the nucleus around which new mountain belts have developed. They lack active *volcanoes* or *earthquake* activity and are severely eroded forming extensive areas of low plateau. Examples include the Canadian Shield, the Baltic Shield and the Brazilian Shield.

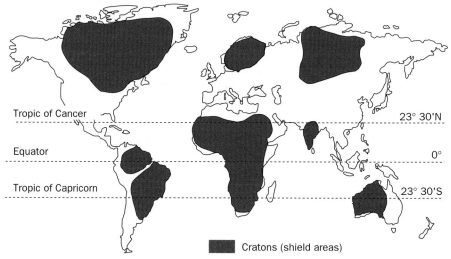

Distribution of cratons

creep is the slow downhill movement of soil and other material such as *scree* (*talus*). It operates on slopes steeper than 6°. A number of processes, each of which are capable of producing only slight movements, combine to cause it:

- raindrop impact – the hitting of raindrops on to a loose surface during a period of intense rainfall
- the expansion and contraction of the soil caused by either seasonal wetting and drying (especially clays) or seasonal and diurnal freezing and thawing of the soil surface. In both cases, expansion causes the surface of the slope to heave at right-angles to the original surface, but it then falls back under gravity in a more perpendicular manner
- the swaying of vegetation during windy spells
- the treading of animals both wild and domesticated.

The rates of creep vary in differing environments. In temperate humid environments such as the British Isles rates are commonly 1–2 mm a year, but in moist tropical areas the rate can be 3–6 mm a year. Common forms of evidence of creep are tilted telephone poles, terracettes and the build-up of soil on one side of a stone wall causing the wall to lean to one side.

cross-profile is the view of a feature or landform from one side to another. For example, a river valley in an upland area has a typical V-shaped cross-profile with steep sides and a narrow bottom. A glaciated valley has a U-shaped cross-profile, with steep sides and a wide flat bottom.

crust is the name given to the outer layer of the Earth, generally thought to be between 6 and 70 km thick. It is separated from the underlying *mantle* by the *Mohorovicic* discontinuity. There are two types of crust:

- continental crust (also known as the *sial*) – with a lighter average density (2.7 gm per cubic cm), consisting largely of granitic rocks
- oceanic crust (also known as the *sima*) – with a relatively higher density (3.0–3.3 gm per cubic cm), consisting of basaltic rocks.

The oceanic crust is continuous around the planet's surface, whereas the continental crust only occurs where there are continental land masses and it rests on top of the basaltic layer. The continental crust is also much older than the oceanic crust. In Greenland the continental crust is older than 3500 million years, whereas the oceanic crust is at no point older than 250 million years.

cryptobiotic soil crusts are communities of cyanobacteria, green algae, *lichens*, fungi and mosses. They form a fragile cover just a few centimetres thick in hot and semi-arid *ecosystems*.

cuesta: a landform with a steep scarp slope and a gentle dip slope, which forms on sedimentary rocks that are gently dipping. They are formed by differential *erosion* of alternating resistant and less resistant strata. *Chalk* and *limestone*, because of their permeability, are left as upstanding *escarpments* while weaker sands and *impermeable* clays are subjected to greater erosion and produce lower lying vales.

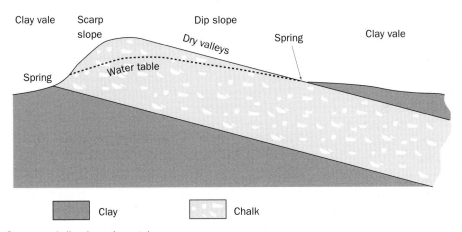

Scarp and dip slope (cuesta)

cultural diversity is the variety of human societies or cultures in a specific region. This is increasingly seen in many areas as the *United Nations* estimates that around 175 million people are living away from the country of their birth.

cultural imperialism is the practice of promoting, distinguishing or artificially injecting the culture or language of one culture on another.

cultural quarter: in many cities, it is possible to find 'quarters' associated with a particular ethnic group such as Chinese (Chinatown) and Italians.

cumulative causation is a process suggested by *Myrdal* in 1957 to explain regional differences in *economic growth*. Where a region has some form of initial advantage such as raw material resources, nearness to a port or is the source of some innovative industrial invention, this may set cumulative causation in motion. The establishment of a leading industry may trigger off other developments; other industries which supply that industry with **inputs** may be attracted to the region. These are termed ancillary industries. Subsidiary firms, which use outputs or products of the initial industry, may also be attracted. These industrial **linkages** can lead to the agglomeration of industry and other **agglomeration** economies. This activity may lead to other **multiplier effects** with in-migration of **labour** and increased employment in construction, transport and services to supply the needs of the larger population. Economic growth leads to a higher local tax yield and improved **infrastructure** which makes the region more attractive to other industries; in this way one development leads to other growth and the process is cumulative. The process can work in reverse following the closure of a major employer leading to a downward spiral in the economy of a region.

cumulative frequency: the data for each class in a distribution is converted into a percentage and each percentage figure is added successively. For example, if 5% of the population is under 5 years, 4% is 5–9, 6% is 10–14, then cumulatively 5% are under 5 years, 5 + 4 = 9% are under 10, and 5 + 4 + 6 = 15% are under 15. This would be continued until all groups are included when the cumulative total percentage would be 100%. This can be displayed on a cumulative frequency curve with cumulative frequency on the vertical y-axis (scale 0–100%) and range of values on the horizontal x-axis. This would allow identification of the **median** (50th percentile) and upper and lower **quartile** positions (75th and 25th percentile).

Worked example: grain size distribution in glacial/fluvioglacial deposits
The **till** contains a wide variety of grain sizes. **Esker** and **kame** deposits contain a small percentage of smaller grains and a large percentage of grains between 0.3 mm and 5 mm.

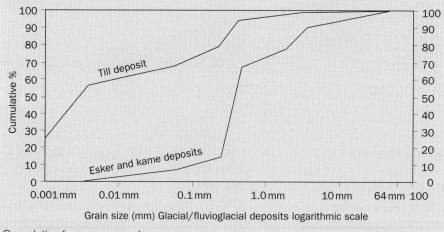

Cumulative frequency graph

cuspate forelands are triangular beach forms, varying in scale from small features, around 500 square metres, to the capes in the Carolinas on the eastern seaboard of the USA, which can cover 150–200 square kilometres. There is no doubt that they are formed in part from coastal **deposition**, but their development is not always easy to follow as they result from the complex interaction of a number of variables including **longshore drift**. The best and largest example in the British Isles is Dungeness on the south Kent coast.

cusps are small arcuate hollows or embayments which form on beaches. They vary in size up to 50 m across and they lie parallel to the water's edge with projecting low ridges pointing towards the sea. The sides of the cusp channel incoming **swash** into the centre of the embayment and this produces stronger **backwash** in the central area which drags material down the beach deepening the embayment. The low projecting ridges contain coarser sediment than the hollows.

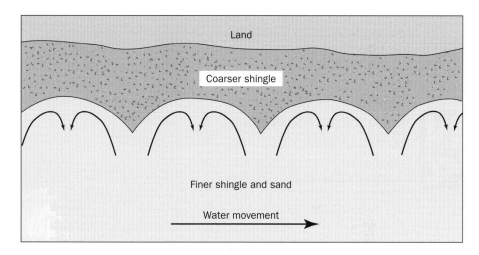

cycle of poverty: a vicious circle that exists within the economy of rural regions of **Third World** countries (represented diagramatically overleaf). Farmers who lack money cannot invest in better seeds, equipment and **fertiliser**. This inevitably results in poor crop yields giving low incomes, which in turn relates back to the lack of investment in the system. Loans and grants from government or international funds will help to break the cycle as long as the farmers are also educated in more efficient agricultural methods. Also helping farmers to set up **cooperatives**, where resources are pooled, will help individuals in breaking the cycle.

cyclogenesis is the sequence of events leading to the formation of **cyclones**, especially the middle-latitude **depressions**. Cyclogenesis is caused by the **convergence** of **air masses** at ground level due to **divergence** of air in the upper **troposphere**. The main areas of cyclogenesis are along the line of the **polar front** in the North Atlantic and North Pacific and in the Mediterranean Basin (mainly in winter).

cyclone: the name given to the low-pressure systems accompanied by severe weather that occur in the western Pacific Ocean and affect areas such as the Philippines, Taiwan, Hong Kong, China and Japan. (See also **hurricane**.)

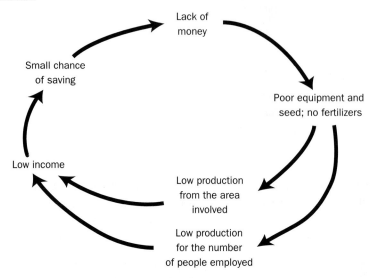

Cycle of poverty

Dalmatian coast: a drowned coastline where the main relief trends run parallel with the line of the coast. The ridges of upland produce elongated islands separated from the mainland by the flooded valley areas. The name originates from the Adriatic coast of Dalmatia. Alternative names for this type of coastline are concordant and Pacific-type coasts.

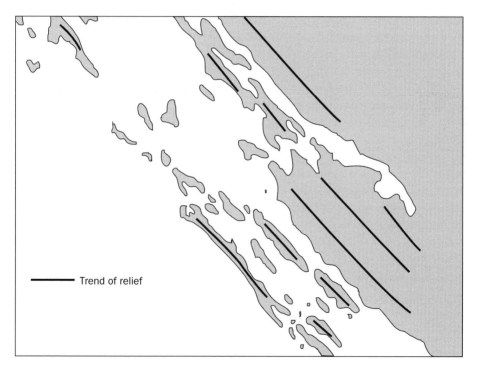

— Trend of relief

Dalmatian coast

DALR: see *dry adiabatic lapse rate*.

death rate: the number of deaths in a year per 1000 of the population. This is referred to as a crude death rate as it takes no account of the differences in the sex/age structure of populations.

debris is the name given to the fragments of loose material which are produced as a result of the breakdown of minerals and rocks. This material may accumulate on slopes as *scree* or *talus*, or may be transported by *mass movements*, *channel processes*, wind action or as *supraglacial*, *englacial* or *subglacial* material.

debt (Third World): there are two main types of international debt: commercial debt, which is owed mainly by government departments, public corporations and private companies, and official debt, which is the repayment and accumulated interest which is due on loans through aid programmes.

During the 1970s increasing amounts of foreign money poured into developing countries, much of this from commercial banking in industrial countries. Governments and their lending agencies were also providing money for aid projects set up with long repayment periods in anticipation of a good return as economies began to expand in developing countries. These recipient countries assumed they would be able to meet their repayments and this led to a rapid increase in debt levels. As debt rose, so did annual interest charges. The *recession* in industrial countries in the 1980s led to a fall in trade and commodity prices and an increase in interest rates. Many countries were using new loans and aid to service existing interest payments on previous debts.

More recently, such debt has been the result of the continued *globalisation* of economies and the *interdependence* of countries through trading. An increasing dependence on imports has exposed all nations to variations in economic stability. Developing countries find themselves without sufficient foreign exchange currency to purchase essential goods from abroad. Their governments approach commercial banks and private investors or allied countries for loans. If these are unwilling to lend to the country, the government approaches organisations such as the *IMF* or *World Bank*. These institutions are financed by affluent nations and influenced by commercial interests. Free market policies control lending and this exposes the borrowing country to interest rates that occur in the affluent countries. The debt crisis has continued to increase leading to calls from some organisations for the cancellation of debt on the grounds that the original loans have been repaid many times over. For the poorest countries (about 60), $550 billion has been paid in capital and interest over the last 30 years on $540 billion loans, but there is still an outstanding debt of $523 billion to pay.

decentralisation: the outward movement from established central areas, for example the movement of population and employment from the inner-city areas towards the suburbs or to smaller urban centres. This may be a natural response to the negative aspects of higher crime, noise, pollution and high land costs which are found in central locations or the process may be encouraged by agencies such as government trying to spread investment and development from the *core* area towards the *periphery*.

deciduous woodland: a woodland comprising of trees that are mainly broad-leaved such as oak, birch, ash and beech, which shed their leaves in the autumn, or 'fall'. Leaves are shed before the onset of lower winter temperatures. As soil temperature falls, the tree roots can only absorb small amounts of water and growth is retarded. Leaf loss reduces *transpiration* and the demand for water. In winter there is also a decrease in the water content of cells, together with a rise in the sugar concentration of the sap. This allows water to enter the cells but prevents its outward passage thus preventing excessive loss of moisture.

decision-making is the process in which alternative strategies towards achieving a goal or different solutions to a problem are evaluated and a decision is taken. A number of theoretical models have used the concept of Economic Man, an optimiser who aims to maximise returns and who has perfect knowledge enabling rational decisions to be taken. In Behavioural Geography it is recognised that in the real world, decision-making is likely to be influenced by a number of factors such as the decision-maker's perception, circumstances, access to relevant information and their ability to interpret information and analyse different alternatives. People who make decisions can be regarded as decision-makers.

decomposition is the breakdown of plant and animal remains which releases energy and nutrients into the soil. *Organisms* such as earthworms, mites and slugs help in the decay and incorporation of leaf litter in the soil and fungi and *bacteria* secrete enzymes which break down organic compounds in this detritus. The organisms which are decomposers are called detritivores.

deflation is the removal of small grained particles by wind action. It is usually a small-scale process which involves the transport of previously loosened sand grains leaving behind small hollows and blow outs.

deforestation is the deliberate clearance of forested land by cutting or burning. It can have a major impact on surface water flows, channel *hydrographs* and *soil erosion* as the *interception* layer of the *canopy* and the soil binding properties of the roots are removed.

Defra is the UK government department responsible for the environment, food and rural affairs. It oversees the health and welfare of animals, protection of the environment, farming and fisheries, plant and seed development, research projects and data analysis, wildlife, countryside and *sustainable development*.

deglaciation is the retreat of an ice margin, either the snout of a valley *glacier* or the edge of an *ice sheet*. It occurs when there is a negative mass balance and *ablation* at the ice margin exceeds the rate of supply of material accumulating in other parts of the glacier or ice sheet. Large amounts of *meltwater* are released during deglaciation which can produce *glacial lakes* and overflow channels, and if the deglaciation is extensive *eustatic* changes of sea level occur. As the weight of the ice is removed during melting, the land mass undergoes *isostatic* uplift and readjustment.

deglomeration: the opposite of *agglomeration*, as firms disperse from a site because of increased costs such as those of *labour*, transport and land. Firms may find expansion difficult in the area and there may also have been a decline in the local market.

deindustrialisation occurs when there is an absolute, or perhaps relative, decline in the importance of manufacturing in the industrial economy of a country and a fall in the contribution of manufacturing to *gross domestic product*. In spatial terms it is most severe in the areas which have traditional heavy industries, such as iron and steel, chemicals, shipbuilding and textiles, which in the UK and USA have suffered from strong overseas *competition* from areas where *new technology* and less unionised practices have been adopted.

delta: a landform produced by the *deposition* of sediment at the mouth of a river as it enters a sea or lake. Deposition occurs as the river's velocity and sediment-carrying *capacity* decrease as it enters the lake or sea, and *bedload* and suspended material are dumped. In addition, clay particles *flocculate* due to the chemical change in sea water and settle on the bed. Deltas form where the rate of deposition exceeds the rate of sediment removal. Unless the sediment *load* is very large, as in the case of the Mississippi Delta, the feature develops in coastal areas which have a small tidal range and limited wave action or offshore currents. Deltas are classified according to their shape as arcuate, bird's-foot or cuspate.

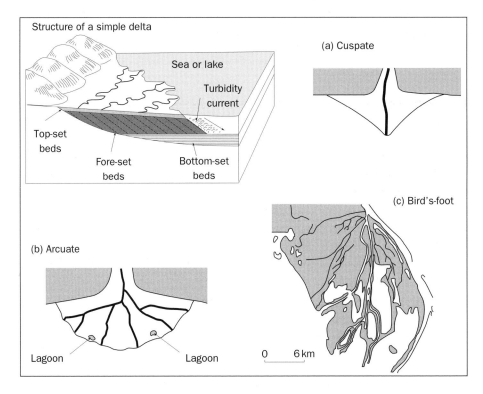

Structure of a simple delta

demographic transition model describes a sequence of changes over a period of time in the relationship between *birth rates* and *death rates* and the overall population change. The model suggests that all countries should eventually pass through similar stages, although the rate of change will vary from country to country. The model is a generalisation and particular countries can be expected to deviate from the pattern suggested; some may even miss out a stage. Some countries have moved into a fifth stage beyond that shown in the model where death rate is consistently higher than birth rate and there is a natural decrease in population. In 2008, these included Germany, Russia, Hungary, Bulgaria and Ukraine.

demography is the study of population numbers and change.

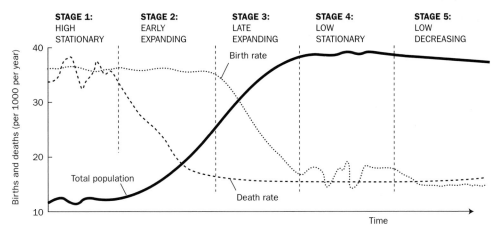

The demographic transition model (five stages)

dendrochronology: the method of using tree rings (by taking a core through the trunk) to find the age of the tree and to use as evidence of past climates. Each year's growth is shown by a single ring, but when the year is warmer and wetter the ring will be larger, showing more growth than if it was cooler and drier. Research has shown, though, that *precipitation* is perhaps more important to growth than temperature. The oldest trees found by this method are over 5000 years in age.

dependency ratio: the relationship between the economically active or working population and the non-economically active population. For ease of calculation, the active population is taken as all those in a population aged 15–65 and the non-active are those aged under 15 and over 65. In the 15–65 age group, for the purposes of calculating the ratio it does not matter if a person is employed or unemployed. Ratios in the *developed* world usually lie between 50 and 75 but in *developing* countries the ratio is high, sometimes over 100. The ratio is calculated as follows:

$$FORMULA: \quad \frac{\text{children } (0–14) + \text{elderly (over 65)}}{\text{all population } 15–65} \times 100$$

Worked example: dependency ratio of UK in 2001 (in thousands)

$$\frac{11,105 + 9,341}{38,343} \times 100 = 53.32$$

dependent variables are those which are directly affected by variations in another variable (the *independent variable*). For example, when considering the relationship between the amount of rainfall and altitude, rainfall varies with altitude, rather than altitude being a function of rainfall. Therefore, rainfall amount is the dependent variable – it 'depends' on altitude. The dependent variable should be placed on the vertical y-axis of *scatter graphs* when attempting to identify a *correlation* between two sets of data.

deposition: the laying down of solid material in the form of *sediment* such as mud and sand, on land, on the bed of a river or on the sea floor. Deposition is carried out by several agencies such as rivers, *glaciers*, wind and by marine processes and usually follows from the *erosion* of the land and the transportation of the resulting *debris*. Deposition is one of the main ways in which the land is built up.

depression: an area of low atmospheric *pressure* with a roughly circular pattern of isobars, that occurs in the mid-latitudes. Most depressions form along the zone of contact between cold, dense *polar* air and the warmer, lighter air from *tropical* latitudes. This zone of contact is known as the polar *front*. Each depression consists of a series of sections: a *warm front*, a *warm sector*, a *cold front* and an *occlusion*.

Theories of origin were first put forward in the 1920s by a group of Norwegian meteorologists, but with improved technology, particularly *satellite photographs*, refinements have been made to the original suggestions. (See also *cyclone* and *cyclogenesis*.)

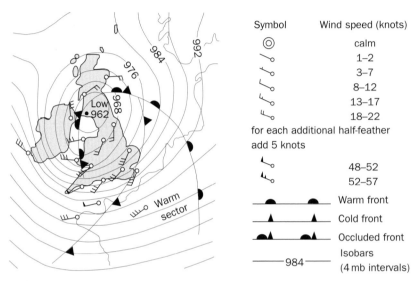

Depression

deprivation: in Britain, deprivation is defined by the Department for Environment, Food and Rural Affairs (Defra) as when 'an individual's *well-being* falls below a level generally regarded as a reasonable minimum for Britain today'. To measure the extent of deprivation, geographers have used a range of indicators from the areas of employment, housing, health and education. In developed countries, such as Britain, the problem is seen to be one which particularly affects the central residential parts of urban areas, the *inner cities*. Some of the indicators that have been used to show deprivation within these areas include: unemployment, mean family income, overcrowding, levels of housing benefit, numbers of children on free school meals, electricity disconnections, *life expectancy*, levels of basic amenities (bathroom, WC, hot water).

deprivation cycle: a *downward spiral* that is particularly seen within *inner cities*. Such cycles prevent the poor from raising their living standards and contribute to the *quality of life* experienced within such areas. The aim of many inner-city policies in the UK was to try to break such cycles and to bring about a rise in the living standards of the families concerned.

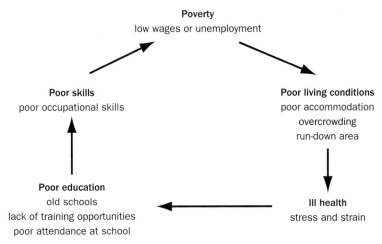

A typical deprivation cycle

deregulation is the removal of government rules, regulations and laws from the workings of business. This could be such moves as ending *monopoly* rights to supply services such as letter deliveries or bus services.

PROS:
- fewer regulators need be employed therefore cutting public spending and theoretically, taxation
- encourages more *competition*

CONS:
- rules were set up in the first place to give protection to workers; should they be removed?
- the effects of deregulation may be to provide only that which is profitable, and not what the public want, e.g. buses running only at peak periods but not at other times, such as late at night.

derelict land: previously used land which is now abandoned such as old dock areas, factory buildings, areas where inner-city terraces have been cleared, waste heaps from mineral working. These areas are all in need of some form of reclamation.

desalination is the separation of water from salts dissolved in it. Desalination can solve water shortages, improve water quality and be an economical supply alternative. Salts can be removed in two ways:
- *Evaporation* – water is heated to cause evaporation, leaving the salts as a residue. The vapour is then condensed to produce desalinated water. This process uses large amounts of energy and is only viable in areas that have cheap energy supplies. This method is used widely throughout the Middle East
- Reverse osmosis (a membrane process) – salt water is forced through a membrane that prevents the passage of salts. This is more energy efficient as the water does not need to be heated.

Thames Water is constructing a four-stage reverse osmosis plant at Beckton in east London. Planning permission was granted in July 2007 (approved by *Defra*) to build a £200 million plant to supply water for 1 million people.

desertification is the spread of desert-like conditions into neighbouring semi-arid regions of bush, grassland or woodland. Since the early 1970s there has been increased frequency of **drought** in the **Sahel** giving severe **famine** in Sudan, Mali and Niger. Although low levels of rainfall, and unreliability of rainfall, do influence soil and land degradation which leads to desertification, drought problems have also been aggravated by human activity. Increased population pressure in relation to the **carrying capacity** of the resource base has led to overuse of the land through **overgrazing** and overcultivation. The scarcity of fuel supplies has also led to the widespread removal of woodland for firewood. This results in less **interception** of rainfall, reduced **infiltration**, faster **runoff** and greater **soil erosion**. Vegetation removal also exposes the soil to greater **wind erosion**.

desire line is a line on a map that represents the movement of people from their homes to certain destinations such as schools, workplaces, shopping centres, holiday resorts and other such places. The thickness of the line is proportionate to the number of people who are moving. A suitable scale should be identified before the map is drawn so as to avoid excessive clustering at the destination point. A desire line is drawn in a straight line between the home or home area and the destination, with no account being taken of the actual route.

destructive plate margin: in **plate tectonics**, when two plates are converging or colliding this leads to the destruction of plate material. **Subduction** zones occur when an oceanic plate subducts beneath another oceanic or continental plate. When two oceanic plates converge subduction produces a deep **ocean trench** and an **island arc**, a line of volcanic islands. If the convergence involves an oceanic plate and a continental plate, the subduction produces a line of mountains along the edge of the overriding continental plate, as in the case of the Andes. These are mountains produced by the folding and faulting of **sediments** which accumulated on the **continental shelf** on the margin of the continental plate and the melting of the subducting oceanic lithosphere which produces rising volcanic **magma**. If subduction leads to the closing of an ocean so that two continental plates collide, complex fold and fault structures are produced, for example, the Himalayas. The two crustal plates are of similar density which is much lower than the underlying **asthenosphere**; as a result there is little subduction and the collision causes buckling of the plates. These margins are also called convergent margins.

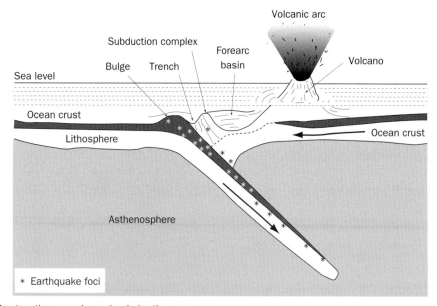

Destructive margin and subduction zone

destructive waves transport sediment back down the beach producing an overall loss of material. They break at a high frequency of 13–15 per minute and they are steep waves. As the wave crest breaks, water plunges downwards at a steep angle. This produces very little *swash* which reduces the transport of sediment up the beach. The *backwash* is stronger and is more effective in dragging material down the beach.

developed is a term that is usually applied to a country or region that has a high standard of living and an advanced economy based on the effective utilisation of resources. Such areas are usually more dependent on manufacturing and service industrial activity and have much lower levels of primary employment. They may be referred to as *economically more developed countries (EMDCs)*, which acknowledges that those countries that are not 'developed' in the sense of this definition may be developed in other non-economic ways, such as with regard to cultural, religious or social conditions.

developing is a term that is usually applied to a relatively poor country or region which has a low standard of living but which is beginning to achieve some economic and social development. Such areas tend to be more dependent on primary activities and some are significant producers of minerals (Brazil) or oil (Nigeria). Some countries have developed industrial growth and are emerging as *newly industrialised countries (NICs)*.

development is the use of resources and the application of available technology to bring about an increase in the standard of living within a country. Earlier views on 'development' emphasised economic expansion and increased output, but the current view is much broader involving social and cultural advancement as well as technological change and *economic growth*.

development continuum is the term used to describe the progressive changes in level of development. *Boserup* proposed a multi-stage progression of agricultural change based on five levels of increasing intensity reflecting increases in population pressure. *Rostow* suggested that economic change passed through five stages from 'traditional society' to 'age of high mass consumption'. It is perhaps more appropriate to view changes as a continuum, for example, a transition from low population density hunter-gatherer systems to high population density agriculture-based societies. Given the wide range of factors that can influence development and rate of change, it may be over-simplistic to describe these as fixed stages.

development gap: the difference between the level of development of the economically more developed countries of the *North* and that of the poorer countries of the *South*. To many, this gap is widening and is seen as a direct result of policies and trading systems pursued by the richer countries. There are many standardised economic and social indicators that can be used to measure the extent of this *inequality*, such as *per capita* income, *calorie intake* per capita, number of doctors per 1000 population, percentage of population with access to clean water and percentage of population educated to high-school level.

development models describe the varying ways in which nations or regions have changed their use of resources, discovered new uses for them and exploited their own human skills. They all concern the manner in which manufacturing and service industries have emerged from an agricultural base. There are a number of development models, of which the main ones are:
- the *Clarke–Fisher model*
- the *Rostow* model.

dew is the deposition of water droplets on to the surface of grass and the leaves of plants. It forms under clear, calm, anticyclonic conditions when there is rapid heat loss at night. The **dew point** is reached as the air cools, and the moisture in the air condenses on to these surfaces.

dew point is the temperature at which a body of air at a given atmospheric pressure becomes fully saturated. If an unsaturated body of air is cooled, a critical temperature will be reached where its **relative humidity** becomes 100% (i.e. saturated). This temperature is the dew point, and further cooling results in **condensation** of excess water vapour.

diastrophism: a general term for the action of those movements which produce relative or absolute changes in the position, level or attitude of the rocks forming the Earth's **crust**. It has been usual to classify diastrophism into two groups:

- **orogenic** movements, or mountain building, which involves intense **folding**, **faulting**, and thrusting, causing much deformation to the rock strata
- **epeirogenic** movements, which are less intense and may only involve uplift and at the most some gentle folding with associated faulting.

Some authorities include within diastrophism **isostatic** and **eustatic** movements and also that of molten rock (igneous movement).

diatoms: minute single-celled algae with hard shell-like skeletons composed of silica. These skeletons are distinctive for each diatom species and are preserved in the **sediments** on lake bottoms when the **organism** dies. They can, therefore, be used to assess the history of the lake's development when cores from the sediment are analysed.

diet refers to the average food intake of people, and is usually measured in megajoules **per capita** per day. The **United Nations** regards 10.8 megajoules per capita per day as being appropriate for healthy living. However, the quality of diets varies within the world, in terms of both the amount of food and the quality of food consumed. A balanced diet should contain:

- proteins – in meat, milk, eggs – to build and renew body tissues
- carbohydrates – in cereals, sugar, fats, potatoes – to provide energy
- minerals and vitamins – in dairy produce, fruit and vegetables – to prevent many diseases.

People who do not eat enough food are said to be undernourished, whereas those who do not eat the right diet suffer from **malnutrition**. Both of these conditions may lead to a variety of diet deficiency diseases.

diffluence happens when a small **glacier** breaks away from a main glacier and crosses over a drainage divide by means of a low col. In some cases, this can mean that the smaller glacier is forced upwards and over the divide.

diffusion is the process whereby an **innovation** is gradually adopted by more and more people through time and across space. At first there are the initial innovators, the leaders, followed by the first adopters of the idea. With time more people will adopt the innovation and its use will expand across an area.

disaster: an individual hazard event (or set of interrelated events) that has an effect on the human, built or economic environment resulting in injury/death together with property damage/economic loss. Disasters occur when the scale of the hazard event exceeds the capacity of the area to cope with the effects. Natural disasters result from physical hazards such as **earthquakes, tsunamis, hurricanes** and **volcanoes**. Manmade disasters are often caused by technical or social hazards such as fire, **radiation** contamination, accidents such as rail collisions and structural collapse. The scale of the disaster is influenced by the severity of the hazard event and the level of **vulnerability** of the locality.

discharge is the volume of water in a river passing a measuring point in a given time. It is calculated by multiplying the velocity of the river by the cross-sectional area of the river at the measuring point. It is measured in cubic metres per second, or cumecs.

discrete variable: in statistics, this refers to a variable which can only take a particular whole number value, such as the number of cars in an urban area or the number of people in a country.

disease of affluence: a number of diseases such as cancer, heart disease, strokes, type 2 diabetes and allergic reactions are considered to be caused by the lifestyle of Western or advanced societies. These are classed as non-communicable diseases (NCDs) in that they are not passed from one individual to another. It is argued that they reflect a number of factors: less exercise, high fat and sugar content in the **diet**, processed foods and chemical preservatives in food, and a longer life span. As health systems reduce the impact of **infectious diseases**, there is a greater proportion of deaths from 'affluent' causes. Recent medical research suggests that there are also high levels of incidence of these diseases in less well developed countries, but it may be that as these become more developed and more people move to work in urban areas, the population develops some of the traits of advanced urban societies.

diseconomies of scale: factors causing higher costs per unit when the scale of output is greater, i.e. causes inefficiency in large organisations. The major causes relate to the increased costs of internal communications and decision-making within large companies, which may not be as efficient as those of a smaller company.

dispersion measures: in statistics, these are the methods of calculating or displaying the distribution of a set of numbers around the **central tendency** (**mean**, **median**). The main statistical measures are the **range**, the **interquartile range** and the **standard deviation**. Visually, data can be represented on a dispersion diagram, **histogram**, frequency polygon and a **cumulative frequency** curve.

Worked example: annual rainfall totals at two widely separated stations over a period of 25 years

Station A	71	93	63	74	71	82	65	79	55	86
	84	70	79	98	75	89	71	79	68	75
	81	83	61	74	84					

$\bar{x} = 76.4$ cm $\quad \sigma\ 9.9$

Station B	59	90	86	36	53	123	90	43	111	68
	74	79	99	58	73	38	120	60	68	77
	89	80	78	87	48					

$\bar{x} = 75.5$ cm $\quad \sigma\ 22.9$

$\bar{x}$ = mean

σ = standard deviation

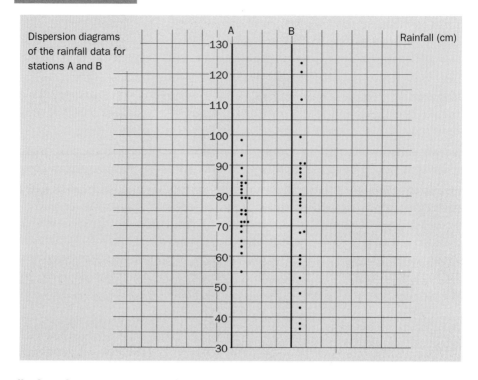

Dispersion diagrams of the rainfall data for stations A and B

displaced person: someone who is forced by circumstances beyond their control to leave their home but who remains within the borders of their home country. Individuals may be forced to leave their home region as a consequence of a natural *disaster* or war. This is a result of forced *migration*.

disposable income: the proportion of a person's income that is left after essentials have been paid for (e.g. housing, food, clothing, heating, taxes). People can then decide how they wish to spend this income. It is one of the factors that has been put forward to explain the increase in international tourism.

distance decay: the fall in the amount of movement or spatial interaction between two centres, the greater the distance that they are apart. This is seen in the *gravity model* where flows between places are inverse to the distance which separates them, e.g. the number of people travelling to a market centre to shop or to obtain services declines as the distance away from that centre increases.

distributary: a branch from a river that does not return to the main stream after leaving it. Distributaries are common in *deltas* such as those of the Nile and the Mississippi.

distribution: this can have two meanings:
- the pattern made by the occurrence of a feature within a given area, such as the distribution of doctors' surgeries within a city
- the entire process of getting products from the producer to the consumer.

distribution channels: the stages of ownership that take place as a product moves from the manufacturer or producer to the consumer. (See also *retailing* and *wholesaling*.)

Main channels of distribution

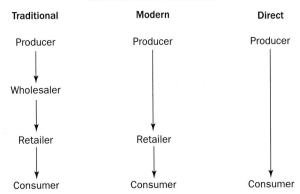

Main channels of distribution

divergence is where air flows out of an area. Divergence near the ground results from/in air sinking (**subsidence**) and the creation of high pressure systems such as the subtropical high pressure zone. (See also **convergence** and **atmospheric circulation models**.)

diversification is the spreading of business risks by reducing dependence on one product or market. It is an important objective for companies of all scales, from small ice-cream manufacturers wanting security from cool summers to large **transnationals** wanting to move into new growth markets.

PROS: • improves prospects for long-term company survival
  • can enable companies within saturated markets to find new growth opportunities
  • provides new outlets for a company's skills and resources

CONS: • with expansion, companies can find that **diseconomies of scale** could arise
  • companies may lack expertise within new markets which can lead to disappointing results
  • the company's main business could be weakened as resources are redirected towards new opportunities.

doldrums: the old name for the area of the **inter-tropical convergence zone** where there is a belt of generally light winds coupled with high temperatures and humidities. In the days before steam, sailing ships would often find themselves becalmed in these areas.

do nothing is an approach to **coastal management** in which no coastal defence systems are implemented except those needed as safety measures. Otherwise known as 'no active intervention', this approach lets natural processes run their course. The increasing cost of building and maintaining sea defences has led the government and organisations such as the **National Trust** to question the viability and effectiveness of physical structures. If the cost of protection is greater than the value of the land being protected, a 'do nothing' strategy is proposed. Expensive installations can be destroyed by marine processes within a few years of construction. In 2008, the National Trust adopted a 'do nothing' policy for Studland Beach in Dorset.

dormitory settlement refers to a rural village which has become increasingly urbanised in recent years, and is largely occupied by people who work in nearby towns and cities (commuters). It has new estates of detached or semi-detached houses, frequently occupied by families or retired people. Most people own one or more cars, and shop as well as work in the nearby towns and cities. Consequently, the provision of services such as shops and bus services is low. However, local schools are often enlarged.

doubling time is the number of years that it has taken for the world's population to double in size. It may also be used for the similar growth of population of a smaller area such as a continent or country. Doubling time has progressively become lower, and this is evidence for the rapid population growth which has taken place over the last 200 years. The world's population is expected to double over the next 40 years.

downward spiral: where decline occurs within a region it may be irreversible, as the region loses its more motivated people, less investment is attracted, which in turn means that more people will leave leading to even less investment and so on. Also known as a *vicious circle*.

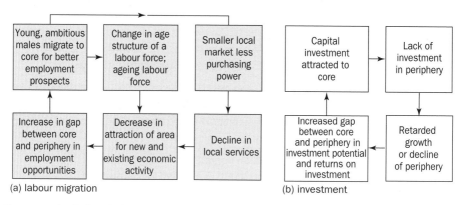

(a) labour migration   (b) investment

Downward spiral typical of the periphery in a developing country

draa: an extensive desert dune often having smaller *dunes* on the sand surface. They may be produced by the amalgamation of a series of smaller dunes. These dunes can extend over 5000 m in length and reach heights of 400 m.

drainage basin is the catchment area from which a river system obtains its supplies of water. *Precipitation* falls over the area bounded by the major *watershed*, and water makes its way either over the ground surface or by underground routes to the various streams which then converge to form the main river.

dredging is the removal of sand, silt and mud from the bottom of the sea or of a river in order to make it easier for navigation. In the case of river management, dredging is usually a short-term measure as it increases the cross-sectional area, and hence lowers the velocity, which increases the tendency for the river to deposit its *load*.

drift is the collective name for all the materials (boulders, gravels, sands and clays) deposited under glacial and *fluvioglacial* conditions. These deposits may be subdivided into:
- *till* – which is deposited directly by a *glacier*
- fluvioglacial *debris* – which is deposited by *meltwater* streams from a glacier.

drought is a lack of rainfall over a long period of time. Droughts occur in many parts of the world but they are especially likely in places where the climate is dry and variable, such as desert margins and **monsoon** areas. One of the worst droughts in recent years has been in the **Sahel** region of Africa. Rainfall has been very low for up to 20 years in some parts, and human activities such as **deforestation** have made the climate even drier. The drought in this area has caused one of the worst **famines** this century. Drought can now be detected from space. **Satellite photographs** can show the 'greenness' of an area and therefore the level of vegetation cover. A negative change in this vegetation cover can indicate a lower amount of rainfall, which can then alert national governments and relief agencies to possible future food shortages.

drumlins are smooth elongated mounds of unsorted boulders, sands and clays (*till*) deposited by the action of ice. They may be over 50 m in height, 1000 m in length and 500 m wide, with their long axis parallel to the direction of ice movement. The steep, blunted stoss end faces the direction of the ice movement, while the downstream lee end is much more streamlined. Their length is invariably greater than their width, and they usually occur in large numbers or 'swarms'. The formation of drumlins is still the subject of debate. One explanation is that they represent a period when a **glacier** became overloaded with **debris**, and was forced to shed some as it moved along. This could have taken place at a point where the ice movement slowed due to valley widening. Subsequent streamlining and moulding by later movements of cleaner ice would have then taken place. Some drumlins have a rock core which may have impeded the movement of the ice and resulted in deposition. Another theory is that they represent fluctuations in the extent of an **ice sheet** or glacier. Thus, they are the product of ice readvancing over its own deposits on more than one occasion.

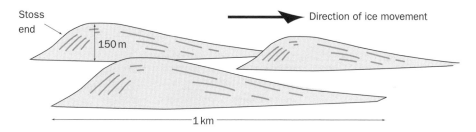

dry adiabatic lapse rate (DALR) refers to the rate at which the temperature of an unsaturated body of air falls as it is forced to rise. The temperature fall is caused by heat loss due to the expansion of the body of air. The DALR remains constant at 9.8°C per 1000 m or approximately 1°C per 100 m. (See also **adiabatic** and **saturated adiabatic lapse rate**.)

dry valleys have the typical V-shape and winding pattern of a river valley, but do not have a stream in their bottom. They are common features of **chalk** and **limestone** landscapes. However, during wet weather many dry valleys do have seasonal streams called bournes in them, particularly in their lower stretches. There are a number of possible explanations for their existence, of which the main ones are:

- during **interglacial** and **post-glacial** times the climate was wetter than present-day times, and this caused the **water table** to be higher. Consequently, streams could flow in the valleys that are now dry
- following the end of the last **ice age**, the ground would have been frozen and therefore **impermeable**. The valleys were therefore cut by **meltwater** streams flowing over this frozen ground

- chalk areas were once covered by impermeable rocks on which rivers flowed and excavated valleys. These were then superimposed on the present landscape.

dual economies: in the *developing* world this refers to a country which contains one or two economically developed areas (*core* and sub-cores) with regions that surround them being economically poorly developed (*periphery*). Good examples are Brazil and Nigeria.

dumping describes the selling of a good in another country at less than its cost price. A country may dump for a number of reasons:
- to earn foreign exchange – this was a common practice among countries formerly under communist control which were attempting to earn Western currency
- to get rid of excess production – European steelmakers in the late 1980s and early 1990s tried to dump steel in the USA
- to try and destroy foreign industry in order to create a market for the high price goods that are being produced.

dunes are ridges of sand found both in *arid* areas of the world, and along coastlines. Examples of sand dunes in arid areas are the *barchan* and *seif* dunes. Coastal dunes are formed by the wind on the landward side of a beach. They form best where there is a wide foreshore that dries out between the tides. Strong onshore winds dry out the beach and remove large quantities of sand with which they build dunes. At first embryo dunes develop which become colonised and stabilised by *marram grass*. These then join up to create larger foredunes which frequently lie parallel to the shoreline. With time, other grasses, fescues and heath plants then colonise the dunes producing a *psammosere*. Exposed areas of sand may become eroded by the wind to create blow outs. A cross-section across coastal dunes is shown below.

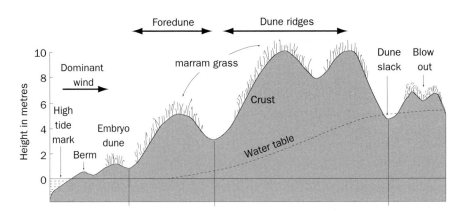

duricrust refers to the hard crust formed on the surface of the ground resulting from the accumulation and cementation of salts. It is a common feature of *arid* and semi-arid areas where high temperatures bring *groundwater* rich in dissolved salts to the surface by *capillary action*, the water then being evaporated. Duricrusts can be classified according to the nature of their chemical composition:
- calcretes – rich in calcium carbonate, they are the most widespread and can be several metres thick
- silcretes – rich in silica, and commonly occurring in southern Africa and Australia
- gypcretes – formed of calcium sulphate, most common in very arid areas.

dust bowl: a region of environmental degradation brought about by inefficient farming methods that covered an area across the High Plains of the USA including the Dakotas, Nebraska, Kansas, Oklahoma and the west of Texas. Farmers initially ploughed the treeless grasslands mainly for cereals, but a long continuation of this with an over-exploitation of the soil led to disaster in the early 1930s when after years of *drought*, high winds stripped off the topsoil. Many farmers were forced off the land and migrated westwards to California. The story of one Oklahoma family is vividly told in John Steinbeck's novel *The Grapes of Wrath*. *Soil conservation* methods have now been successfully applied to the region. The term can be used to cover other parts of the world which have been similarly affected.

dykes are vertical intrusions of *magma* that cut across the *bedding planes* of *sedimentary rocks*. The magma cools slowly although those parts which come into contact with the surrounding rock cool more rapidly to produce a chilled margin. Most dykes are more resistant to *erosion* than the surrounding rocks and therefore tend to stand out as ridges across an area. Dykes may also refer to ditches or natural watercourses in fenland areas, and also to embankments used to prevent *flooding*, as in the Netherlands.

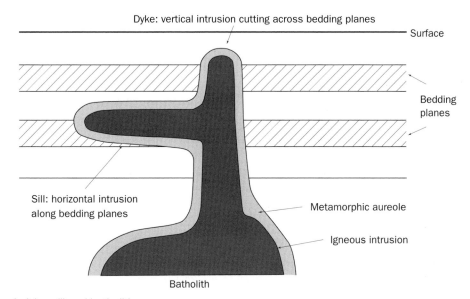

A dyke, sill and batholith

dynamic equilibrium refers to the balanced state of a system when opposing forces, or *inputs* and outputs, are equal. If one element in the system changes because of some outside influence, then this upsets the equilibrium and affects other components. This is known as *feedback*, which may be *negative feedback* or *positive feedback*.

earthflow: a form of *mass movement* in which *debris* moves downslope. When weathered material becomes saturated, internal friction between the particles is reduced and stress can cause the debris to move under gravity. This can occur on slopes as gentle as 5° once mobile, but usually needs a slope of about 10° to initiate movement. Flows are generally faster than *creep*.

earthquake: a series of vibrations and shock waves initiated by volcanic eruptions or movements along the boundaries of oceanic and continental plates and can occur along *constructive*, *destructive* and *conservative plate margins*. The point of origin of an earthquake in the *crust* is called the focus, represented on the surface as the *epicentre*. Shallow focus earthquakes are associated with both constructive and destructive margins but intermediate and deep foci earthquakes only occur where there is *subduction* beneath a destructive margin. There are three types of shock wave: primary (P) waves, which travel fastest and are compressional, vibrating in the direction that they are travelling; secondary (S) waves, which travel at half the speed of P waves and shear rock by vibrating at right angles to the direction of travel; and surface (L) waves which are the slowest, travel near to the ground surface and may have a rolling motion that produces vertical ground movement. *Seismic* recording of the time-lag between the arrival of P and S waves enables the identification of an earthquake epicentre. The energy released by an earthquake, the *magnitude*, is measured on the 10-point *Richter scale*. The severity of ground movement, the intensity of the earthquake, is measured on the 12-point *Mercalli scale*.

ecological footprint: a measurement of the area of land or water required to provide a person (or society) with the energy, food and other resources he/she consumes and render the waste produced harmless. Using this assessment, it is possible to estimate how much of the Earth (or how many planet Earths) it would take to support humanity if everybody lived a given lifestyle. Although the term ecological footprint is widely used, methods of measurement vary.

ecology is the study of the interrelationships between plants and animals and their environment and is concerned with the processes at work in the *ecosystem*. Ecologies can also be used to describe areas of distinctive *habitats* within an ecosystem. For example in urban areas, ecologies refer to exotic species of plants, animals and insects that may exist alongside a railway line, or a road, or in parks and gardens. As with the above, it is the slight variation in process that makes the ecology distinctive.

economically less developed country (ELDC) is a term used to cover those countries whose *economic growth* and therefore development is not as great as the *economically more developed countries* of the world. Originally, these countries were known as *developing*, but a general distaste of this term led to its replacement by the *Third World* and, after the *Brandt*

Report (1980), the **South**. The nations grouped together under this blanket term have shown a widening spread of wealth and living standards. There are the **LDCs**, the poorest countries of the world (mainly to be found in sub-Saharan Africa); the oil-rich states of north Africa and the Middle East; and the **newly industrialised countries (NICs)** particularly to be found in southeast Asia and Latin America. In the twenty-first century, the **BRICs** (Brazil, Russia, India and China) have emerged. All these definitions are based on economic growth and, therefore, wealth, but there are an increasing numbers of geographers who are looking for other ways of making definitions of such areas based on criteria other than economic.

To group so many nations within one category is also difficult as they vary so much on the following features:

- size of area and population
- historic and colonial background
- different physical and human resources
- employment structures
- dependence on external economic and political forces
- political structures
- relative importance of **public sectors** and **private sectors**.

economically more developed countries (EMDCs) are the richest countries in the world with economies that have grown, usually through industrialisation. They were formerly known as the 'developed countries' and later as the **First World/Second World**, and, after the **Brandt Report**, the **North**. Within this group there is a large difference in economic development and, therefore, wealth, between the industrialised countries of North America, Western Europe and Japan (First World) and those countries that were formerly under communist control with **centrally planned economies** – the countries of eastern Europe and the former **USSR** (Second World).

economic growth describes the way real incomes per head increase over time, a nation's growth being measured in terms of **gross national product**. The reasons for growth are complex, although it can be explained in terms of increasingly efficient uses of the **factors of production** to provide more and more goods and services.

economic migrant is a person who moves from one country to another for the sole purpose of raising his/her standard of living, mainly through gaining employment. There are many examples of such movement around the world and through time. The opening of the **European Union**'s borders has allowed many Eastern Europeans to come to the UK to gain work in a variety of industries. Similarly, the economic expansion of the Gulf States has led to many migrants moving to that region from countries in southeast Asia such as Bangladesh.

economies of scale are the factors that cause average costs to be lower in large-scale operations than in small-scale ones, therefore doubling the output results in a less than double increase in costs. In such a case, the cost of producing each unit falls because **inputs** can be utilised more efficiently. Economies of scale fall into two groups, internal and external. Internal economies are:

- specialisation – with a large workforce it is possible to divide up the work processes and recruit people whose skills exactly match the job requirements. The workforce is generally then more effective
- fixed costs of equipment – can be spread over more units of production

- purchasing economies with the benefits of bulk buying – obtaining *raw materials* and components at lower unit costs
- financial economies stem from the lower cost of capital charged to large firms by the providers of finance.

External economies of scale are the advantages of scale that benefit the whole industry, not just individual firms. If an industry is concentrated in one geographical area:
- a pool of *labour* will be attracted to that area and trained to gain the specialised skills useful to the whole industry
- a large grouping of firms will attract a large network of suppliers whose own scale of operations should also yield lower costs.

ecosystem: a community of plants and animals within a physical environment or *habitat*. An ecosystem can exist at any scale as a natural unit from a single tree or a freshwater pond to a climatic *climax vegetation* or the entire Earth. The various components of the ecosystem such as soils, vegetation, climate and animals are interrelated; energy flows occur in the system and materials such as nutrients are recycled and circulated between the soil, plants and leaf litter and as part of the *food chain*.

ecotone: a zone of transition from one major vegetation community to another, the width of the zone depending on the rate at which the environmental conditions change. For example, an ecotone exists between the *deciduous woodland* areas and the *temperate grasslands*. There is usually a greater number of species in the ecotone than in the neighbouring communities because the most tolerant species of each community can withstand the changes in environmental conditions and in addition, species which are particularly favoured by the conditions in the ecotone also develop.

ecotourism is an environmentally friendly alternative form of tourism. The tourist industry is rapidly expanding at a rate faster than the overall world economy and around 500 million people travel each year for tourism purposes. An increasing number are seeking more exotic destinations with different cultures; international tourist visits to developing countries have more than doubled in the last 20 years. Tourism is being increasingly blamed for massive environmental, cultural and social damage: polluted beaches, degraded *coral reefs*, displacement of local population, low financial return to host country, abandonment of traditional economic activity. Although other terms such as 'responsible tourism', 'sustainable tourism' and 'low impact tourism' are also used, ecotourism is commonly used, referring to a niche market for environmentally aware tourists and this has become the fastest expanding sector in the industry. While these approaches may be more beneficial than mass tourism, they do have problems; low impact schemes are likely to become more damaging as they become more successful and it is difficult for such small operations to compete with the cheaper deals of the large tour operators. Ecotourism is an option for the wealthy, therefore attempts to limit the effects of tourism must be aimed at improving mass tourism rather than developing alternatives for minority groups.

ecumene: a term used to denote the most densely populated part of the Earth's surface. It has been estimated that about 60% of the surface is ecumene, the rest being described as nonecumene, the sparsely or intermittently populated or uninhabited area. Delimitation of such areas is difficult because high density areas merge into lower density and within the ecumene there are open areas of parkland, forests and agricultural land.

edaphic: a term which relates to the soil properties and factors which affect plant growth and distribution. These include *soil texture*, *soil structure*, organic content, acidity, soil moisture, nutrients and *organisms*. Although major vegetation regions largely reflect climatic conditions, within these regions local variations are strongly influenced by edaphic factors.

effective precipitation is that part of the total *precipitation* which is of use to plants. The effectiveness of precipitation depends on its pattern and the temperatures and rates of *evapotranspiration* present at the time when the precipitation is falling. The type of precipitation is also important as long steady periods of rain allow the moisture to *infiltrate* the soil, whereas short, heavy downpours can lead to a lot of surface *runoff* and are therefore less effective for plants.

efficiency: the shape of a stream or river channel in cross-section affects its efficiency as a conveyor of water and sediment. The most efficient channel shape is the one that minimises the ratio between the cross-sectional area of the channel and the length of the *wetted perimeter*. This ratio is known as the *hydraulic radius* and is widely used as a measure of *channel efficiency*.

ELDC: see *economically less developed country*.

el Niño: a warm ocean current that occasionally replaces the normal cold Peru current off the Pacific coast of South America. The *Hadley cell* transfers heat energy between the equator and the sub tropical high pressure zone and there are east to west circulations called *Walker cells*. The normal Walker circulation produces easterly *trade winds* between South America and Indonesia–Australia, taking warm water in a westerly direction and allowing cold water to move north along the Pacific coast of Peru. Every few years this Walker circulation breaks down, the easterly trade winds decline, and warm water moves eastwards across the Pacific to shut off the cold Peru current and produce a warm ocean current (el Niño, 'the Christ child') off the South American coast. This change to the normal pattern is referred to as an 'el Niño–southern oscillation event' (ENSO). The high pressure which forms over the cold ocean is now replaced by low pressure over the warmer ocean, which can be 6–10° above normal. This produces heavy rainfall on the usually *arid* coastline and brings natural hazards such as *flooding*. The change in circulation and the effect on the route of the *jet stream* can produce *drought* in other areas and even affect climatic events in the northern hemisphere, bringing increased snow and floods to North America. Conversely, the number of *hurricane* events in the Caribbean tends to be reduced in el Niño years.

Under the effects of the normal Walker circulation the easterly trade winds can cause warm water to build up in the area of the Coral Sea (between Australia and Indonesia). Sea levels rise and cause ecological and economic damage to low lying islands within the region. Recent research in Australia indicates that even a well developed Walker circulation can cause extreme problems. With unusually strong easterlies, the cool water off South America can be moved northwards across the central Pacific, lowering ocean temperatures and changing atmospheric pressure patterns to produce higher than average rainfall and *cyclone* intensities in a zone extending from Australia through Indonesia to Bangladesh. This is termed a la Niña event.

These changes to the normal circulation are not annual events; they occur at intervals of 2 to 7 years, although el Niño seems to have operated more regularly in recent years, with several in the early 1990s and a particularly catastrophic event in 1997–98.

ELR: see *environmental lapse rate*.

eluviation is a general term for the washing out or removal of any material from a soil horizon. It usually involves the downward movement of clay and other fine particles in suspension.

embargo: an order prohibiting trade with a particular country, perhaps imposed because the country has broken international law or conventions.

EMDC: see *economically more developed country*.

emergent coast: one that results from a fall in sea level and/or uplift of the land. Features of such a coastline include *raised beaches*, relict cliffs and exposed estuarine mudflats.

emigration is the movement of people away from an area or country to live in another. (See also *migration*.)

employment structure refers to the relative proportions of employment in an area in each of the main sectors of employment. The main sectors are:
- the *primary sector*
- the *secondary sector*
- the *tertiary sector*
- the *quaternary sector*.

There are broad differences in the employment structure of *economically less developed countries (ELDCs)* as compared to that of *economically more developed countries (EMDCs)*. ELDCs have a much greater proportion of people employed in the primary sector, and few in the secondary and tertiary sectors. EMDCs have very low proportions in the primary sector, more employed in the secondary sector, and high proportions in the tertiary sector. The quaternary sector has become increasingly important in EMDCs in recent years.

endemic refers to a disease that is habitually prevalent in a certain area or country and due to permanent local causes.

endemism is the ecological state of being unique to a place. *Endemic* species are not naturally found elsewhere. The place must be a discrete geographical unit such as an island, *habitat*, nation or other defined area or zone. For example, the orange-breasted sunbird is endemic to Fynbos, meaning it is exclusively found in the Fynbos vegetation type of southwestern South Africa.

endogenetic factors and processes are those that are from within the Earth; for example in landform development, slope formation is affected by factors such as rock type, chemical composition, structure, degree of permeability. These endogenetic factors interact with *exogenetic* processes to produce particular landforms.

endoreic rivers are those that flow to an inland lake rather than to the sea. One example is the River Jordan, which flows into the Dead Sea.

energy budget (the Earth): this refers to the balance between the incoming solar *radiation* (*insolation*) and the outgoing radiation from the planet. Since the Earth's geological record shows that it is neither cooling nor warming to any appreciable extent, the two types of radiation given above must be relatively equal. However, there are variations between the Earth's surface and the atmosphere. The Earth's surface has a net gain of energy, receiving incoming radiation from the sun and downwards from the atmosphere. On the other hand, the atmosphere has a net deficit of energy. To compensate for this difference, heat is transferred from the Earth's surface to the atmosphere by radiation, conduction and by the release of *latent heat*. There are also variations between the different latitudes on the surface. Low latitudes receive a net surplus of energy, whereas high latitudes (polewards of 40°N and S) have a net deficit. Since the poles are not becoming colder, and the equatorial regions are not getting hotter, heat is transferred between the two by means of air movements (winds) and water movements (*ocean currents*).

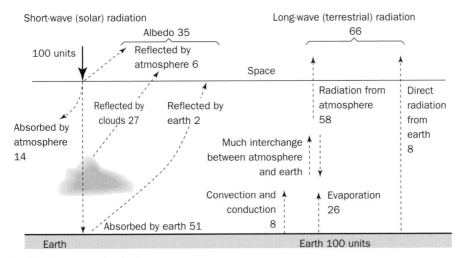

The Earth's energy budget

energy conservation refers to the variety of methods by which the use of all types of energy, but particularly electricity and motor vehicle fuels, is limited or reduced. It may be achieved by:
- greater efficiency – for example, more economic fuel consumption in motor vehicles, cavity wall and roof insulation, low-energy light bulbs in the home
- the use of alternative sources of energy which are less wasteful in their methods.

energy poverty is a term for a lack of access to electricity, heat or other forms of power. Often referring to the situation of peoples in the *developing* world, the term also implies any *quality of life* issues relating to this lack of access.

energy security: access to cheap energy has become essential to the functioning of modern economies. However, the uneven distribution of energy supplies among countries and the critical need for energy has led to significant vulnerabilities. Threats to an individual person's and national energy security include the political instability of several energy-producing countries, the manipulation of energy supplies, the *competition* over energy sources, attacks on supply *infrastructure*, as well as accidents and natural *disasters*. It is also the limited supplies of the most common forms of primary energy, i.e. oil and natural gas that changes perceptions on this topic. Although plenty of coal (up to 150 years' worth) is readily available, coal is not the *fossil fuel* of choice for many *economically more developed countries (EMDCs)* because of its highly polluting nature. The potential need to change our primary energy sources in the foreseeable future is the crux of the energy security question, leading to higher prices, more limited access to sources of energy, competition and political troubles, which in turn make the threat even larger.

energy sources provide heat and motive power for all form of human activities. There are a wide range of energy sources, and they may be classified as being:

- *non-renewable* – coal, oil, natural gas, *fuelwood* and *nuclear energy*
- *renewable – hydroelectric power*, *wind power*, *solar energy*, *tidal energy*, *geothermal energy*, *biogas*, *biofuel* and wave power.

englacial: within a *glacier*.

English Heritage is the organisation responsible for the *conservation* and *preservation* of over 400 historic buildings and other monuments within England. It inherited the responsibilities of the former Ministry of Works, but is now a private body that both encourages membership subscriptions and charges entry fees. Its main aim is to save Grade I and II listed buildings that are at risk from neglect and decay. Buildings for which it has responsibility include Dover Castle, Osborne House, Mount Grace Priory and Brodsworth Hall. Monuments include the bridge at Ironbridge, Stonehenge, Hadrian's Wall and the Wellington Arch at Hyde Park, London.

enterprise culture: a social climate that applauds the profit motive in general and starting a small business in particular. The term was widely used in the 1980s and formed a key element in *Thatcherism*.

entrainment: the process by which individual or groups of particles are removed from the bed of a river channel for transport as either *bedload* or *suspended load*.

entrepreneur: an individual with a flair for business opportunities and risk trading. The term is often used to describe a person with the entrepreneurial spirit to set up a new business.

environment is the natural or physical surroundings where people, plants and animals live. It is very complex and many factors are involved in its evolution and character, all of which interact with each other. It is also important to recognise that the environment is in a state of constant flux because of the range of forces acting on it.

environmental impact assessment (EIA) is a process whereby the significant effects of a development project are identified, predicted and evaluated. Its purpose is also to ensure that the findings are taken into account during the planning, design and authorisation of the scheme. EIA cannot prevent environmental damage, but it informs those who make decisions of the environmentally damaging consequences, so that they may be avoided or reduced. EIA is now a requirement within the *European Union* for schemes such as major motorways, airports, power stations and oil refineries. It may also be necessary for smaller schemes in protected or politically sensitive areas. An important aspect of any EIA involves the canvassing of public opinion, the main aim of which is to identify the range of potential problems, and then to identify priorities.

environmentalism refers to the growing political concern over a range of issues affecting the world's environment. These include *global warming*, *ozone depletion*, *acid rain*, *rainforest* removal and others. 'Green' parties and environmental pressure groups are becoming more influential, both at a national scale (e.g. the Green Party in Germany) and at an international scale (e.g. *Greenpeace*).

environmental lapse rate (ELR) is the change in temperature with height above a particular place at a given time. It is that which would be measured by a thermometer moving vertically through a still atmosphere. The average value for the ELR is a decrease of 6.5°C per 1000 m. However this value varies both with height, being lower near ground level and, with time, being higher in the summer season.

environmentally sensitive areas (ESAs) are rural areas of national environmental significance that are worthy of being conserved. The ESAs cover a wide range of farming and landscape types, from the granite moorlands of West Cornwall, and the hills of the Welsh borders, to the marshes of the Norfolk Broads and the Pennine Dales of Yorkshire. In many of these areas the main aim is to preserve the farming landscape and other features such as walls, hedges and barns. The participation of farmers is voluntary, and all farmers in such areas are invited to participate in the scheme for a period of 5 years. They agree to adopt 'environmentally friendly' farming practices for all or part of their farm and they receive payment for doing so. The details of the type of farming required, and the payments for them, vary for each ESA. In the Breckland, for example, farmers receive payment if they leave a strip around their *arable* fields permanently uncropped for the benefit of wildlife. (See also *environmental stewardship*.)

environmental stewardship: in the UK, it is recognised that farmers have an important part to play in protecting and managing the *environment*. *Defra* therefore set up an Entry Level Stewardship Scheme (ELS), which aims to encourage a large number of farmers to deliver simple yet effective management. Farmers receive £30 per hectare per year (£8 if over 15 hectares are involved) and there are over 50 options to choose from. These include hedgerow management, stone wall maintenance, creating buffer strips, ditch and pond management, infield tree protection, management of rush pastures, stubble control, management of archaeological sites and bird and flower *conservation*. There is a similar scheme for *organic farmers* as Defra recognises that they are able to deliver greater environmental benefits.

epeirogenic movement involves relatively gentle raising or lowering of parts of the Earth's *crust*. These usually operate on a large scale and may be referred to as continent building. Such movements do not produce strong folding of rocks, but some tilting may occur.

ephemeral rivers flow intermittently, or seasonally, after rainstorms and are features of desert landscapes. Despite being short-lived they can generate very high *discharges*. This is due to the torrential nature of the rain which falls in such areas, which cannot easily infiltrate into the ground. This is because the presence of *duricrusts* in deserts creates an *impermeable* surface which inhibits *infiltration*. The lack of vegetation in such areas also means that the rain reaches the ground without being intercepted. Some desert lakes (*playas*) are also ephemeral.

epicentre is the point on the Earth's surface directly above the origin or focus of an *earthquake*.

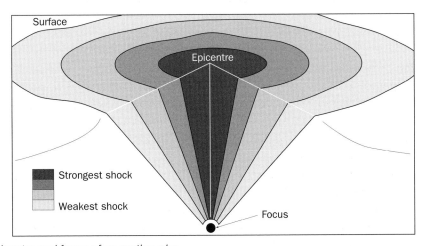

Epicentre and focus of an earthquake

epidemic: a disease that affects many people simultaneously in a community or area.

epidemiological transition is a model leading on from the simpler *demographic transition model* which shows how countries go through different stages of diseases and causes of death as they develop. The four stages are:

- age of *epidemic* diseases and *famine* – mainly *infectious diseases*, smallpox, *malaria*, *cholera*, typhoid
- receding *pandemics* due to vaccines, immunisation, improved living conditions
- degenerative and human-induced diseases due to unhealthy *diets*, stress, smoking, general unfitness (lack of exercise) – cancer, heart disease, respiratory diseases
- degenerative diseases – cancers, Alzheimer's.

The value of such a model is:

- it sets up a framework within which to set healthcare strategies
- it can help planners to project future healthcare needs both nationally and within international health agencies and to decide where best to allocate funds
- it can help manufacturers of medicine and health equipment to project future demand
- it draws attention to the problem of the world's ageing population.

epiphytes are plants which develop on the branches and trunks of trees within the tropical *rainforest*. They are mainly small shrubs and herbs that have escaped the shade conditions on the forest floor by germinating in the tree layer and living on host plants. Many of these epiphytes have a mass of tangled roots that catch falling leaves and *debris*, which as it decays supplies nutrients and acts as a sponge to hold and provide a water supply.

Equatorial climate occurs in lowland areas within 5° or 10° latitude of the equator and is sometimes referred to as the tropical rainy climate. The annual temperature range is small, normally as low as 3°C, reflecting the fact that there is little seasonal variation in the angle of inclination of the sun. Mean monthly temperatures are of the order of 26°–28°C. Diurnal range is also small, about 10°–12°C; night temperatures rarely fall below 20°C and day temperatures may rise to 30°–32°C. With its high *humidity* and monotonous temperatures this climate can be oppressive. Rainfall is high, over 2000 mm per year, and the daily pattern is rather repetitive and predictable; the morning tends to be hazy and as this clears *convection* currents develop and produce cumulus *clouds* which intensify during the afternoon heat. Heavy rain occurs in late afternoon or early evening. Convectional uplift is related to the position of the *inter-tropical convergence zone* and therefore a double maxima of rainfall can occur when the sun is overhead at the spring and autumn equinox.

erosion involves the removal of weathered material by the action of gravity, water, wind or ice. These agents of erosion transport material that has already been attacked by *weathering*, but they also use the fragments of *debris* to wear away, or erode, other material through the processes of *abrasion*.

erratic is the name given to a rock which has been transported by a *glacier* or an *ice sheet* and deposited in an area of different geology to that of its source. Therefore it can be used as evidence to indicate the direction of ice movement. Rocks from Ailsa Craig, in the Firth of Clyde, have been found in southwest Lancashire, 240 km from their source. The Norber erratics, near Austwick in the Yorkshire Dales, are large Silurian boulders which have been carried about 2 km from their original outcrop before being dumped on the surface of the Great Scar Limestone.

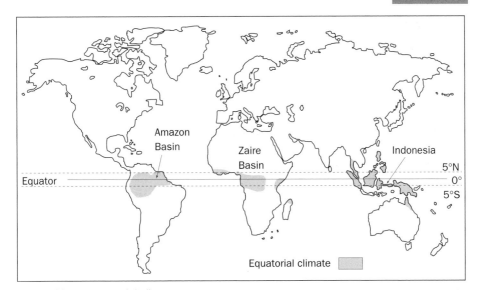

Areas with an equatorial climate

escarpment: the steep slope that forms a more or less continuous line along the margins of an upland area such as a plateau or a *cuesta*. The term is often abbreviated to *scarp*. These scarp faces or scarp slopes are the result of differential *erosion* on horizontal or gently dipping rock strata.

esker: a sinuous ridge of material deposited by *meltwater* flowing in *subglacial* channels or through tunnels within the ice. These are termed ice contact *fluvioglacial* features. They are formed at right angles to the ice front and consist of silt, sand and gravel deposits which often display some degree of sorting, both from the centre of the ridge outwards and in the downstream direction, although this may be disrupted by the *slumping* which occurs as the ice which forms the channel side melts during deglaciation.

estancia: a large farm or estate. It is a term that is used in the Spanish-speaking areas of South America, particularly in relation to the extensive commercial cattle-rearing stations.

estuary: the area of a lower river, or mouth, which is affected by tidal change. Estuaries are produced by the *post-glacial* rise of sea level and drowning of former lower valley areas.

ethnic cleansing is a euphemism for the actions of one ethnic or religious group forcing people of another such group to flee their homes, either through eviction or as a consequence of fear and intimidation. It became a feature of the civil wars in the former Yugoslavia where ethnic cleansing has ensued in Bosnia-Herzegovina and Kosovo. The civil strife in Rwanda in the mid-1990s also illustrated ethnic cleansing when the actions of Hutu militias led to the displacement and death of thousands of Tutsi people.

ethnicity refers to the grouping of people according to their ethnic origins or characteristics. In narrow terms it refers to the racial make-up of a population – whether people are Caucasoid, Mongoloid, Negroid or Polynesian. More recently the term has broadened in meaning to refer to groups of people classed according to one or more of common racial, national, tribal, religious, linguistic or cultural origins or backgrounds.

ethnic segregation is the clustering together of people with similar ethnic or cultural characteristics into separate residential areas in a town or city. There are numerous cases of this in many UK towns and cities. For instance, in London, there are concentrations of:
- Jewish people in the area around Golders Green
- Indian people in Southall
- African-Caribbean people in Brixton.

Once a concentration of one ethnic type has become marked, it attracts others of a similar *ethnicity*. Sadly, such segregation has been beset with problems resulting partly from prejudice, ignorance and intolerance on behalf of the host population. In some inner-city areas, deprivation due to low rates of employment has also helped to create many social problems and tensions, which have surfaced as riots in St Pauls (Bristol), Manningham (Bradford) and Tottenham (London).

However, groupings of such people have also contributed positively to host cultures. There has been enrichment by different foods, clothes, music and festivals, as in the Notting Hill (London) and Chapeltown (Leeds) carnivals.

EU: see *European Union*.

European Central Bank (ECB): one of the world's most important central banks, responsible for monetary policy covering the 15 member states of the *eurozone*. It was established by the *European Union* in 1998 with its headquarters in Frankfurt, Germany.

European Commission: the executive of the *European Union.* The body is responsible for proposing legislation, implementing decisions, upholding the EU's treaties and the general day-to-day running of the union. The Commission is headed by a president and 27 commissioners appointed for a 5-year term. Proposals for any new EU regulations or directives pass from the Commission to the *European Parliament* for debate and possible modification, before going to the European *Council of Ministers* for approval or rejection.

European Court of Justice is the judicial arm of the *European Union*'s legal system. The court makes judgments on European law and treaties when they are in dispute. The Court has the power to fine firms, but it can only apply moral pressure on governments. This has often resulted in the slow implementation of European law by some member states.

European Parliament: the elected chamber of the *European Union*. The members sit each month alternately in Strasburg, France or in Brussels, Belgium. They debate and amend proposals put forward by the *European Commission* before they are passed to the European *Council of Ministers* for approval or rejection. The Parliament is very restricted in its powers and can only modify or delay decisions proposed by the Commission. As the only directly elected European institution, the Parliament feels it ought to be given more powers, and indeed it is slowly having greater influence. The Parliament is composed of 785 MEPs (Members of the European Parliament), who serve the second largest democratic electorate in the world (after India) and the largest transnational democratic electorate in the world (342 million eligible voters in 2004).

European Union (EU): known as the European Community prior to 1 November 1993 and before that as the European Economic Community. It currently consists of the following members: France, Germany, the Netherlands, Belgium, Luxembourg, Italy (the first six members), the UK, Denmark, Ireland (joined 1973), Greece (1981), Portugal and Spain (1986), Austria, Finland and Sweden (1995), Czech Republic, Cyprus, Estonia, Hungary, Latvia,

Lithuania, Malta, Poland, Slovakia and Slovenia (2004), and Romania and Bulgaria (2007). The EU was established under the Treaty of Rome in 1957 with the objective of removing all trade barriers between member states. The background to this was the desire to form a political and economic union which would prevent the possibility of another war in Europe. The EU has a number of institutions:

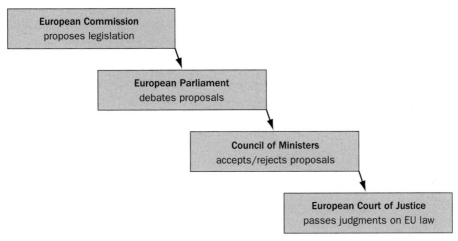

The institutions of the EU

Arguments for and against the EU include:

PROS:
- huge potential market of around 500 million people
- the combined strength of the members forms a powerful **trade bloc**
- inward investment is encouraged, from which the UK has already gained huge benefits
- some income and wealth redistribution throughout Europe
- greater freedom for workers within a wider job market

CONS:
- poor distribution of Union income, particularly as the **Common Agricultural Policy** takes so much of the budget
- overproduction within agriculture with some corrupt practices
- over-bureaucracy within the **European Commission** has brought into question the centralisation of law-making
- individual states, in some cases, still put their own interests before the wider community. European laws, therefore, have sometimes been sporadically applied.

eurozone: the Treaty of Maastricht (1991) paved the way for monetary union, which came about in 2002 with the adoption of a common currency (the euro) by the existing members of the **European Union** at that time, with the exception of the UK, Denmark and Sweden. Since then, Cyprus, Malta and Slovenia have been accepted into the eurozone and it is hoped that all the nations that joined the EU in 2004 and 2007 will eventually adopt the euro. Slovakia joined the eurozone in 2009.

eustatic adjustment refers to a worldwide change in sea level. For example, the **ice ages** created a universal fall in sea level since water was stored as ice on the land. Similarly, the melting of the *ice sheets* at the end of the ice ages caused a worldwide rise in sea level as large quantities of water returned to the sea.

eutrophication is the nutrient enrichment of water in rivers and lakes by the accumulation of chemicals from *fertiliser* and/or slurry from farms and farmland. Farmers use fertiliser to produce healthy crops and to increase yields, but if too much nitrogenous fertiliser is used, or too much animal manure is added to the soil, then some remains unabsorbed and may be leached to contaminate rivers and lakes. Due to the nutrient enrichment in the water, algae and other *phytoplankton* multiply rapidly to produce *algal blooms*, using up oxygen in the water and blocking out the light. Further *bacteria* multiply to decompose these algae and use up even more oxygen. Consequently other *organisms* such as fish are starved of oxygen and die.

evaporation is the process by which liquid water is transformed into water vapour, which is a gas. A large amount of energy is required, and this is usually provided by heat or by the movement of air.

evapotranspiration is the total amount of moisture removed by *evaporation* and *transpiration* from a vegetated land surface. (See also *potential evapotranspiration*.)

exfoliation: a process of *physical weathering* where the outer layers of the rock peel away like the layers on an onion. Some texts refer to this feature as onion-skin weathering. The process was thought to be the result of the heating of the outer surface of an exposed rock surface, which would heat faster than the inner areas leading to a greater expansion and contraction rate. Stresses were therefore set up between the outer and inner parts of the rock which led to cracking and the outer layers peeling away. A number of laboratory experiments, however, cast doubt on this process, and it would now appear that water has to be present before the rock will behave in this way. The process is therefore probably connected with *chemical weathering*, one likely explanation being that water causes the minerals in the rock to swell (*hydration*) and then the outer layers gradually peel away when rapid heating and cooling occur.

exogenetic: this is the term that covers all those processes that are at work on the surface of the Earth shaping the land. These are the agencies that are generally responsible for the wearing away of parts of the surface and the subsequent *deposition* of the *debris* that is created. Such processes include *weathering*, *mass movement* and *erosion*.

exogenous rivers are perennial rivers (flow throughout the year) that rise in areas beyond deserts and then flow through the *arid* area. The best-known example is the Nile, which has its origins in the Ethiopian Highlands and central Africa and flows across the arid regions of north Africa to reach the Mediterranean. In North America, the River Colorado rises in the Rocky Mountains and then flows across the dry regions of southwest USA in order to reach the Pacific Ocean. All of these rivers are highly managed with dams, reservoirs and irrigation schemes reflecting their great importance to the areas across which they flow.

exotic river: a river which maintains its course through an area which has insufficient rainfall to support the *channel flow*. The river receives the majority of its *discharge* from outside the immediate area through which it is flowing, for example, where rivers maintain their flow through desert or semi-desert regions they are supplied with water from their source regions which begin in high rainfall, or snow-fed, mountain areas. The River Nile is fed by the White Nile (from Lake Victoria) and the Blue Nile (swollen by heavy rain in the Ethiopian uplands).

export processing zone: an industrial area, often on the coast, where favourable conditions are created to attract foreign **transnationals**. These conditions include low tax rates and exemption from **tariffs** and export duties.

extensional flow: a form of **glacial movement** that occurs when the valley gradient becomes steeper. The ice accelerates and becomes thinner, leading to reduced **erosion**.

extensive agriculture is where a relatively small amount of agricultural produce is obtained from a large area. **Inputs** per unit of land are low. Extensive can apply to both **arable** and **pastoral** agriculture and can be found at a **commercial** as well as a **subsistence** level. Pastoral types which are extensive include **nomadic** pastoralism, beef cattle ranching and hill sheep farming. In arable farming, extensive agriculture occurs in the Amazon Basin (**shifting cultivation**) and in areas where wheat is grown on a large scale, the Canadian Prairies for example. (See also **intensive agriculture**.)

external economies of scale are the advantages of scale that benefit a whole industry and not just an individual firms. (See also **economies of scale**.)

extractive industry refers to that part of the **primary sector** in which minerals are removed from the ground. It therefore covers the **mining** and **quarrying** industries, e.g. coal, oil, natural gas, iron and other mineral ores, salt, **limestone**, granite.

extrapolation: in forecasting the near future it can be assumed that the recent past will be a good guide. This is known as extrapolating from past to future. When trends have been established, they can be plotted on a graph and extrapolation by eye or by mathematical means can be undertaken.

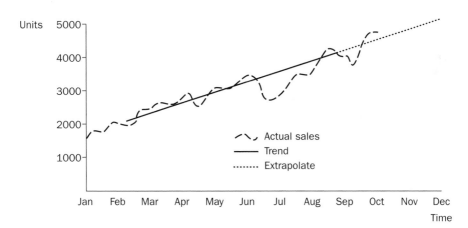

Extrapolation of a sales trend

extrusive: when molten rock is forced to the surface it is known as **lava**, which is said to be extrusive (along with associated material such as **pyroclastics**). This gives rise to a number of extrusive landforms such as lava flows and **volcanoes**. Minor extrusive features include:

- solfatara – small volcanic areas without cones, produced by gases (mainly sulphurous) escaping to the surface, for example in the Bay of Naples area in Italy
- geysers – these occur when water, heated by volcanic activity, explodes on to the surface, for example Old Faithful, Yellowstone National Park, USA

- hot springs/boiling mud – sometimes the water, heated below, does not explode on to the surface. If this water mixes with surface deposits, boiling mud is formed. Such features are very common in Iceland. There are hot springs at Bath in the west of England.

(See also *intrusive*.)

eye: an area in the centre of a *hurricane* that develops only when the system has reached maturity. This is an area 25–50 km across, with subsiding air producing clear skies, light winds and a temperature higher than the surrounding windy areas.

Aiming for a grade A*?

Don't forget to log on to **www.philipallan.co.uk/a-zonline** for advice.

factors of production include land, *labour*, capital and enterprise (*entrepreneurship*). These are the factors that combine in order to produce profit.

factory farming is the use of *production line* techniques to ensure maximum output from farm *livestock* with a minimum of *inputs*. This usually means that the animals live in isolation in very confined spaces and are sometimes fattened up with the use of hormones. It is not without controversy as the 1995 protests over the use of veal crates have demonstrated.

fair trade: under fair trade agreements, agricultural products such as bananas, coffee, nuts and tea are sold to companies in the *developed* world at a rate above the market price. This extra money is used to provide a better standard of living for the producer's family. In 2006, fair trade coffee accounted for 5% of all the coffee consumed in the UK.

famine: it seems rather obvious that a famine is a chronic shortage of food in which many people die from starvation. In recent years, however, this traditional view of famine has come to be challenged by many leading authorities. Questions that have been asked include:

- Why have some famines occurred when there has been an increase in food availability?
- How is it that famines affect some groups in society but not others?
- Why is it that food can sometimes be found in markets in famine-affected areas?
- How is it that food is sometimes exported from famine areas?

A now widely accepted view is that most famines result from a combination of natural events and human mismanagement. Many authorities refer to famine as not being the result of not enough food, but rather caused by a decline in the access to food. Famines are thus the result of a decline in access to food but not a decline in its availability. People are seen as having endowments of goods and resources and using these to obtain what they want. 'Entitlement' failure occurs where an individual's endowment is too small to secure sufficient food for survival. Different groups in society have different endowments, therefore there must be differences in peoples' *vulnerability* to famine. Definitions here are important, because if famines are simply instances of widespread starvation, then food *aid* is the obvious relief. But if they are long-term processes of asset deprivation, then apart from the immediate provision of food aid, the solutions should be directed towards improving the asset situation.

The 1974 famine in Bangladesh was not simply a shortage of food resulting from *flooding*. The rice crop was not completely destroyed but fears of poor harvests led to large increases in price. Combined with inflationary pressures in the economy, this led to a decline in rural employment, decreased rural wages, leading to a collapse in the entitlement of the rural workforce and their families.

Northeastern Africa is an area that in the 1990s was subject to food shortages. *Drought*, *desertification* and *overpopulation* have all been put forward as factors that could explain the famines that have occurred in such countries as Ethiopia and Sudan, but observers have increasingly pointed out that civil unrest and deliberate misman-agement were at the heart of famine in this area. During the famine that affected the Bahr El Ghazal region of the Sudan in 1998, some observers accused the country's government of deliberately preventing food reaching the area because of internal politi-cal differences.

FAO: see *Food and Agriculture Organization*.

Farm Diversification Scheme: since 1988, British farmers were able to apply for grants for diversifying into non-agricultural activities such as timber, golf courses, horse riding cen-tres, tourist accommodation and farm visitor centres. The scheme also gave help to farmers wishing to establish direct marketing such as farm shops and pick-your-own (*PYO*) facilities. The main aims of this scheme were:
- to reduce agricultural surpluses
- to maintain farm incomes
- to prevent rural depopulation.

The scheme has now been replaced by a number of other ways of helping to maintain, or even increase, farm incomes.

farmers' markets are where farmers, growers and producers from a defined local area are present in person to sell their own produce direct to the public.

Farm Woodland Scheme: another scheme introduced in Britain in 1988 by the Ministry of Agriculture, Fisheries and Food (MAFF), in which farmers were encouraged to plant woodlands on land currently in agricultural use. The scheme has four main aims:
- to reduce agricultural surpluses
- to enhance the landscape, create new *habitats* for wildlife and to encourage recreational use
- to maintain farm incomes and help rural employment
- to encourage greater interest in timber among farmers, which in the longer term will contribute to Britain's timber requirements (and save the cost of imports).

fatalism regards hazards as acts of God about which nothing can be done. It is a belief that all such events are predetermined, and the consequences are inevitable.

faulting involves the fracturing of the Earth's *crust* along which the rocks have been displaced either vertically, horizontally or at some intermediate angle. The movement takes place along a fault plane, and the angle between it and the horizontal is called the dip. The vertical displacement is known as the throw, whereas the lateral displacement is known as the heave. There are different kinds of fault:
- a normal fault – the result of tension, the rocks being displaced in the direction of the fault plane
- a reverse fault – the product of compression, the rocks of one side of the fault plane being thrust over those on the other side
- a tear fault – where the movement is in a horizontal direction.

(a) Types of faulting

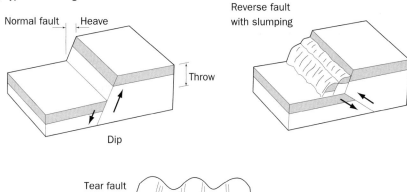

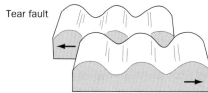

(b) Landforms resulting from faulting

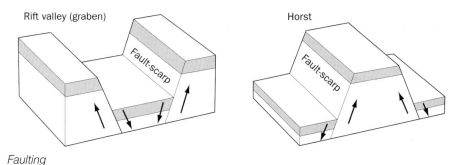

Faulting

FDI: see *foreign direct investment*.

federalism occurs when a number of states, which have some *autonomy* or responsibility for governing their own area, combine into a federation for the control of some government functions. National issues such as defence or foreign policy would be dealt with at the national level while decisions on education, housing and health provision may be taken at state level. The USA and Australia operate as a union of federal states.

feedback occurs where one element of a system changes because of some outside influence. This will upset the *dynamic equilibrium*, or state of balance, and affect other components in the *system*. (See also *negative feedback* and *positive feedback*.)

fermentation refers to one of the surface layers in a *soil profile* where the decay of freshly deposited plant litter is active. It lies immediately below the litter layer where leaves, twigs and cones have fallen, and the form of some of the plant remains is still visible.

ferralitic soils are found in the humid tropical areas of the world, and are deep and intensively weathered. They result from the high annual temperatures and levels of rainfall which cause rapid *weathering* and growth and subsequent *decomposition* of luxuriant vegetation. Continuous leaf fall creates a thick litter layer, but the underlying humus layer is thin due to the rapid rates of decomposition by intensive *biotic factors*. The heavy rainfall causes severe *leaching*. The silica in the soil becomes more mobile than the iron and aluminium *sesquioxides*, and is removed. This leaves a soil rich in iron and aluminium which gives the soil a deep red colour. The continual leaching and abundance of mixing fauna inhibits the development of soil horizons. These soils are very sensitive to environmental change. Once the supply of nutrients is removed, by *deforestation*, the soils soon lose their *fertility*. The soils have a loose structure, and, when exposed, are easily eroded by heavy rainfall.

Ferrel's law states that a body moving over the surface of the Earth will be deflected to the right in the northern hemisphere and to its left in the southern hemisphere. This is due to the Earth's rotation and the *coriolis force*, and is particularly related to *general atmospheric circulation*.

ferruginous soils are found in the tropical regions of the world that experience marked wet and dry seasons (regions known as the *savanna*). As the grasses of these areas die back during the dry season they provide organic matter which is decomposed to give a thin layer of humus. With high rates of *evaporation*, *capillary action* also brings bases and salts in solution to the surface. During the wet season, *leaching* removes the silica from the upper horizons, leaving behind iron and aluminium *sesquioxides* which give the soil a red colour. This alternating pattern often produces a hard cemented layer called *laterite* just below the surface of the soil which impedes drainage, root penetration and ploughing. These soils are not particularly suitable for cultivation, and often form grazing lands. They contain few nutrients, and are vulnerable to *erosion* both by the heavy rains of the wet season, and by the action of wind during the dry season.

fertiliser is used by farmers to produce healthy crops and to increase yields. It is added to a soil in order to replace the nutrients that have been removed by the growth of crops, and which are not replaced when they are harvested. Fertiliser may be:
- organic – farmyard manure which is then ploughed into the soil. This not only returns nutrients to the soil, it also improves the structure of the soil
- inorganic – chemicals, especially compounds of nitrogen, potassium and phosphorous.

fertility (population) is the average number of children each woman in a population will bear. It is usual to refer to women between the ages of 15 and 50 in such calculations. If fertility is 2.1 children, then it is likely that a population will replace itself.

fertility (soil) is a statement of how suitable a soil is for the growth of crops. If a soil is fertile, plant growth is rapid and yields are high. Such a soil provides water, air and nutrients in the correct amounts and proportions. A fertile soil therefore has all or some of the following characteristics:
- a neutral to slightly acid *pH* value
- a loamy texture giving good drainage and aeration
- a crumb structure associated with adequate amounts of humus
- a good supply of nutrients either from the decay of organic matter or the addition of chemical *fertiliser*.

fetch is the length of open sea over which a wind blows to generate waves. The longer the fetch then the greater is the potential for large waves. In Cornwall the fetch is to the southwest, and the largest waves approach from this direction. It is possible therefore that some waves may have originated several thousand kilometres away.

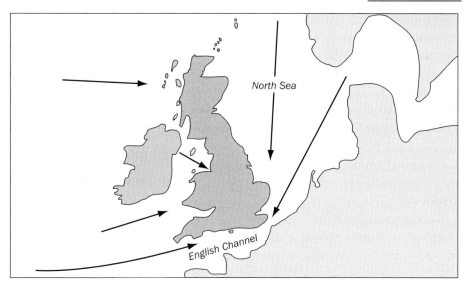

Varying directions of maximum fetch on the British coastline

fieldwork is the first-hand experience of going out of a classroom to observe and record what is present in an area. It may be undertaken within the grounds of a school, in a local area, on day excursions further afield, or on residential visits in a country and abroad. Fieldwork enables a student to explore and understand the different environments that exist in the world. It also provides the basis for students to analyse their results and to formulate conclusions. In addition, fieldwork develops generic skills such as problem-solving, decision-making, communication, leadership and cooperation.

finite is a term which is applied to resources when they are ***non-renewable***. Once they have been used up they cannot be replaced; minerals and ***fossil fuels*** are examples of this type of resource which is being consumed at an increasing rate as more countries progress towards industrial development.

fire hazard: this occurs where there is some threat to the natural, built or human environment as a result of fire, whether this is due to natural or human causes. The term tends to relate to natural fires rather than those which are the result of domestic or industrial accidents and as such can be regarded as an ecological hazard. The brushwood and coniferous vegetation of the ***Mediterranean*** regions of France, combined with the long, hot dry summers, provide ideal conditions for fires, which once ignited are often fanned southwards by the Mistral, a dry wind which blows down the Rhône Valley towards the coast. This threatens the vegetation, the urban areas, the people and the economy of the Riviera coast.

firn: the term generally applied to snow that has survived summer melting and has not yet become ***glacier*** ice. Névé is another term for the same feature.

First World: the economically more advanced countries which have some form of free ***market economy***. This would include countries in North America, Western Europe, Australia, New Zealand, Japan and Israel.

fiscal policy is the government policy towards the raising of revenue and its level of public spending. These are taken in line with the economic objectives of the government and the balance it wishes to achieve between revenue and spending. Governments will also wish to maximise political advantage through its decisions. Fiscal policy is a task undertaken by the Chancellor of the Exchequer.

fish farming involves the deliberate feeding, breeding and harvesting of fish in manmade pools or tanks, as well as the use of enclosures in river channels. Careful monitoring is needed to avoid losses from disease in such confined environments; regular application of chemicals and antibiotics has enabled effective control.

fissure: a line of weakness, such as a joint or a fault, in crustal rocks through which volcanic lava may erupt. Such action usually involves more basic, fluid lava which is accompanied by little explosive activity. (See also *faulting*.)

fjord: a long, narrow, steep-sided coastal feature which is formed by the drowning of a glaciated valley. They usually result from valley *glaciers* eroding below the level of the sea producing an overdeepened U-shaped valley. This valley has been inundated by the *post-glacial eustatic* rise of sea level and in *long profile* often reveals a series of glacial basins which become shallower towards the seaward end of the fjord where a submerged threshold is found. This can occur as a major rock bar, sometimes with *moraine*, and it marks the point where the lower section of the glacier reached the sea and began to float thus reducing the vertical erosive power of the ice. Evidence for *isostatic* uplift in post-glacial times may be seen in *raised beaches* at the head of some fjords. These coastal features are well developed in Norway, but are also found in British Columbia, South Island, New Zealand and South Chile.

flagship developments are usually large-scale, high-profile and high-investment projects, such as major new museums, art galleries, theatres and sports facilities, which are a key means of regenerating the economy of city centres. Iconic 'signature' buildings designed by well-known architects may also be part of the new image. The idea is that these projects improve the image and reputation of an area and help to attract people and other activities.

flash flood: a large but temporary increase in channel *discharge*. The *hydrograph* is modified and is characterised by a steep rising limb, a short *lag time*, a brief period of *peak flow* and a gentle recession limb. These short-lived floods are particularly associated with deserts, but they can occur in other environments when heavy rainfall follows an extended period of wet weather which has saturated the soil. With reduced *infiltration*, faster *overland flow* transports water into channels more quickly giving a rapid increase in discharge.

flexible manufacturing systems make a wider range of specialised products than traditional (*Fordist*) industries. They are able to adjust to alterations in market demand and in order to achieve this they rely on many different suppliers to provide materials and components as they are required. These firms also use computerised machinery which can be adapted to make a variety of products, unlike traditional firms where there is mass production for a mass market. *Labour* is also more flexible and less unionised, and, unlike the demarcation and division of traditional factory working, a smaller workforce operates multi-tasking where one person is capable of undertaking a number of jobs as production targets demand.

flocculation is the process by which a river's *load* carried in suspension is deposited more easily on its meeting with sodium chloride in sea water. The meeting of fresh and salt water in river estuaries and *deltas* causes a clustering or coagulation effect on silt and clay particles, and these larger particles sink more rapidly.

flood control seeks to reduce the frequency and *magnitude* of *flooding* that takes place and therefore limit the damage that floods cause. There are two major approaches to flood control:

1 Flood protection. Achieved by:
  - modifications to the banks and/or channel to enable the river channel to carry a larger volume of water. Artificially raised and strengthened banks form a significant part of this strategy. In some cases parallel lines of such floodbanks act as a double form of protection – if the river overtops the first barrier, then it has difficulty rising over the second bank some distance behind. The removal of large boulders from the bed of the river reduces roughness, therefore increasing the velocity of flow
  - the building of dams to regulate the rate at which water passes down a river
  - diverting rivers away from vulnerable areas
  - increasing the height of the *floodplain* by dumping material on it.

2 Flood abatement. Achieved by:
  - *afforestation*, which slows down the rate at which water reaches a river as well as reducing the amount that actually does reach it
  - contour ploughing and strip farming in semi-*arid* areas, which reduce the amount of surface *runoff* and therefore reduce the liability to flooding.

flooding occurs when a river's *discharge* exceeds the capacity of its channel to carry that discharge. The river overflows its banks. Flooding may be caused by a number of factors:
  - excessive levels of *precipitation*
  - the melting of snow
  - the failure of manmade dams and/or embankments
  - the changing of land use in catchment areas such as *deforestation* and urbanisation.

Flooding may also take place in coastal areas due either to rising sea levels or tidal surges caused by storms. Bangladesh, being located at the mouth of the rivers Ganges and Jumana, and at the head of the Bay of Bengal which suffers from tropical *cyclones*, is often trapped between two sets of floods – one caused by rivers and the other by the sea.

floodplain: that part of a valley floor which a river may flood from time to time. As the river floods it deposits a layer of silt which gradually builds up the height of the floodplain. The edge of the plain is often marked by a prominent slope known as a *bluff* line. Flood-plains may be found throughout a river valley, even extending in restricted fashion into upland areas.

flora and fauna: flora is the name given to the plant species which make up the vegetation of an area and fauna is the name given to the animal species.

flow line: a technique for presenting data in which the width of the line between two points in the network is proportional to the volume of movement. The route followed by the line represents the line of movement of the people, traffic, goods, information or other data being presented whereas a *desire line* simply shows the origin and destination of the movement.

flow production is the manufacture of an item in a continually moving process. Each stage is linked with the next by a conveyor belt or in liquid form, so that the production time is minimised and production efficiency is maximised.

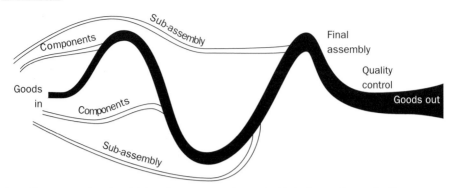

A continuous system in which sub-components are fed into the main production

fluvial is a term applied to the action of rivers. Fluvial *erosion* includes mechanical *abrasion*, where fragments of *debris* are rolled and dragged along the bed of the channel, slowly wearing away other material. The sheer power of water flow may lead to *hydraulic action* eroding material. Fluvial transport involves traction, where sediment moves along the channel bed, *saltation*, where pressure differences across particles cause them to be pro-jected temporarily into the main velocity flow which moves them downstream, and suspen-sion where small particles are carried along in the water flow. As stream velocity decreases fluvial *deposition* occurs; particles settle according to their grain size with the largest diam-eter material being deposited first. The relationship between stream velocity and particle size in relation to fluvial erosion, transport and deposition is displayed in the *Hjulström curve*.

fluvioglacial (glaciofluvial) is a term that is used to describe processes and landforms resulting from the action of *meltwater* streams associated with glacial environments.

fluvioglacial landforms are produced by the action of *meltwater* streams transporting and depositing material in a glacial environment. These streams may flow on the ice (*supraglacial*), within the ice (*englacial*) or beneath the ice (*subglacial*) and they can carry material beyond the snout of the *glacier* or the edge of the *ice sheet*. The resulting landforms can be divided into two groups: proglacial and ice-contact features. Proglacial landforms include *outwash* plains and valley trains which are deposited beyond the ice margin. Ice-contact landforms are produced under or on the margins of the ice and include *eskers*, *kames* and kame terraces. Fluvioglacial landforms consist of stratified deposits which distin-guishes them from the unstratified materials produced as a direct result of glacial *deposition*.

fluvioglacial processes include the transport and *deposition* of material already eroded by glacial action. When *meltwater* streams are flowing within or under the ice they are sub-jected to greater hydrostatic pressure which increases their *capacity* to transport *sediment*. As the *discharge* of streams varies seasonally, some material is deposited along the bed and banks of these stream channels. When streams emerge at the ice margin the decrease in pressure, and resulting fall in velocity, leads to a reduction in the stream's capacity to carry material and deposition occurs. This results in graded deposits with larger particles nearer to the ice front and progressively smaller particles at greater distance.

fog is a term which is applied to an atmospheric condition when visibility is less than 1 km. It is caused by the cooling of the air to *dew point* and the resulting *condensation* of water vapour in the atmosphere at ground level. This cooling can result from three different modes of formation giving three types of fog: *radiation*, *advection* and frontal. Hill fog, which can occur on high ground in all seasons in the British Isles, is in effect low **cloud** which forms below the summit of hills.

fohn: a warm, dry wind that descends from the Alps. Temperature rises of between 15°C and 20°C may be experienced and this can cause rapid snowmelt and **avalanche** problems. The fohn effect, which is similar to the changes caused by the Chinook wind in the Canadian Rockies, is caused by low pressure systems to the north of the Alps drawing moist air from the Mediterranean. As this air rises over the Alps it is cooled at the **dry adiabatic lapse rate (DALR)** until it reaches its **dew point**. At this temperature, when **relative humidity** is 100%, the air becomes saturated and on rising higher it will cool at the **saturated adiabatic lapse rate (SALR)**. Condensation and **precipitation** occurs on the windward side of the mountain barrier and as the air begins to descend on the leeward side, it warms up at the dry rate. Because of the difference between the dry and the saturated rates, there is a net increase in temperature as the air crosses the mountains.

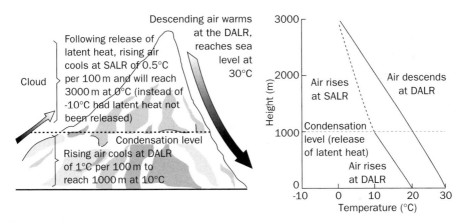

The fohn effect

folding is the bending and crumpling of sedimentary layers and is caused by compressive forces in the Earth's **crust**. When rigid blocks move towards each other, the **sediments** in between are crumpled and may be contorted into folds of varying intensity and complexity. These can range from simple monoclines, anticlines and synclines, and more complex recumbent folds to extreme forms such as nappes. The forces involved are so powerful that strata may be metamorphosed leading to the transformation of rocks. For example, clays may be changed into slates. Fracturing associated with the folding may lead to **intrusive** and **extrusive** volcanic activity. Fold mountain development is related to the movement of crustal plates at **destructive plate margins**, when ocean crust subducts beneath continental crust or when two continental plates collide.

Food and Agriculture Organization (FAO) is the oldest permanent specialised agency of the **United Nations**, established at the end of the Second World War with the objective of eliminating hunger and improving nutrition. The FAO seeks to coordinate the efforts of governments and technical agencies in programmes for developing agriculture, forestry and fisheries. The headquarters of the organisation are in Rome. The work of the FAO includes:
- research into new methods, crops (particularly **HYVs**), pest control etc.
- providing technical assistance on various projects
- running programmes to promote more rural employment

- promoting agricultural exports
- running an educational programme
- maintaining statistics on world production, trade and the consumption of agricultural commodities
- publishing a number of periodicals, yearbooks and research bulletins.

food chain: the flow of energy through an *ecosystem*. Each link in the chain feeds on and obtains energy from the one preceding it and in turn is consumed by and provides energy for the following link. Each link in the chain is known as a *trophic level*. It is usual to recognise several trophic levels in a chain. At its simplest, this could consist of plants, then herbivores, followed by carnivores and finally omnivores (that feed on both animals and plants and include humans in the group). At each stage in the chain material is lost to decomposer *organisms*, mainly *bacteria* and fungi, which break down this matter into its constituent parts, making it available for reuse by plants.

food insecurity: this results either from a lack of available food, due to adverse physical factors (e.g. *drought*) or when there is adequate food available but the people are too poor to be able to access it. Chronic food insecurity translates into a high degree of vulnerability to *famine* and hunger.

food marketing channel refers to the institutions and routes involved in the production, distribution and consumption of food. The 'traditional' marketing channel for food has been from the farmer to a wholesaler and then to a retailer. For some goods, such as milk, government-backed marketing boards were also available. However, in recent years there have been a number of major changes to these channels:

- the growth of *contract farming*
- producer cooperatives – an alternative to marketing boards, for example frequently used by wine producers in France
- direct marketing through farm shops, *farmers' markets*, roadside stalls and pick your own (*PYO*).

food miles is the distance food travels from the producer to the consumer. The term can also be expressed as food kilometres.

food processing involves the preservation of food so that it can be sold either further away from the point of production or longer after its production. The main forms of food processing are freezing, refrigeration and canning. The creation of pre-packaged convenience foods forms an increasing proportion of food processing activities.

There are benefits resulting from food processing for both the farmer and the consumer:
- farmers receive a greater degree of income security
- consumers receive regular supplies of food, as well as a greater variety of foods.

food retailers are the organisations through which food is sold. They exist at varying scales from the small local shop to the larger supermarket and *hypermarket*. Much food used to go from farms to wholesale markets and then on to the smaller retail outlet. Nowadays, a larger proportion goes directly to the supermarket chain as a result of *contract farming*. Farmers are also retailing their produce themselves with the development of farm shops and pick your own (*PYO*).

food security: this means that a population has access to enough food for an active healthy life.

food surpluses occur when production has exceeded demand. They have been a common feature of the *European Union*'s *Common Agricultural Policy* due to its system of guaranteed prices which has encouraged farmers to overproduce. 'Mountains' of food have been created in grain, beef, butter and sultanas, in addition to 'lakes' of milk, wine and olive oil. Many of these surpluses are either placed in storage at great expense or sold at cheaper prices to other countries, such as those of eastern Europe.

food web: a complex series of *food chains* where *organisms* operate at more than one feeding or *trophic level*. Any animal's diet is usually quite varied and one species of animal or plant may be part of the food of a wide range of different animals. Consequently, complex interactions of feeding are set up. Humans operate at several trophic levels by being both herbivores and carnivores and therefore complicate any simple food chain.

footloose industries are those that have a relatively free choice of location. They are not tied to the location of *raw materials*. Many are the newer industries which either provide a service for people or produce light goods of a high value, and are therefore market-orientated. Examples include *high-tech industries* and electronics industries, *food processing* and distributive companies. Footloose industries are frequently located on large new *industrial estates* on the edges of towns and cities, or alongside major motorways to utilise the efficient road transport system.

Fordism: the application of Henry Ford's faith in mass production, run by an autocratic management. This involved the division of *labour*, i.e. the breaking down of a job into small, repetitive fragments, each of which could be done at speed by workers with little formal training. The benefits for production were outstanding and costs dramatically declined. By the 1960s, however, such methods were coming under increased scrutiny as people demanded more satisfaction from their employment.

foreign direct investment (FDI) is investment made by foreign *transnationals* into a nation's economy, either to build new facilities such as factories or to acquire or merge with an existing company in that country.

foreset beds are layers of sediment that are laid down at the seaward edge of a *delta*. They consist of mainly clays and silts deposited at an angle as the delta builds outwards. The coarser sands and silts are deposited first in the *topset beds* and even finer clays are carried furthest to form the *bottomset beds*.

forest clearance: the removal of woodland in order to establish land which could be used for *arable* production, pasture or village settlement. In medieval England, much of the land was still wooded, even in areas of well-established settlement. On lowland clays this was mainly oak-ash forest and on the drier *chalk* and *limestone* it was largely beech. By the twelfth and thirteenth centuries, much of this woodland had been cleared for economic reasons.

forest management is the deliberate preparation, planting and cutting of forest areas. It is a planned process of felling and replanting; as one section of the forest is cut new areas are ploughed over and planted. The land is prepared by ditching to lower the *water table*, and *fertiliser* is used to encourage growth. The trees are protected from pests and disease and fire-breaks are left between sections of the forest. Dead or diseased trees are removed and sections are thinned out to provide optimum conditions so that fully mature trees can be felled.

forest parks are areas of forest which may be used for recreational as well as economic purposes. They are often provided with facilities for visitors such as car parks and picnic sites and may include signposted tracks for walking, cycling or nature trails. Some areas may have permanent sites for camping and these can be landscaped within the forest environment. With increases in leisure time some forests may be more important for recreation than timber supply.

Forestry Commission: a body established in 1919 to combat the effects of overfelling of timber during the First World War and to build up stocks for the future. The Commission controls over 800,000 ha of land, largely under coniferous *plantations*. Private owners also planted conifers in the 1980s as a part of tax reduction schemes; this ended with the 1988 budget. Grants to plant woodland still exist under the woodland grant scheme, but this is aimed at broad-leaved woodland (*deciduous woodland*).

The decision to plant fast growing species not native to Britain has led to criticism and conflict:

- it has altered the natural appearance and replaced it with an artificial habitat, less variety and richness
- it has destroyed moorlands and upland bogs
- upland bird communities have been damaged
- clearance and pre-*afforestation* ditching have increased drainage density and the initial level of surface *runoff*, which has raised the flood risk and the level of sediment in streams
- coniferous foliage tends to trap, rather than process, atmospheric pollutants; these are eventually washed into water courses
- *fertiliser* and *pesticides* used in the plantations are washed into streams and this affects aquatic life
- many areas are closed to public access, including some of the most attractive upland moorland.

Some areas have been opened as *forest parks* with facilities to encourage visitors, such as car parks, picnic sites, signposted tracks and walking, cycling or nature trails. These areas therefore have both an economic and recreational function.

forward integration: a type of vertical integration where a company gets control of activities downstream of it, i.e. those firms that deal with the finished product of the firm in question. For example, car manufacturers could seek to get hold of car distributors and retailers. (See also *vertical integration*.)

fossil fuels are those fuels, consisting of hydrocarbons, that were laid down in past geological periods. They are coal, oil and natural gas and are classified as *non-renewable resources*. The bulk of the world's energy is produced from them and this reliance is likely to last well into the next century.

PROS:
- widespread in their occurrence and therefore accessible to many nations
- have been converted economically for a long time into forms of energy
- readily converted from one form to another, i.e. from liquid to gas, from solid to gas
- excellent fuels for transport
- there are several hundred years of recoverable coal deposits

CONS:
- non-renewable
- oil and gas reserves are far more limited than coal
- a valuable source of raw materials for a wide range of chemical products and many feel that they should not be 'squandered' by being used as fuels
- burning fossil fuels contributes to atmospheric pollution such as smog
- a major contributor to acid rain
- the release of carbon dioxide is a major contributor to the greenhouse effect and thus global warming.

fragile environments are those where any disruption to the **ecosystem**, however slight, can have serious effects. The **tundra** is considered to be fragile because of its climate and limited productivity. The slow rate of plant growth means that any disruption to the ecosystem takes a long time to be corrected. The low productivity and limited species diversity mean that plants are very specialised and any disruption causes difficulty when it comes to regeneration. In such circumstances, species have great difficulty in adapting to a changed environment.

fragmentation is where agricultural land is broken up so that one farm may consist of numerous small and scattered fields. This is usually the result of inheritance laws where the land is divided up between the sons or land is given in the dowry of a bride. Fragmentation wastes a lot of the farmer's time as he moves between fields and does not promote an efficient farming system. **Mechanisation**, for example, is seriously hindered.

freeport: an area set aside for commercial activities where costs are saved as the area is not considered to be within the country for taxation purposes. Therefore **raw materials** can be imported, made into finished products, which are then exported, all without attracting any taxation or duty of any kind. Singapore, for example, has seven such zones, six based around its docks and the seventh at its airport. Goods can be manufactured or simply assembled, with no import or export duties being paid. Profits can be sent to the parent country of the companies involved, again free of taxation. The main advantage to Singapore is the employment that is created by these zones together with the economic stimulus this gives to Singapore's economy through the increased spending power of those employed within the zones.

free trade exists when trade between countries is not restricted in any way by **tariffs**, **quotas** or other barriers. It is based on the theory that every country will be better off if it specialises in producing goods at which it is comparatively more efficient. The **General Agreement on Tariffs and Trade** accepts the notion that more trade benefits everyone, and through successive 'rounds' has sought to reduce trade barriers around the world.

freeze–thaw: see **frost shattering**.

frequency is how often an event occurs. For example, a flood of 1 m in height may occur, on average, on a particular river every year, whereas a flood of 5 m may occur only once in 100 years. Low-**magnitude** hazard events therefore have a more frequent recurrence level.

frequency distribution: see **dispersion measures**.

friable: a description of a soil that is easily broken up.

frictional unemployment occurs in the time delay between losing one job and finding another. By its nature it is temporary, as opposed to **structural unemployment** which is more fundamental and therefore longer-term. If a government wished to reduce frictional unemployment it could improve the quality of service in Job Centres, so that the newly unemployed are able to find work more quickly.

friction of distance refers to the lesser likelihood of people using a service, the greater the distance away that they live from it. Distance is perceived to be a disadvantage, due either to the time, cost or effort involved. The effect of the friction of distance is to create a *distance decay* in the use of a service.

Friedmann, J. produced a model of the economic development of a country, with particular reference to the changing spatial relationships within that country. The model progresses through four stages of development for a country:

1 A number of relatively independent local centres exist, each of which serves a small region, with no settlement hierarchy.
2 The development of a single strong *core* during the initial phases of industrialisation, with an underdeveloped *periphery* in the remainder of the country. Development occurs in the core region which has a specific advantage over the rest of the country, for example a *natural resource* or dense population. The initial advantage is maintained by *cumulative causation* as more capital, entrepreneurs and *labour* move to the core.
3 The core-periphery structure becomes transformed into a multi-nuclear structure with the national core and a number of peripheral sub-cores. These may develop due to large regional markets or important natural resources.
4 A functional interdependent system of cities resulting in national integration and maximum growth potential.

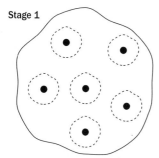

Relatively independent centres, no hierarchy. Each town lies at centre of a small region.

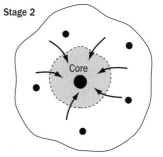

A single strong core, with a periphery. Labour and capital move to the core.

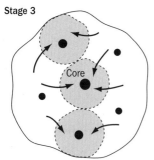

A single national core, with a number of peripheral sub-cores

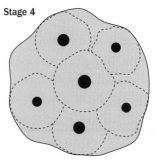

An interdependent system of cities, with a maximum growth potential for the whole country

Friedmann's development model

From this model four types of areas can be designated:
- the core region – the focus of the national market and seedbed of new industry and *innovations*
- upward transitional areas – regions with some form of natural endowment characterised by inward *migration* of people and investment
- downward transitional areas – regions with unfavourable locations and resource bases, characterised by outward migration of people and investment
- resource frontiers – areas where new resources are discovered and exploited.

fringing reef: an area of coral which is attached to the coast and extends out to sea. It represents an early stage in the sequence of subsidence or sea level rise which leads to the formation of *coral reefs*. As relative sea level rises the coral continues to grow upward and away from the subsiding land and this produces a barrier reef and then an atoll.

front: a boundary separating two *air masses* with different temperatures and different densities. The term dates from the First World War when meteorologists compared such a meeting of contrasting air masses to the trench warfare of that time. There are two main types of front:
- a *warm front* – when warmer air is advancing and rising up and over a wedge of cold air ahead of it. Although the gradient of the incline is gradual (1°), the rising air cools and the water vapour within it condenses to produce *cloud* and *precipitation*
- a *cold front*.

Fronts and their passage over an area are significant features of a *depression*.

frontogenesis: the activity which takes place along a *front* and results in the growth of waves along the front into fully fledged *depressions*. Contact on the front between air masses of different temperatures, *humidity*, density, speed and direction of movement, creates friction which in turn sets up waves. *Convergence* and uplift at the apex of the wave allows surface pressure to fall creating the centre of the depression. *Divergence* aloft allows the rising air to be moved on (by *jet stream*) and this is fundamental in deepening the depression and bringing maturity to the system.

frost is the deposit of fine ice crystals on to a surface of grass, plant leaves and walls. It forms under clear, calm, anticyclonic conditions in winter when there has been a rapid heat loss at night. Water vapour condenses directly by *sublimation* on to these surfaces. Glazed frost is formed by raindrops falling through a layer near the surface with temperatures below freezing, and on to a very cold surface, producing a solid sheet of ice. On a road surface, this is known as 'black ice'.

frost heave: in *periglacial* regions, as the *active layer* starts to refreeze in winter, ice crystals begin to develop. They increase the volume of the soil and cause an upward expansion of the soil surface. Frost heave is most significant in fine-grained material.

frost hollow: a natural depression or valley where *frosts* occur more regularly than in surrounding areas. They frequently occur as a consequence of a *temperature inversion* in winter associated with anticyclonic conditions. They may also occur due to the influence of *mountain winds* (*katabatic* winds).

frost shattering: a form of *physical weathering* in rocks that contain crevices and *joints* and where temperatures fluctuate around 0°C. Water enters the joints and, during cold nights, freezes. As water occupies around 9% more volume than water, it exerts pressure within the joint. This alternating freeze–thaw process slowly widens the joints, eventually causing bits to break off from the main body of rock. It leads to the formation of such features as *scree* slopes and *blockfield*.

fuelwood: the use of wood as a fuel for cooking and heating for many families in *economically less developed countries*. In many parts of the world, the collection of fuelwood is a very time-consuming occupation for the women and children, with them having to travel many kilometres to find sufficient amounts. As the demand for fuelwood increases, more trees are cut down with damaging environmental effects such as *soil erosion*.

full employment is the level of employment which provides jobs for all those who wish to work apart from those *frictionally unemployed*. What level of unemployment in the UK now represents full employment is a matter of some debate. Structural change, which includes an increasing rate of technological advance, appears to be creating higher and higher levels of unemployment at each trough in the *trade cycle*.

function is the main reason why a settlement was built or continues to exist. Different settlements have differing functions, for example:
- some are market towns, where people can buy and sell produce
- some are ports where goods are either imported or exported
- some are industrial where the main form of employment is in *manufacturing industry*
- some are resorts, where tourism is a significant factor.

Settlements usually have a combination of functions and their functions may change over time. Consequently, the type of classification given above is seen by many as being too simplistic.

functional interdependence represents the final and ideal stage of *Friedmann*'s development model. All of the cities in a country are fully integrated into a complex system of economic production and communication. The country should be able to adapt to any change in circumstances, as it has the most efficient locational pattern of cities and has the maximum potential for further growth.

functional zone: district of a town or city where one type of *land use* is dominant. The principal functional zones of a town or city are:
- the *central business district*
- industrial areas
- residential areas.

Each of these zones can be further subdivided into sub-zones. For example, residential areas could be subdivided into areas of inner-city housing, high-rise tower blocks, inter-war semi-detached housing, post-1960s suburban estates, private or council.

fungicide: toxic chemicals used in *arable* farming in order to eradicate fungal diseases that attack plants.

G8: see *group of eight*.

gabion: an example of a *hard engineering* solution in coastal defence. Gabions consist of large boulders contained in a steel wire-mesh cage placed in front of a cliff or sea wall to take the full force of the waves. Boulders placed in a similar position without the cage are known as rip-rap (rock armour).

Gaia: the theory that the functioning of the planetary *ecosystem* is determined for its own long-term good by the sum total of living and non-living components of the system. The theory was first put forward by James Lovelock, who took the name Gaia from the Greeks' name for the goddess of Earth. Lovelock maintained that the planet is not an inert cinder beyond the influence of its organic passengers. Just as it determines their fortunes, they serve to shape its make-up. Lovelock did not set out to show that all this was done willingly, he maintained that *organisms* work in accord with their surroundings, so that they do not have to know what they are about for them to work beneficially with each other. Ideas on Gaia are beginning to gain some measure of acceptance, particularly because of the way in which man is increasingly responsible for polluting the atmosphere and the oceans, deforesting vast areas and expanding the deserts.

garden city movement: the ideas of Ebenezer Howard (1850–1928) of completely planned settlements with much open space, first put forward by him in 1898. Howard suggested a housing density of ten units per acre or 35 persons, which meant that the same population occupied an area 20 times greater than it would have done in the back-to-back housing present in British cities of the late nineteenth century. The first garden city developed was Letchworth in 1903 followed by Welwyn Garden City in 1920. Howard's ideas were slow to gain recognition until after 1945 when they became the basis of the development of *new towns*.

garrigue: the vegetation of the *Mediterranean climatic regions* that occurs on drier and more *permeable* rocks such as *limestone*. It is lower and less dense than the more typical Mediterranean vegetation of *maquis*. The more common plants are gorse and a range of aromatic shrubs such as thyme, lavender and rosemary. It is a good example of a *plagioclimax* as much of the vegetation is the result of people's interference.

GATT: see *General Agreement on Tariffs and Trade*.

gavelkind is a system of land inheritance in which, on the death of the landowner, the property is divided equally between the sons. The land may be passed on to daughters if there are no surviving male heirs. Over a few generations this system leads to *fragmentation* of land holdings which can produce plots too small to be economically viable. It contrasts with *primogeniture* in which land is passed to the eldest son.

GDP: see *gross domestic product*.

gender: the socially or culturally defined difference between men and women, while sex is the biologically defined difference. Gender differences are based on socially defined roles of men and women and the relations between them, such as the degree of power they exert over one another. They have different degrees of access, for example, to certain jobs, higher education and some social and recreational organisations.

General Agreement on Tariffs and Trade (GATT) was established after the Second World War within the *United Nations* to encourage the growth of international trade by removing or reducing *tariff* and non-tariff barriers. Agreements are reached after what are known as 'rounds' of negotiations, which often last many years. The latest was the Uruguay round, which started in 1986, but talks were repeatedly stalled because of disagreements between the *EU* and the USA over agricultural *subsidies*. Negotiations came to a conclusion on 15 April 1994 when the agreement was signed in Marrakesh, Morocco. From January 1995 GATT was replaced by a new organisation called the *World Trade Organization (WTO)*.

general atmospheric circulation is the pattern made within the atmosphere of winds and *pressure* belts. Although the circulation is extremely complex, there are certain movements that occur regularly enough for us to recognise patterns of air pressure distribution and winds.

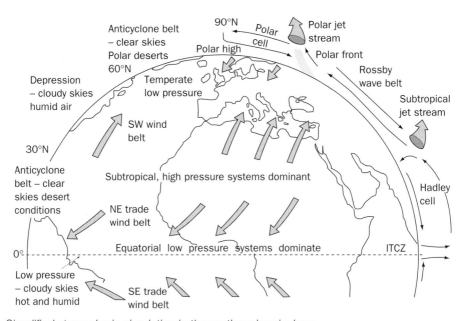

Simplified atmospheric circulation in the northern hemisphere

Our modern understanding began with the work of Halley (1686) followed by Hadley in 1735, who recognised that a global *convection* system was set in motion by a tropical heat source. That the system was fundamentally a three-celled one was discovered by Ferrel in 1856 and in more modern times refined by Rossby in 1941.

genetically modified crops: also known as GM crops, genetic modification involves taking a gene from one *organism* and inserting it into the genetic material of another one, so the recipient now produces a protein that the first one normally produces. Major crops that have been modified, and are now grown commercially, include maize, oilseed rape, tomatoes and potatoes.

Advantages that have been claimed for modified crops include:
- it enables agriculture to become more profitable through greater intensity and cheaper crops
- yields are boosted by making crops resistant to pests, which also removes the need for *pesticides* etc. (This could also allow more food to be produced for the *Third World*.)
- crop quality can be improved
- it could allow crops to be grown in new areas such as those with *drought* conditions
- it allows farming to be more intensive and thus concentrate on smaller areas, sparing more land for wildlife
- foods can be designed to be more convenient (such as melons without pips) and plants could be engineered to contain medicinal properties.

Opponents of GM crops have put forward these arguments:
- GM crops damage *biodiversity*, rather than making agriculture more sustainable
- genes can escape from GM crops, via insects or wind-borne pollen, to nearby wild plants which may then take on some of the properties of the GM crops. This cross-pollination with wild relatives could also create 'superweeds', leading to the use of even more chemicals
- not enough is known about the effects of such modifications on people's health
- some areas already produce enough food and GM crops will just encourage a greater *overproduction*
- to some people there is the ethical question of gene manipulation – whether it consti-tutes 'playing God'.

gentrification: this is a process of housing improvement associated with a change in neigh-bourhood composition when lower income groups are displaced by more affluent people, usually in professional or managerial occupations. This is one of the processes that can regenerate the *inner cities*. It essentially involves the rehabilitation of old houses and streets, often in areas originally developed in the early part of the nineteenth century. Gentrification often depends on the cooperation of property developers, estate agents, building societies and local authorities.

PROS:
- contributes to the social balance of inner-city areas
- increases residents' purchasing power and therefore the general prosperity of the area
- creates employment in inner cities

CONS:
- lower income people are driven into poorer quality accommodation as the privately rented sector shrinks, therefore increasing urban deprivation
- decline in the social quality of neighbourhood life for the original residents
- friction often occurs between the gentrifiers and the original residents.

Geographic Information System (GIS): a structured framework for acquisition, storage, retrieval and analysis of data within a common geographically referenced spatial system. Such systems are made up of data on various topics (e.g. soil, vegetation) and from a range of sources (e.g. maps) processed by computer. For example, information obtained from satellites can be placed in a GIS where maps are produced of an area and layers of different information may be combined so that relationships become apparent. GIS databases are frequently used in environmental and resource management.

GIS can therefore:
- help with questioning and understanding data
- enable multiple interrogation of data
- illustrate difficult abstract concepts in a dynamic visual way
- make use of three-dimensional representations
- provide opportunities for modelling and decision-making.

geological survey: an investigation into the nature and structure of rocks at a particular location. It may also seek to identify evidence of ancient life forms in the form of fossils.

geophysics: the branches of physics which are concerned with the Earth and its atmosphere. Meteorology, geomagnetism and seismology are all geophysical subjects.

geopolitics is the study of the relationships between a country and the rest of the world. Each nation has a sphere of influence it exerts over surrounding nations in areas such as trade, economic aid and military intervention.

geostationary satellites maintain a fixed position over a point on the Earth's surface and therefore monitor only that part of the surface. The meteorological satellite METEOSAT 3, launched by the European Space Agency in 1988, is in geostationary orbit over the equator at 0° longitude. It observes the Atlantic, Europe and Africa, producing an image every 30 minutes. Other such satellites are fixed at longitudinal intervals of 70° around the equator and together monitor the whole globe.

geostrophic winds blow parallel to the isobars in the upper atmosphere. They are the result of the balance between the *pressure gradient* force and the *coriolis force* that exist in air movement at such high altitudes. They are unaffected by friction from the Earth's surface. Their existence gives rise to the maxim: 'If, in the northern hemisphere, you stand with your back to the wind, low pressure is to your left and high pressure is to your right.' (See also *Buys Ballot*.)

geosyncline: the name given to a *linear* depression in the sea floor in which vast amounts of *sediments* would have accumulated prior to being pushed together and upwards to form fold mountains. Such a depression would have had to subside at the same rate at which it was being infilled by sediments. A geosyncline called the Sea of Tethys is said to have existed in Eurasia. The sediments in it were pushed upwards to create the Himalayan mountain range when the Indian plate moved north and collided with the Eurasian plate.

geothermal energy is derived from the heated rocks beneath the surface of the Earth. Cold water is pumped down bore holes, heated by contact with the hot underlying rocks and turned to steam. It is then returned to the surface, where it is used to generate electricity. Geothermal power is used in a variety of locations around the world such as Iceland,

New Zealand and El Salvador. In Iceland, it has enabled heat to be provided to homes, open-air swimming pools and greenhouses. Due to such heat, Iceland is the only European country that is self-sufficient in bananas!

ghetto: originally referred to a concentration of Jews within a city. However, particularly in the USA, the term is now used to refer to a highly segregated area where a minority, often ethnic, group is dominant. The most common forms of ghettos in American cities are those dominated by African-Caribbeans and Hispanics. Such areas are characterised by low incomes, high unemployment, substandard housing and high levels of delinquency and crime.

Gini coefficient is a statistical measure of the degree of similarity between two sets of percentage data. It is calculated using:

FORMULA: Gini (G) $= \frac{1}{2} \Sigma(X_i - Y_i)$

where X_i and Y_i represent the two sets of percentage data. The coefficient ranges from 0 to 100; a value of 0 indicates that the two sets of percentages are identical, i.e. $X_i - Y_i = 0$. A value of 100 indicates that the data is as different as it is possible to be. Because the Gini coefficient has a fixed range of values, it provides a simple and consistent framework for interpreting results.

For example, differences in the variation of manufacturing employment in a region could be compared with the pattern of manufacturing employment on the national scale.

Worked example: manufacturing employment for four industries in two regions and national pattern

Industry	Region A %		Region B %		National % (Y_i)
W	$(X)_A$	20	$(X)_B$	10	25
X		10		20	10
Y		40		10	25
Z		30		60	40

For Region A:

$G = \frac{1}{2}$ [(20 − 25) + (10 − 10) + (40 − 25) + (30 − 40)] $= \frac{1}{2}$ [5 + 0 + 15 + 10] = 15

For Region B:

$G = \frac{1}{2}$ [(10 − 25) + (20 − 10) + (10 − 25) + (60 − 40)] $= \frac{1}{2}$ [15 + 10 + 15 + 20] = 30

In calculating the coefficient it is not necessary to take account of the + or − answer to each X − Y value; the formula allows for this by halving the final sum of X − Y.

Region A has a pattern of manufacturing employment which is more similar to the national pattern than Region B.

GIS: see *Geographic Information System*.

glacial budget refers to the net balance between *accumulation* and *ablation* within a *glacier*'s system. With reference to a temperate glacier:

- if accumulation is greater than ablation, then the snout of the glacier will advance
- if ablation is greater than accumulation, then the snout of the glacier will retreat
- when accumulation and ablation are equal, the snout's position is stationary.

(Note that ice continues to move downhill within the glacier in both of the latter two circumstances.)

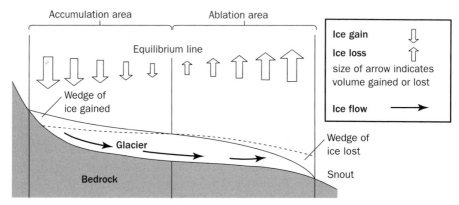

Annual budget model of a valley glacier

glacial control theory: this was put forward by Daly to explain the growth of *coral reefs* in relation to fluctuating sea levels. During the glacial periods of the *Pleistocene*, sea level was lower as a result of *eustatic* changes and sea temperatures would have fallen. As a result any corals would have been killed and erosion would have levelled off existing reefs to produce platforms of dead coral. At the end of the *ice age*, as temperatures rose, sea level would rise eustatically and the sea would have been warm enough for coral growth. New reefs would be built up on the eroded platforms and these would continue to grow upwards as sea level continued to rise. Although this theory explains a number of observed facts it does not fit all situations. Borings through coral reveal thicknesses up to 340 m and *seismic* evidence suggests that some corals are over 760 m thick. If only glacial control had been involved coral thickness would not exceed 90 m, i.e. the base of the coral would be related to the low glacial sea level. This evidence tends to support the subsidence theory and act against Daly's theory. However the subsidence theory cannot explain all features and it may be the case that both approaches are applicable to particular coral sites.

glacial diversion: see *diffluence*.

glacial lake refers to any lake created largely through the actions of a *glacier* or *ice sheet*. There are a wide variety of such lakes:

- tarn (*cirque* lake) – a small circular lake located in the base of a cirque. It occupies the natural bowl depression of the cirque, although some are dammed behind a ridge of *moraine*
- proglacial lake – created adjacent to a glacier or ice sheet by *meltwater* issuing from it. Proglacial lakes existed at a variety of scales in the British Isles towards

the end of the last *ice age*. Several are thought to have existed in North Yorkshire. The Vale of Pickering was covered by a huge lake that eventually overflowed down Kirkham Abbey *gorge*. The meltwater was provided by the ice sheet to the east being dammed by the North Yorkshire Moors to the north, the Yorkshire Wolds to the south and the Howardian Hills to the west. Several smaller such lakes existed in the western end of the Esk valley. A number of overspill channels were created from these lakes

- *ribbon lake*.

(See also *kettlehole*.)

glacial landform refers to a feature created largely through the actions of a *glacier* or *ice sheet* (*glacial processes*). (See also *arête*, *cirque*, *drumlin*, *erratic*, *fjord*, *glacial lake*, *glacial valley*, *hanging valley*, *lodgement till*, *moraine*, *nunatak*, *ribbon lake*, *roche moutonnée*, *terminal moraine*, *till* and *truncated spur*.)

glacial movement refers to the manner in which a *glacier* or *ice sheet* flows from one area to another. Ice is a relatively rigid substance which behaves more as a plastic as more stress is applied to it. The greater the pressure applied to it, the more it will deform in the direction of that pressure. The movement of ice has two main components:

- basal slippage – the sliding effect of a glacier over the bedrock surface. There is frequently an increase in pressure and friction between the bedrock and the ice causing a slight rise in temperatures which results in the melting of some ice. The resultant *meltwater* acts as a lubricant enabling the glacier to move more easily. (See also *basal slipping* and *pressure melting point*.)
- internal flow – movements within a body of ice resulting from the stresses applied by the force of gravity. Such movements often result in tears or crevasses within and at the surface of the ice. These may extend several metres down into an ice mass.

Glacial movement is greatest in temperate areas where more meltwater is available, and in areas with steep gradients. For temperate glaciers, the rate of movement is largely determined by the gradient of the valley in which it is located. Where a glacier moves into a section of the valley which has a reduction in its gradient (i.e. it is flatter), the ice will decelerate and become thicker. This is called compressing flow. Conversely, if the gradient becomes steeper, the glacier will accelerate and become thinner. This is called extending flow.

glacial processes are the actions of a *glacier* or *ice sheet* either within itself or on the area immediately around it. Such processes include *erosion*, *glacial movement* and *deposition*. The outcome of many processes will be to create distinctive *glacial landforms*.

glacial valley: also known as a glacial trough, this is a steep-sided, wide and flat-bottomed valley that was previously occupied by a *glacier*. The glacier filled the whole valley floor and performed *erosion* over its entire width. Subsequent *deposition* of sediment may have flattened the floor even more. The lower parts of the valley sides have often been smoothed by the passage of the glacier, whereas the upper parts are more jagged. The *long profiles* may be stepped, with rock basins occupied by *ribbon lakes* alternating with slightly raised sections. They are usually straight in their plan, with pre-existing interlocking spurs removed by the moving ice to form *truncated spurs*. Other typical glacial landforms such as *roche moutonnées* and *hanging valleys* may be found.

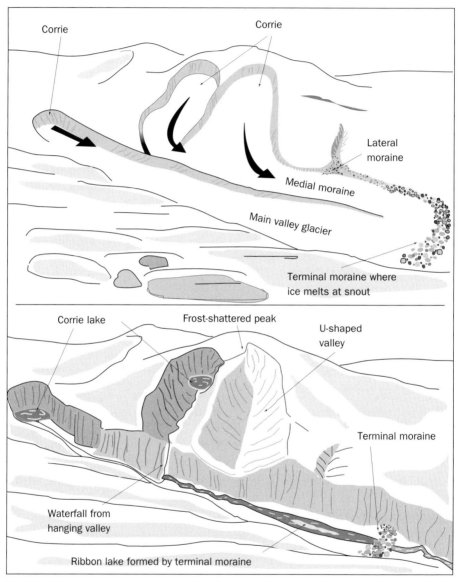

Glacial valley

glacier: a tongue-shaped mass of ice moving slowly down a valley. Processes and landforms associated with such a feature are said to be glacial.

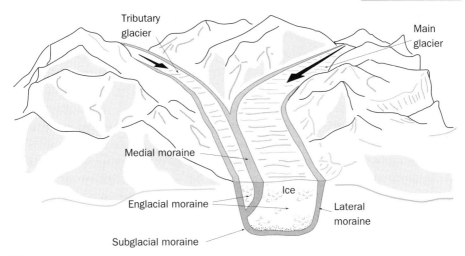

Valley glacier

gleying: a process of soil formation that takes place when the conditions are waterlogged or *anaerobic*. Under such conditions the pore spaces are filled with stagnant water which becomes de-oxygenised, causing the reddish coloured oxidised iron (ferric) to be chemically reduced to the blue-grey ferrous iron. Re-oxygenising causes red-orange patches to appear within the blue-grey soil, which is known as mottling. Many of the soils of upland Britain show evidence of gleying.

global brands are branded products, such as Coca-Cola, McDonald's and Bacardi, that have been marketed worldwide, the key to which is standardisation. In other words, a Big Mac should taste the same in Manchester, Milan and Melbourne.

global dimming is the gradual reduction in the amount of global direct irradiance at the Earth's surface. It is thought to have been caused by an increase in particulates, especially from aerosols. The vapour trails of aircraft (contrails) are believed to be another cause.

globalisation: the increasing interconnectedness of the world economically, culturally and politically.

global shift: a term that refers to the locational movement of manufacturing production on the global scale. In the late nineteenth century and for much of the twentieth century, manufacturing activity was concentrated in North America and Western Europe but in the 1970s and 1980s, *transnationals* began to establish plants in the *newly industrialised countries*. The traditional areas became more oriented towards tertiary activity and research and development. The focal point of manufacturing activity has to some extent shifted to the countries around the Pacific Ocean (Japan, Korea, Taiwan, Australia) and the west coast of the USA and Canada (the area referred to as the *Pacific Rim*).

global village: a term used to convey the idea that the world, in terms of transport cost and time, is 'shrinking'. Transportation of *raw materials* and products, transfer of capital, and communication of information and ideas have all become much faster as a result of technological developments. Events taking place in almost any part of the world can be transmitted by satellite and shown on television as they happen and the emergence of the internet and the superhighway further emphasise the trend. This space–time convergence in terms of a global village is to some extent exaggerated; many of the new forms of communication simply

117

follow established networks and therefore access to **new technology** and the instant world are not universally available.

global warming is the gradual warming of the Earth's atmosphere due largely to human activities. Evidence for this warming comes from climatic data and the recession of **glaciers** and ice margins. The atmospheric concentration of **carbon dioxide** has increased by about 15% in the last 100 years and the current increase in carbon dioxide is considered to be approximately 0.3% per year. Carbon dioxide allows incoming short-wave **radiation** to pass through but it absorbs some of the long-wave radiation emitted from the Earth to space. This produces a warming of the atmosphere, the **greenhouse effect**. It is generally agreed that this change will cause a rise in surface temperatures, but it is not yet possible to predict the extent or speed of the increase. A further doubling of carbon dioxide levels could perhaps raise average surface temperatures by 2 or 3°C, with greater warming in higher latitudes, perhaps by 7 or 8°C. At current rates of increase this would be reached towards the middle of the twenty-first century. One of the major reasons for the increase in carbon dioxide has been the burning of fuels that contain hydrocarbons – coal, natural gas and oil. Large-scale **deforestation** of areas such as the Amazon Basin have also contributed as trees are a major store of non-atmospheric carbon dioxide. The changes to the Earth's climatic regions and the melting of **ice cap** areas have serious implications for major areas of food production and low lying coastal zones, which would be inundated by a rise in sea level.

glocalisation is the local sourcing of products by **transnationals** in places where they assemble their 'global products' close to markets. At the same time, they are able to customise their products to meet local tastes or laws.

GNP: see *gross national product*.

Gondwanaland is the name given to the former supercontinent comprising the present-day continental areas of South America, Africa, South India, Australasia and Antarctica. *Wegener* suggested that about 200 million years ago all the continental areas were joined as one land mass which he called **Pangaea**, but in late Mesozoic times the land mass broke up and **continental drift** produced two supercontinents, the northern **Laurasia** being separated from Gondwanaland by the Tethys Sea.

gorge: a deep, steep-sided and narrow valley usually occupied by a river. Gorges are produced when vertical **erosion** is much faster than lateral erosion. This may result from excessively high river flow as when glacial **meltwater** discharges through a former valley and undertakes rapid downcutting. It may also occur when an existing river course is interrupted by uplift of land as in **antecedent drainage** when the river cuts vertically in order to main-tain its course. If the downcutting is fast enough the river will continue on its original course through a newly cut gorge.

government policies is a term that covers all aspects of government intervention to influence the natural, built or human environments. This includes agricultural measures such as grants, subsidies and **set-aside**; industrial incentives through **regional policy** for the attraction or relocation of industry; schemes to regenerate inner-city areas; general policies on housing, education and health which influence the development and distribution of these services.

government spending is the amount of public expenditure that a government undertakes with the income it has generated from taxes and all other sources.

graben: the downfaulted section of a *rift valley* system. When extending forces operate in an area of crustal rocks, faults develop and the mass between two parallel faults will move downwards. The blocks which are left upstanding on either side are called horsts.

graded profile: this is the long section of a river, which some researchers have suggested displays an even and progressive decrease in gradient down valley. This idea was originally put forward by W. M. Davis, who argued that irregularities in the profile, which could reflect changes in underlying geology, are worn away by *river erosion* to give a smooth graded profile. This is also referred to as the profile of equilibrium and it is linked to the concept of a graded river which has been defined by some geographers as being attained when the river uses up all of its energy in the movement of water and sediment so that no free energy is left to undertake further *erosion*. Such a balance between energy and work cannot occur at a particular point in time, but is suggested as an average position over a long time. Recent approaches consider rivers as open systems with *inputs* and outputs of energy, water and sediment. These systems can achieve equilibrium when inputs and outputs are balanced, but changes in the system bring adjustments in the profile as the river attempts to counter the change. In this way it regulates the system. Modern studies have shown that even irregular profiles may be 'graded' in that they are in equilibrium but the concave form of long section is very common and can be explained in relation to the controlling factors. The increased efficiency of the channel, with a larger *hydraulic radius* and increased *discharge* downstream enables the river to transport sediment over a progressively shallower gradient.

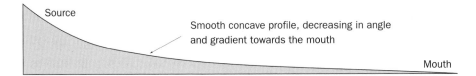

Source

Smooth concave profile, decreasing in angle and gradient towards the mouth

Mouth

gravity model: this can be used to predict the amount of interaction between two places in relation to their distance apart. The predicted interaction is calculated using:

FORMULA: $\quad I_{ij} = \dfrac{P_i \times P_j}{(d_{ij})^k}$

where P_i and P_j are the populations of the two settlements, d_{ij} is the distance between the two places and k represents a mathematical power, e.g. 1, 1.5, 2, 2.5, 3. If k = 1, the bottom line of the equation would be (d); if k = 2, the bottom line would be $(d)^2$ and so on. A low value of k represents a landscape that has good transport networks or is easy to cross because it is a flat area. A high value of k represents an area over which interaction would be more difficult; a high value of k indicates greater resistance to movement.

By predicting interaction between places and comparing this with actual or observed levels of movement, it is possible to calculate the value of k. If I, P_i, P_j, and d are known values, then by substitution and rearrangement of the formula it is possible to calculate k. Studies suggest that k normally equals 2 and the formula is usually written in the form:

FORMULA: $\quad I_{ij} = \dfrac{P_i \times P_j}{(d_{ij})^2}$

Interaction should refer to the movement of people, or goods, or the volume of transport services between places. It could also be applied to the interaction of people in terms of telephone calls between points in the network. Because distance is not always a good indicator of the separation between places, interaction could be predicted using time or cost of movement in the equation.

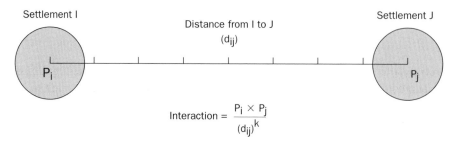

Gravity model

green belt: an attempt to prevent urban sprawl through the establishment of areas around the major cities in Britain in which planning restrictions were enforced to curb housing and other buildings and to conserve areas for agriculture and recreation. Green belts were first created following the Town and Country Planning Act 1947. Since their establishment, residential development in particular has leapfrogged the belts. This, in part, shows their success as relatively little development has taken place within them. In the 1980s they came under renewed pressure, particularly the green belt around London, as the demand grew for building land within southeast England.

greenfield sites: the sites for new developments (such as industry, housing and retailing) which have not previously been built on and are therefore found outside of present urban areas. With the predicted expansion of housing in the UK in the late twentieth and early twenty-first centuries, the argument has centred on the proportion of housing which will be constructed on *brownfield sites* within existing urban areas and that to be put on greenfield sites, many of which will be taken from the present *green belt*.

greenhouse effect: see *global warming*.

greenhouse gases are gases in the atmosphere, both natural and anthropogenic (caused by humans), which absorb outgoing terrestrial *radiation*. They are essential to helping determine the temperature of the Earth. The main greenhouse gases are *carbon dioxide*, *methane*, nitrous oxide, ozone and *chlorofluorocarbons (CFCs)*.

green movement: the environment, or green, movement has been well established in the USA for over 100 years. In recent times it has emerged as a strong political force, with parties such as the Die Grünen (Green) party of Germany taking 8.3% of the vote in the 1987 elections. In the 1989 European elections, green parties received high levels of support, nearly 15% for example in the UK. Since the late 1960s there has also been the establishment of a number of green pressure groups such as Friends of the Earth (1969) and *Greenpeace* (1971).

Greenpeace: a green organisation established in 1971 by activists protesting about American nuclear tests in the Aleutian Islands. The organisation has been particularly involved in the following campaigns:
● control of the whaling industry
● protection of seals, dolphins and porpoises

- reduction of trade in endangered species
- closure of nuclear power stations and nuclear reprocessing plants, and stopping the disposal of radioactive waste (and other chemicals) at sea
- stopping tests of nuclear weapons
- stopping **acid rain** and other atmospheric pollution
- keeping Antarctica free from exploitation
- controlling the disposal of North Sea oil platforms.

green revolution: in its narrowest sense, this refers to the breakthrough in plant breeding that produced **HYVs** of grain, but it has come to mean the application of modern farming techniques to **Third World** countries. This package consists of technology (including **fertiliser** and **pesticides**), water control, and **mechanisation**. The impact of the green revolution has been large in India where the government in the early 1960s, faced with food shortages, encouraged the introduction of new methods. India has therefore seen the ways in which the green revolution can benefit farmers, but it has also seen some of the drawbacks:

PROS:
- wheat and rice yields have more than doubled (in some areas up to a four-fold increase)
- an extra crop can sometimes be fitted into the year
- farmers' living standards have risen
- farmers can now afford tractors, better seeds, fertiliser and pesticides
- there is now a much more varied **diet** among rural dwellers
- new jobs have been created in such areas as fertiliser manufacture
- local **infrastructures** have improved, along with markets
- areas under **irrigation** have increased

CONS:
- HYVs require heavy fertiliser application
- HYVs need more weed control, are more susceptible to pests and diseases and are not as well adapted to **drought**
- some environmental damage through increased use of fertiliser and pesticides
- some farmers have certainly become richer, but many have grown poorer, increasing the differentiation between farmers and therefore disrupting the rural fabric
- this, in turn, has led to more rural-to-urban **migration**
- rural indebtedness has increased as farmers borrow money to pay for these **inputs**
- mechanisation has increased rural unemployment
- some irrigation schemes have resulted in **salinisation**
- some HYVs have an inferior taste.

In recent years there have been suggestions that a second green revolution is beginning to appear, based on sustainable agricultural practices and generally geared towards small-scale farmers with limited financial resources.

green tourism: see *ecotourism*.

greying population is a population structure in which the proportion of people aged 65 and over is high and rising.

grey water: non-industrial wastewater generated from domestic processes such as bathing, dish washing and laundry. This consists of 50–65% of all wastewater generated by domestic premises. There are a number of devices on the market that can recycle such water.

gross domestic product (GDP) is the sum total of the value of a country's output over the course of a year. It differs from *gross national product* because it does not include net income from abroad.

gross national product (GNP) is calculated by adding the value of all the production of a country plus the net income from abroad. Net income from abroad is the income earned on overseas investments less the income earned by foreigners investing in the domestic economy. In most countries, the growth in real GNP per head of population is the main measure of *economic growth*.

groundwater: water that collects underground in the pore spaces in soil and rock. When it fills all the pore spaces available, the rock or soil is said to be saturated. This water can be transferred slowly through rock as groundwater flow or *base flow*.

group of eight (G8) is the collective name given to the richest nations in the world, whose leaders meet at regular intervals. The original members (Japan, the USA, Germany, France, Canada, Italy and the UK) were joined in 1998 by Russia. The *European Union* has observer status at these meetings. Because of their economic power, members carry considerable political 'muscle' and consequently their meetings are widely reported.

growing season: that part of a year when the average temperatures are above 6°C. It may also refer to the length of time between the last severe *frost* of spring and the first of the autumn. This is therefore linked to the number of frost-free days required for the growth of a particular crop.

growth pole: a relatively small area within a country in which new economic development is targeted. The aim is to increase employment in such an area, and so increase the spending power of the local population. This then creates additional jobs in the service and construction industries as well as attracting more firms linked to the original industry. Growth poles therefore attract migrants, entrepreneurs and capital. Their creation is an integral part of *Myrdal*'s model of economic development. Growth poles have been adopted by many governments as part of an effort to reduce regional imbalance. Investment is concentrated in selected towns in the *periphery*, which will then, it is hoped, set off the *cumulative causation* processes. Incentives are often offered to attract industries to them, and away from the *core* region. An example of such a policy was the French government's creation of métropoles d'équilibres to act as counter-attractions to Paris. Investment was directed towards cities such as Lyon, Rennes and Grenoble. One of the main problems with a growth pole policy has been that few industries have relocated themselves entirely to such areas. More often, *branch plants* have been established, which are often the first to close during times of economic *recession*. Also, it is difficult to determine to what extent growth poles stimulate the local economy. Firms often retain their links with their former areas rather than create new ones.

groyne: a wooden construction built across a beach designed to trap and then stabilise sand and shingle. Sediment brought by *longshore drift* piles up on the updrift side until it is high enough to overtop it, or can drift around the end. They are usually placed at 5° or 10° to the perpendicular in order to prevent scouring on the downdrift side, although the exact angle will depend on the direction of the prevailing waves. Groynes can have disastrous effects for areas on their downdrift side. Deprived of their sediment, beaches are actively eroded, allowing waves to attack the cliffs behind. Tourism may also be affected by the loss of beach material.

guest worker (or Gastarbeiter) is a euphemism for an *economic migrant* from a poorer part of the world who has been attracted to provide cheap *labour* in an economically more developed country, such as Germany. Such migrants usually take up jobs which are not taken by locals because the jobs are dirty, unskilled and demand long and unsociable hours. Host countries hope that guest workers will only stay for a short period of time, say, 2 to 3 years, but in many cases they have not only remained longer, they have brought their families to live with them.

guided trail: a signposted and/or mapped and documented walkway through a rural or urban area designed to present features of the landscape which are both attractive and of interest. Guided trails are increasingly being used by local authorities to illustrate evidence of *conservation* and preservation schemes within their rural and/or urban landscapes. They have been common features of towns and cities with an interest in exploiting their historic heritage, such as Chester and York, but they are now forming part of the tourist provision of many less well-known areas.

gullying: an extreme form of *soil erosion* that produces steep-sided water courses subject to *ephemeral flash floods* during rainstorms. The *United Nation*'s *Food and Agriculture Organization* defines a gully as a stream channel which is more than 0.5 m deep and which prevents the normal use of farm machinery. Gullying commonly occurs on steep slopes which have little or no vegetation. Heavy rains are not intercepted and are unable to infiltrate into the ground. Water flows over the surface of the ground as sheet-flow, but small channels or rills are then formed on the surface which in time develop into gullies. (See also *rilling*.)

Gutenberg channel: a layer in the upper part of the Earth's *mantle* in which the speed of *earthquake* waves is reduced. It is a low velocity channel at varying depths between 100 and 200 km which is made up of less rigid material. The waves from earthquakes originating in this zone take longer to reach a given *seismic* recording centre. The existence of this zone was suspected by the American seismologist Gutenberg as early as 1926, but it took a further 30 years before worldwide recordings of waves from underground nuclear blasts provided confirmation.

guyot: a type of seamount or flat-topped hill rising above the ocean floor. The top of the feature is below sea level and has been planed off by *erosion* and many guyots appear to be the remnants of volcanic peaks. They are often referred to as tablemounts.

habitat: the specific environment in which plants and animals live. It may be regarded as the 'home' of an *organism* or group of organisms.

Hadley cell: the circulation of air taking place on either side of the thermal equator resulting from *convection* and subsequent subsidence. Solar heating close to the equator causes masses of warm air to rise creating surface low pressure. This air rises to the *tropopause* where it spreads out polewards at high altitudes. Further cooling of this air eventually causes it to subside back to the surface at approximately latitudes 30°N and 30°S to create surface areas of high pressure. Some of this air then returns across the surface to the equator as the *trade winds*. These are deflected by the effect of the *coriolis force* to create the northeast and southeast trades in the northern and southern hemispheres respectively. Hadley cells are to be found on either side of the thermal equator, although their position is not always constant due to movement of the overhead sun. They are named after George Hadley, who in 1735 formulated a model to explain why the so-called trade winds were so reliable for mariners in the tropics. (See also *inter-tropical convergence zone*.)

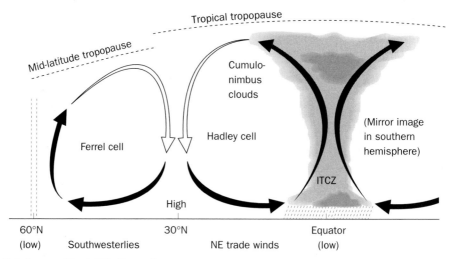

Relative position of Hadley cells

hail is a form of *precipitation* which is usually formed under conditions of extreme atmospheric *instability*, in cumulus or cumulonimbus *cloud* above the squall line of a steep *cold front*. Rapidly ascending draughts of air carry raindrops up into higher sections of cloud where the cold causes freezing. This produces a pellet of ice which, due to the turbulence in the air, collides with droplets of supercooled water. These freeze on impact with the hailstone and form a layer of opaque ice (rime). As the hailstone falls, the outer layer melts but with further uplift this layer refreezes to form an outer layer (glaze). The concentric, or banded, structure of the hailstone reflects the repeated up and down movement in the uplifting air. It will fall as hail when the weight and fall speed is sufficient to overcome the updraught of air. Hailstones have diameters up to 5 mm but can grow larger in lower latitudes, where deeper clouds form as the *tropopause* is higher above the Earth's surface.

halophyte: a plant which can tolerate saline conditions. An example of a halophyte is sea couch grass which is a common feature of coastal sand *dunes*.

halosere: a plant *succession* which takes place in areas where conditions are naturally saline. For example, in a British river *estuary* bare mud is initially colonised by green algae which can tolerate submergence for most of the 12-hour cycle. This traps further accumulations of mud which are then colonised by *halophytes* such as salicornia and spartina. The latter's long roots allow it to trap even more mud which forms the habitat for plants which are less tolerant of inundation and salinity such as sea aster, sea lavender and some grasses. Trapping of sediment is now greatly accelerated, the level of the land is raised and the vegetation cover becomes more dense. A *salt marsh* is created where the frequency of tidal inundation is reduced. Further *accretion* tends to be uneven, and so hollows and creeks are produced where the salinity may be too high, or the inundation too frequent for plants to exist. Further inland, juncus and other rushes and reeds grow, eventually being replaced by non-halophytic shrubs and small trees of rowan and alder.

hanging valley: a tributary glaciated valley perched above the level of the main glaciated valley floor. Streams flowing down these tributary valleys often descend to the main valley floor by *waterfalls*. They were formed by smaller *glaciers* joining a main glacier, but the smaller glacier had less erosive power and so could not erode as deeply.

hard engineering: a form of management of river *flooding* or coastal *erosion* that involves the construction of manmade features. For rivers it involves features such as dams, weirs, artificial and strengthened banks and diversion channels. In coastal areas, hard engineering refers to sea walls, *gabions*, revetments, *groynes* and *barrages*. It can also refer to cliff-face strategies such as drainage, fixing and regrading. (See also *flood control* and *coastal management*.)

hazard: an event that is perceived to be a threat to people, property and nature. It may take place in a variety of contexts:

- in the physical environment, for example, *earthquakes*, *volcanoes*, floods, storms and *drought*
- in the built environment, for example, the disposal of *chemical waste*
- in the human environment, for example, the spread of a *transmittable disease*.

Other hazards are said to be context hazards – these are widespread or global and are due to environmental factors such as climate change. For example, sea level rise increases the risk of coastal floods. Some key features of environmental hazards make them a huge threat:

- the warning time is normally short and onset is rapid
- humans are exposed to hazards because people live in hazardous areas through perceived economic advantage or over-confidence about safety

- most direct losses to life or property occur within days or weeks of the event, unless there is a secondary hazard
- the resulting *disaster* often justifies an emergency response, sometimes on the scale of international humanitarian aid.

Some socioeconomic characteristics, such as high population density, high poverty level and corrupt and inefficient government, increase people's *vulnerability* and amplify the risks, particularly of death from environmental hazards.

HDI: see *human development index*.

head: the local name for *solifluction* deposits in southwest England. Such deposits consist of poorly sorted *debris* usually located along the edge of a valley floor.

headland: a coastal promontory resulting from the existence of a more resistant rock. A headland is usually where there is a concentration of *erosion* by the sea, and erosional landforms such as caves, arches and *stacks* are common features.

heat budget: see *energy budget (the Earth)*.

heathlands are areas of rough pasture consisting of coarse grasses and shrubs such as heather. They can be found in a variety of locations within the British Isles:

- wetter upland areas where dwarf shrubs of heather, bilberry and crowberry exist as well as gorse, rushes and sedges; poorly drained areas within these are characterised by sphagnum moss and cotton grass
- drier lowland sites based on *fluvioglacial* sands and gravels or coarse sandstones; here purple moor grass and bracken exist alongside heathers.

Many heathlands, particularly in the moorland areas of Britain, owe their origin to the interference of humans – they are an example of a *plagioclimax*. The clearance of upland forests over the last few centuries has allowed heather to take advantage of the newly created exposed nutrient-deficient sites. Subsequent heavy grazing and burning have prevented the regeneration of forests and have allowed the heathlands to continue to exist. Indeed, many heather moorlands are now managed to maintain their cover in order to satisfy the growing demands for grouse shooting. Other areas, however, have been afforested with conifer *plantations*.

heavy industry is the large-scale production of goods which are either the basis for other products, or are large in size. Such industries are very reliant on the availability of coal and other bulky *raw materials*. Examples of heavy industry include the iron and steel industry, shipbuilding and chemicals.

helical flow: also called helicoidal flow, this is a downstream spiralling of water flow associated with meandering channels. The main current tends to flow in a straight line and therefore strikes the outer bank of the channel where the river bends. This leads to a return flow of water towards the inner bank close to the bed of the channel. The outward surface flow and inward bed flow produce a circular motion when viewed in cross-section. As the main flow of the channel is downstream, this circular pattern is superimposed producing a spiral movement or helical flow. This water movement may help in the transport of sediment, moving material from the undercut outer bank towards the inner bend where it is deposited as a *point bar*.

HEP: see *hydroelectric power*.

herbicides are chemicals applied to crops to control weeds.

hierarchy: the arrangement of features in an order of importance. In a geographical sense, one of the most significant hierarchies is that of settlements. In simple terms, a hierarchy based on size is: hamlet, village, town, city, *conurbation*. (See also *central place theory*.)

high-order goods and services have a large trade area and require a large population to support them. They tend to be goods that are purchased less frequently such as furniture and electrical equipment, or services such as accountants or barristers. Customers are willing to travel greater distances to obtain these more specialised high-order functions. These goods have a high *range* and high *threshold*. *Central places*, or settlements, which provide such high-order goods tend to be further apart in the landscape than service centres which sell low-order goods.

high-tech industry: refers to those industries that have developed within the last 20–30 years and whose processes often involve micro-electronics. Also included are industries such as *biotechnology*, pharmaceuticals and the manufacture of medical equipment. They could also be defined as industries that have high *inputs* of scientific *research and development* and produce new, innovative and technologically advanced products. In the UK, high-tech industries, although well distributed, have formed major concentrations in such places as:
- the M4 corridor, west of London, towards Reading and Newbury – known as the 'Sunrise Belt'
- central Scotland – 'Silicon Valley'
- Bristol and west into south Wales
- the M11 corridor to the Cambridge *science park*.

high-yielding varieties (HYVs): applies mainly to grain crops where developments in plant breeding produced seeds that could more than double the yields over existing varieties. Initial development took place in Mexico and later at the International Rice Research Institute (IRRI) in the Philippines where the so-called 'miracle rice', IR-8, was produced. Recent work, particularly in Malaysia, has seen a whole range of 'designer' seeds produced for individual farmers' needs. From work on wheat and rice, research has now gone into other crops such as maize and potatoes. HYVs therefore formed one of the major developments that was part of the *green revolution*.

PROS:
- can give up to four times the yield of former varieties
- some are more responsive to fertiliser
- shorter growing period which allows double or triple cropping to occur
- many varieties can be grown in a range of environments

CONS:
- require heavy doses of *fertiliser*
- less adaptable to *drought*
- many varieties are more susceptible to pests and diseases therefore requiring large *inputs* of *pesticides* and *fungicides*
- need more careful weed control
- new seeds have to be bought each year to ensure the purity of the strain
- some varieties have an inferior taste.

HIPC programme: the Heavily Indebted Poor Countries programme provides debt relief and low-interest loans to reduce external debt repayments to sustainable levels. However, countries do have to meet a range of economic management and performance targets to take part.

Hirschman, A.: an American economist who established some of the principles of the *core-periphery* model or the theory of polarised growth. Hirschman maintained that the movements of people, *entrepreneurs* and *capital* to the core was a process of *polarisation*. Growth is then supposed to *trickle down* to the periphery in order for it to develop. Such a feature would be the demand from the core for food and resources from the periphery which should stimulate economic activity in that area.

histogram: a method of illustrating a *dispersion measure* consisting of a set of vertical bars rising from a graph's horizontal x-axis. The vertical y-axis is labelled 'frequency'. The data is arranged into classes before plotting.

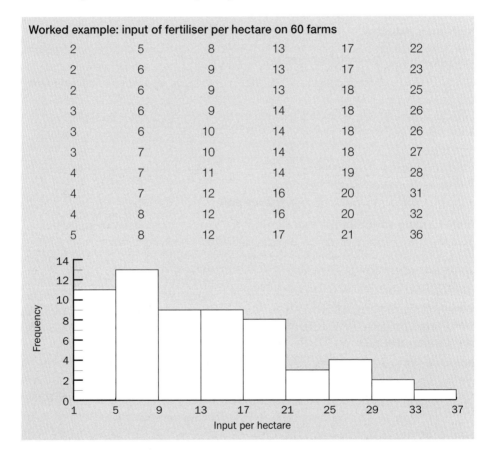

Worked example: input of fertiliser per hectare on 60 farms

2	5	8	13	17	22
2	6	9	13	17	23
2	6	9	13	18	25
3	6	9	14	18	26
3	6	10	14	18	26
3	7	10	14	18	27
4	7	11	14	19	28
4	7	12	16	20	31
4	8	12	16	20	32
5	8	12	17	21	36

Hjulström curve: a graph that shows the relationship between stream velocity and the size of sediment which is picked up, transported and deposited. Velocity increases as *discharge* rises and this enables the stream to pick up particles from the bed or banks of the channel. Hjulström's research showed that lower pick-up or critical *erosion* velocities were needed to move sand particles than for finer silts and clays or coarser gravels. The small-sized clay particles are difficult to pick up because of their cohesive properties; they do not occur as single particles. Silts and clays also lie on the channel floor and because of their smaller size they offer less resistance to water flow than larger particles and therefore a more powerful flow is needed to lift them. Once picked up, particles can be transported at velocities lower than the critical erosion velocity. As discharge and velocity decrease, sediment is deposited following the sequence shown by the fall line or settling velocity curve. Larger grains are deposited first and there is a progressive reduction in grain size as settling velocity decreases. Clays and smaller silt particles, once picked up, are only deposited at very low velocities; this explains the transportation of these fine materials into the lower reaches and estuaries of rivers where they may form mudflats.

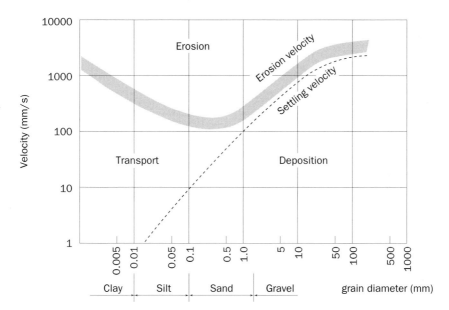

holding the line: a form of *coastal management* where the line of the existing coastline is retained by maintaining current defences or building new ones, such as a sea wall, where existing structures no longer provide sufficient protection.

Holocene: the geological period covering the last 10,000 years since the end of the *Pleistocene* glaciation. Within this period there have been a number of major fluctuations in climate in the British Isles which have resulted in significant vegetation changes.

homeworking: earning income from work undertaken at home. Traditionally homework-ers have completed *labour-intensive*, very low-paid jobs such as hand sewing and packing. Modern technology offers the possibility that more professional employees could work from home, armed with communication links such as electronic mail, fax and telephone.

honeypot sites: places with special interest or appeal which become very popular with tourists and which tend to become overcrowded at peak times. In the English Lake District, for example, Tarn Hows would be considered a honeypot site. This can give rise to a number of problems associated with overuse, particularly with congestion on access routes. In some popular areas it has become necessary to limit vehicular access at certain times such as in the Peak District *National Park* in England and the Yellowstone National Park in the USA. Where planners wish to concentrate tourists in order to protect other areas, it is not unusual for honeypot sites to be developed. This will ensure that these sites have adequate visitor amenities (toilets, car parks and picnic facilities).

horizontal integration occurs where a firm takes over or merges with another firm at the same stage of production. The production process starts with *raw materials* being processed, then moves through manufacture and assembly, followed by sales to wholesalers and then retailers. Horizontal integration can occur at any of these stages. The merger of two breweries would be an example of horizontal integration and would have the following advantages and disadvantages:

PROS:
- increases market power over the next or the previous link in the process
- enables greater *economies of scale* to occur

CONS:
- may restrict customer choice
- could possibly be seen as a *monopoly* and thus be referred to the Competition Commission.

horticulture: strictly speaking, the art of gardening, but in practice it is taken to include the production of fruit, vegetables and 'ornamentals' (decorative non-edible plants, e.g. flowers). It can be carried out both in the open and under protected structures such as glasshouses. Horticulture is highly *intensive agriculture*, particularly in terms of *labour*, usually carried out on small plots with large investments of capital for technology. Shortages of labour have led to some farms developing as pick-your-own (*PYO*) establishments. (See also *market gardening*.)

hot spots: areas where heat under the Earth's *crust* is localised. At such points, rising *magma* can produce *volcanoes*. The Hawaiian Islands, whose summits are volcanoes, have been formed in this way. This hot spot is stationary, so as the Pacific plate moves over it, a line of volcanoes is created. The one above the hot spot is active and the rest form a chain of islands of extinct volcanoes. The oldest volcanoes have put so much pressure on the crust that subsidence has occurred. This, together with marine *erosion*, has reduced these old volcanoes to seamounts below the level of the ocean.

housing associations are independent not-for-profit bodies that provide low-cost social housing for people in housing need. Any trading surplus is used to maintain existing homes and to help finance new ones. They are now the UK's major providers of new homes for rent, while many also run shared ownership schemes to help people who cannot afford to buy their own homes outright. Housing associations provide a wide range of housing, some managing large estates of housing for families, while the smallest may perhaps manage a single scheme of housing for older people. Much of the supported accommodation in the UK is also provided by housing associations, with specialist projects for people with mental health or learning disabilities, substance misuse problems (drugs or alcohol), the formerly homeless, young people, ex-offenders and women fleeing domestic violence. The day-to-day activities of housing associations are funded by rent and service charges payments made by, or on behalf of, those living in its properties. In this sense, housing associations are run as commercial entities and most do not depend on donations for their general activities.

Huff, D. L.: developed a probability model in 1962, which attempted to predict the probability of people purchasing goods in different shopping centres. The model takes into account the relative attraction of different shopping centres in an area and the distance that would have to be travelled in each case. Huff maintains that shoppers consider distance and attractiveness of towns. The attraction of towns can be measured in terms of such variables as population size, number of shops, number of types of shops and total turnover.

FORMULA:

$$\text{Probability of visiting town A} = \frac{\dfrac{\text{attraction of town A}}{\text{distance to town A}}}{\dfrac{\text{total attraction of all towns under study}}{\text{total distance to all towns under study}}}$$

human development index (HDI): a modern method of trying to define development by using several indicators. This was first proposed by the **United Nations** in 1990 and uses three variables based on income, education and **longevity**:

- income is measured by adjusting it to the purchasing power within the country
- education is taken from the adult **literacy** rate and the average number of years spent in school
- longevity is taken as the **life expectancy** at birth.

For each variable, a country scores according to its position on the list on a scale from one to zero. If a country is top of a list its score will be 1. If it was top of all three lists, its combined score would be 3, which averages out to an index of 1. On this basis, Norway is the most developed country in the world with an HDI of 0.965. The country showing the least development is Niger with an index of 0.311. It also helps you to see how a country stands in terms of the best conditions. Swaziland, with an index of 0.500, could be said to be about half as developed as Norway.

There are, however, a number of criticisms of the HDI:

- quantity says nothing about quality – what standards have been reached, for example, in literacy
- the index is a measure of relative rather than absolute development through its use of rankings
- it is not that far removed from classifying development according to **gross domestic product/gross national product**, as all the variables depend to a large extent on wealth
- it contains no measure that refers to human rights or freedom.

humidity is the amount of gaseous water (or water vapour) content of the atmosphere. This content varies a great deal throughout the atmosphere, being virtually zero in **polar** regions to over 5% in parts of the tropics. The amount of water vapour in the atmosphere is very much dependent on temperature, the warmer the air, the more water vapour that it can hold. Humidity can be measured in three ways:

- absolute humidity – the total amount of water vapour in a given mass of air, usually given as grams per cubic metre
- vapour pressure – as above, but in this case expressed as the amount of pressure exerted by the water vapour content compared with the atmospheric pressure as a whole, e.g. 20 millibars out of a total pressure of 1000 millibars
- **relative humidity** – the most important of the measures, being defined as the actual vapour content in comparison with the amount that the air could hold at the given temperature and pressure, measured as a percentage.

humus: the organic matter in the soil that is derived from the breakdown of plant and animal matter and secretions of living **organisms**. To be classified as humus this material has to be completely broken down. Humus gives the soil a dark colour and in combination with clay particles, the clay-humus complex, is one of the major contributors to **fertility** of the soil as it provides it with a high water and nutrient-holding capacity. It also helps bind the soil together and thus combats *soil erosion*. There are two forms of humus which vary in *soil acidity*: *mull*, a mild form of humus and *mor*, the acid variety. Some soil scientists recognise an intermediate category known as moder.

hurricane: the name given to a tropical *cyclone* that occurs in the Atlantic and Caribbean areas. Hurricanes are systems of intense low pressure (up to about 600–700 km across), formed over tropical sea areas and moving erratically westwards until they reach land where their energy is rapidly dissipated. At their centres they have an area of subsiding air with calm conditions, clear skies and higher temperatures, known as the *eye*. A number of factors encourage the development of hurricanes:

- the heat energy and abundant moisture supply of tropical sea areas
- the release of vast amounts of *latent heat* by *condensation* once *convection* has begun
- the deep layer of humid and unstable air to be found on the western sides of ocean basins
- a circulatory motion of the air, anti-clockwise in the northern hemisphere (encouraged by the *coriolis force*).

The major *hazards* associated with hurricanes are:

- strong wind speeds often over 200 km per hour
- heavy rainfall with associated *flooding* – e.g. 40 cm in 24 hours
- *storm surges* – e.g. hurricane Camille (1969): 7 m surge on Gulf coast (USA)
- *landslides* after heavy rain.

The cost of a hurricane, both in human and economic terms, can be enormous. In *First World* countries such as the USA, there may be adequate warning of the event and preparation for it. Consequently, there often tends to be enormous damage but very little loss of life. Communities will be able to adjust and recover in a relatively short space of time. In *Third World* countries, the loss of life, the damage and the economic hardship can be great, particularly where the economy is dependent on a narrow range of agricultural activities.

hybrid: a plant resulting from a cross between two genetically different plants. Many economically important cultivated plants have resulted through natural hybridisation or hybridisation induced through biological research. Examples include bananas, coffee, peanuts, alfalfa and roses. The process is important for agriculture as hybridisation can produce varieties with certain qualities, such as disease and *drought* resistance, quicker ripening and *high-yielding varieties*.

hydration is a process of *chemical weathering* in which certain types of minerals take up water and expand. This creates physical stresses within the rock which can lead to disintegration and assist other forms of weathering. Calcium sulphate can occur in both an unhydrated state as anhydrite and a hydrated state as gypsum. In chemical terms:

$$CaSo_4 \; + \; H_2O \; \rightarrow \; CaSo_4H_2O$$

anhydrite + water → gypsum

The uptake of water alters the solubility of the mineral as gypsum dissolves more rapidly than anhydrite.

hydraulic action is a process of stream *erosion* which moves loose, unconsolidated material due to the force of moving water and its frictional drag on sediment lying on the bed of the channel. As velocity increases, turbulent flow lifts a larger number of grains, particularly sand-sized particles, from the floor of the channel. Hydraulic action is particularly effective at removing loose material in the banks and this can lead to undercutting and collapse.

hydraulic radius: a measure of the *efficiency* of a channel; it is calculated using the formula:

FORMULA: Hydraulic radius (R) $= \dfrac{A}{P}$

where A is the cross-sectional area of the channel i.e. width × depth and P is the *wetted perimeter* i.e. the distance along the bed and banks which is in contact with the water.

A stream loses energy as a result of friction between the water and the bed and banks of the channel. Therefore a stream channel is more efficient, i.e. it loses proportionally less energy in overcoming friction compared with the *discharge* through the cross-section if the ratio between the area and the wetted perimeter is high. As channels become larger they tend to be more efficient; area increases at a faster rate than wetted perimeter. For example, a channel which is 10 m wide × 2 m deep has a value for R (if at *bankfull*) of 20/14 = 1.43. A channel which is 20 m wide × 4 m deep has R = 80/28 = 2.86. The area has increased fourfold but the wetted perimeter has only doubled.

hydroelectric power (HEP) is a clean and efficient *renewable resource* generated by using the power of running water to turn the turbines. This can occur at natural *waterfalls*, but more usually involves the construction of a dam to impound a lake. This creates a head of water, determined by the height difference between the water intake pipe and the turbine room, which produces the pressure required. Conditions favouring HEP schemes are stable underlying geology, *impermeable drainage basin*, high and reliable rainfall, mountainous areas with narrow *gorge*-like valleys. Access to markets is also important although some activities such as aluminium smelting, electro-chemical and metallurgical industries, which have very high electricity consumption, may be attracted to the HEP site even if it is remote. Although HEP is 80–90% efficient (based on the proportion of input which can be used as energy), compared with oil and coal at 30%, capital costs are high and schemes are expensive to install. Many of the most suitable sites are often remote from markets although recent improvements in technology allow transmission over distances in excess of 800 km. Despite its great potential it only contributes a small proportion of total energy, but it is an important source in parts of the world with few oil, gas and coal reserves, such as South America, Africa and parts of southeast Asia. It is a major power resource for some developed countries such as Norway, Sweden and Switzerland.

hydrograph: a graph which shows variations in river *discharge*, in cubic metres per second, over a period of time. Hydrographs can be plotted for a period ranging from a few hours to several months or even for a year. A storm hydrograph is plotted after a storm event to record the effect of the input of rainfall on the discharge in the river system. The starting and finishing level of discharge is called the *base flow* representing the river flow produced by *groundwater* seeping into the channel bed. As stormwater enters the system, the discharge rises, shown by the rising limb, to reach peak discharge which indicates the highest flow in the channel. The recession limb shows the fall in discharge to a final base level. The time delay between maximum rainfall intensity and peak discharge is called the *lag time*. The rising limb is produced mainly by *overland flow* which reaches the channel quickly; the steepness of the limb depends on the intensity and duration of the storm and the amount of antecedent soil moisture reflecting recent weather conditions. Heavy rain falling on a soil which is saturated as a result of a period of wet weather will produce a steep rising limb. The recession limb is produced mainly by water flowing through the soil into the channel and from overland flow from points at a greater distance from the channel. A small *drainage basin* tends to respond more rapidly to a storm than a larger basin and a short lag time and steep rising limb tend to follow after a period of very wet weather which has saturated the soil. These quick response hydrographs are described as flashy. The shape of the graph reflects the amount, intensity and duration of the rainfall, the seasonal weather conditions and the level of residual moisture in the soil layer, vegetation cover, the size of the drainage basin and the slope angles within the basin.

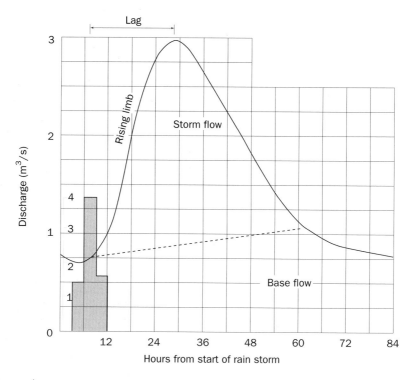

Hydrograph

hydrological cycle: the circulatory system by which water is transferred between the oceans and the land masses. *Evaporation* from the sea, lakes and the land surface and *transpiration* from vegetation produces water vapour in the atmosphere. As this moist air rises or moves laterally to cross land areas, it is cooled and the resulting *condensation* produces *precipitation* which falls on the land and sea. Surface *runoff* in streams or rivers and *percolation* which supplies *groundwater* return water to the sea.

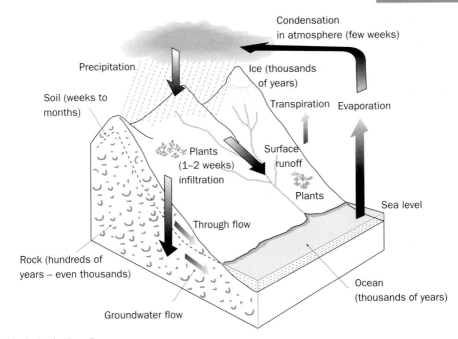

Hydrological cycle

hydrolosis is a type of *chemical weathering* in which there is a chemical reaction between a mineral and water. There is a reaction between the H^+ (hydrogen) ions or OH^- (hydroxyl) ions in the water and the ions of the mineral. It is particularly effective in the breakdown of feldspar which is commonly found in *igneous rocks* such as granite. The potassium feldspar (orthoclase) is broken down into a clay mineral, kaolinite, together with soluble potassium oxide and a soluble silica, or silicic acid, which are both removed in solution. This leaves a residual clay deposit which is considerably less resistant than the original granite. In chemical terms:

$$FORMULA: \quad 2KAlSi_3O_8 + 2H_2O \rightarrow Al_2Si_2O_5(OH)_4 + K_2O + 4SiO_2$$

orthoclase   water   kaolinite   potassium soluble
                                    oxide   silica

hydroponics: a method of growing plants using mineral nutrient solutions without soil.

hydrosere: a type of *primary succession* that begins on a waterlogged or submerged site. These wet environment *vegetation successions* may be divided into hydroseres, which develop in freshwater sites such as lakes or ponds, and *haloseres*, which originate in saltwater sites such as tidal mudflats. In a pond the first plants to develop are submerged aquatics. These help to trap sediment which enables other species, such as aquatics with floating leaves, to move into the area. The next seral stage sees the growth of reed beds and swamp conditions and, as *debris* accumulates with increased thickness of silt and sediment, the surface rises above the water level to produce a carr or fen. This leads to colonisation by alders and ferns which then further modify the environmental conditions to allow the entry of willows and ash. Eventually the climatic *climax vegetation* of *deciduous woodland* (oak or beech) is reached. Throughout the succession there are progressive changes to the soil conditions, the ground level microclimate and the faunal activity as the **ecosystem** is modified through time.

hygroscopic is a term used to describe a substance that will attract water. Impurities in the **atmosphere** act as hygroscopic nuclei allowing **condensation** to take place. Condensation can occur on salt nuclei at **relative humidities** as low as 78%. Large amounts of such condensation nuclei are present in the atmosphere and consist mainly of solid particles which are of varying sizes. The smallest particles are largely produced by combustion, natural and man-made, and these occur in high concentrations near industrial areas and in polluted urban air. Larger nuclei consist of ammonium sulphate and sulphuric acid which are both hygroscopic. These again tend to be the result of industrial processes or power production and are more prevalent in urban industrial areas. The largest nuclei are sea salt particles; these contribute only a small percentage of the nuclei involved in **cloud** formation.

hypabyssal: a name given to a group of **igneous rocks** which cooled and crystallised at intermediate depths in minor intrusions such as **sills** and **dykes**. These rocks have grain sizes which lie between the coarse size of granite which cooled slowly at depth and the fine-grained basalt which cooled quickly as a result of lava extrusion on to the surface. Individual features often display a variety of grain sizes within one rock type and this group of rocks overlaps with those which are coarse or fine-grained. In addition to medium-grained rocks such as dolerite, common in dykes and sills, this group includes volcanic pitchstones and the plutonic granite pegmatites.

hypermarket: a large retailing development usually located at a point that provides access to a large regional market. These centres often provide a wide range of shopper goods and may have other facilities such as restaurants, petrol stations and banks. They have very large, free car parking areas and they cater predominantly for the car-based customer. In Britain they are found both in urban and **urban fringe** locations; in the USA and France they have mostly been constructed on **greenfield sites**. They pose a serious threat to the future success of high street shopping in central urban areas; in many areas large numbers of wealthy car-owning consumers are being drawn to these convenient, large volume and often cheaper centres where there are few traffic hazards for the family and where parking is free. There is the added convenience of easy transfer of shopping from cash out to car space. They are often described as out-of-town shopping centres but the term is often misused describing any sizeable shopping development. Hypermarkets have a floor space in excess of 4500 square metres and although they serve a customer base from beyond the nearby urban area they are not as large as genuine regional shopping centres (RSCs), which typically have a gross area of over 38,000 square metres. These centres can often be distinguished from hypermarkets in that they contain at least one department store, perhaps 50 smaller outlets and leisure facilities such as cinemas or ice rinks. The smaller supermarket and superstore are also distinguished by size and by the fact that they usually consist of one company outlet.

hypothesis: a stated proposition that suggests a relationship between variables. In statistical terms it is something that needs to be tested. Hypothesis testing involves expressing an idea about that relationship in order to provide the basis for the collection of data which can be analysed statistically to prove, or otherwise, the relationship. A hypothesis relating to urban development could be that 'building height decreases with increasing distance from the city centre' or in a **drainage basin** 'discharge increases with distance from the source'. It is more usual to formulate a *null hypothesis*: 'there is no variation in building height as distance from the city centre increases'. If the result rejects the null hypothesis, there is some significant relationship between the two variables.

HYVs see *high-yielding varieties*.

A–Z Online

Log on to A–Z Online to search the database of terms, print revision lists and much more. Go to **www.philipallan.co.uk/a-zonline** to get started.

ice ages: the common term for the periods when there were major cold phases, known as glacials, and *ice sheets* covered large areas of the world. Prior to 1950 it was believed that there were four major glacials during the last 2 million years, but recent studies of *sediments* on the ocean floors now suggest that there have been multiple glaciations, each interspersed by warmer phases or *interglacials*. During the glacials the *ice caps* of the world expanded greatly. In Europe, the Scandinavian ice sheet covered much of Western Europe, including a large part of the British Isles. Smaller centres of ice originated from the Alps, and from the Scottish and Welsh Highlands and the Lake District. The reason for the onset of the ice ages is still the subject of debate. One theory is that there are celestial cycles when there are changes to the Earth's position in space, to its tilt and to its orbit around the sun. These changes combine to create a reduction in insolation. Another theory is that they are related to variations in sunspot activity.

iceberg: a floating mass of fresh water found in the world's cold seas in or near to *polar* latitudes. They are formed by:
- blocks of ice breaking off from floating polar *ice sheets* during the warmer spring weather
- blocks of ice calving away from *glaciers* as they flow into the sea.

Nine-tenths of any iceberg can be found below the level of the sea, and it is this which makes them such a danger to shipping.

ice cap: a large area of ice burying the landscape of up to 50,000 km² of land. Like the larger *ice sheet*, they tend to have a flattened dome-like cross-section.

ice core: obtained by drilling down through an *ice sheet* in order to examine the rings of annual accumulation of snow. From such rings it is possible to identify:
- the overall age of the ice sheet
- the variations in the degree of volcanic deposition during the timescale of the core
- the varying chemical composition of air bubbles at different stages in the past in order to identify changes in global temperature during the timescale of the core.

Ice core can also be applied to the block of ice found in the centre of a *pingo*.

ice lens: an accumulation of ice beneath the surface in *periglacial* and *tundra* areas which causes the ground to dome slightly upwards. They are formed in fine-grained soils such as clays and silts. With the onset of winter, the layer of the ground nearest to the surface freezes first. The ice in the frozen layer attracts water from below by *capillary action*. A layer of ice begins to grow while freezing continues, and while unfrozen *groundwater* is allowed to move upwards. The ice layer forms a lens shape parallel to the ground surface, and the thickening of the ice at its centre causes the ground to heave upwards. Stones resting on the ground surface are rolled sideways by the heaving ground, helping to create landforms such as *stone circles* and other types of *patterned ground*.

ice marginal landform: a feature located and formed at the edge of an *ice sheet* or *glacier*. It may include proglacial lakes, lateral *moraine*, *terminal moraine*, *kame*, *esker* and *outwash*. (See also *glacial lakes*.)

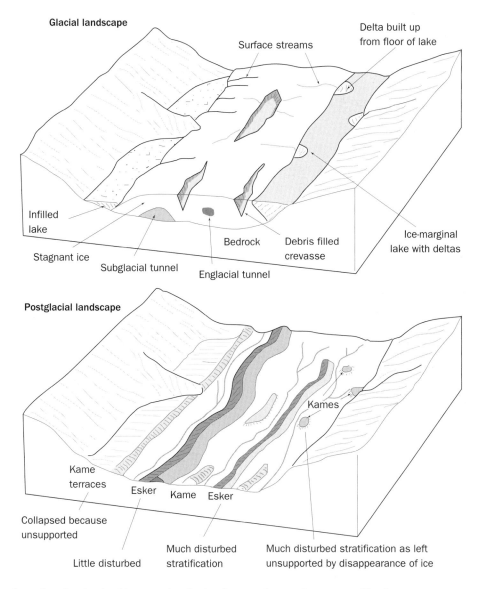

ice sheet: a body of ice covering the landscape of a very large area of land, over 50,000 km². It has a flattened dome-like cross-section, and is hundreds of kilometres in width. The two best known ice sheets are on Antarctica and Greenland.

ice wedge: a deep crack in the ground surface in *periglacial* environments which is broader at the top and may taper down to depths of 3 m or more. In plan, they frequently form a polygonal pattern across an area. In such areas intense cold temperatures (below −15°C) cause a shrinkage in the volume of the ice held within the soil. This leads to the development of cracks in the frozen ground. Early in the following spring, moisture collects in these cracks and freezes in their base since the ground is still frozen at depth (**permafrost**). This prevents the crack from closing again when the ground expands as the temperature rises. Later more moist times will cause the wedge to be filled with stones and gravels. In this way fossilised ice wedges can often be seen in gravel pits in areas which have been previously periglacial. Many such examples can be found in southern England, and are thus indicative of the former extent of periglacial conditions during the **ice ages**.

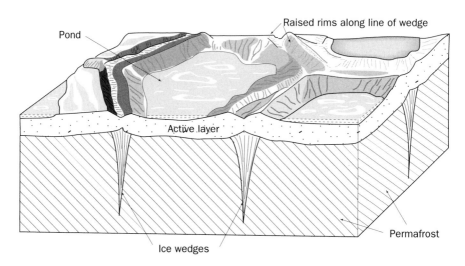

ICT: see *information and communication technology*.

ideology is the science or study of ideas. The term is generally used to describe the shared beliefs of a group of people, such as a nation, a religious sect or a group of economic or political theorists and their idealised concept of how to live in the world. For example, the ideology of **capitalism**, based on market forces, is different from **Marxism**. Ideological differences exist between different factions of Islam, e.g. Sunni and Shi'ite, and this has been significant in conflict within the Middle East. Usually a culture has multiple political ideologies, with some less popular than others, e.g. Conservative, Labour, Liberal and other parties such as the Green party in the UK.

igneous rock: a type of rock composed of crystal minerals that forms from the cooling and solidification of molten **magma**. If cooling takes place rapidly, say on the surface, then fine-grained and glassy rocks form, such as basalt. If cooling is slow, usually at depth, then coarse-grained rocks are created such as granite. They are often very hard, rarely show any layering and never contain fossils.

illegal migrant: someone who has entered a country without the relevant documentation such as a visa or work permit. They may be **economic migrants** seeking a better standard of living who have crossed a national boundary without going through the correct procedures. Such migrants may seek **refugee** or asylum status in that country.

illuviation: the washing in and **deposition** of material from an upper horizon of a **soil profile** to a lower horizon by downward percolating water. Illuvial horizons tend to be enriched by materials such as clays and organic substances.

IMF: see *International Monetary Fund*.

immature soils mainly reflect the properties and characteristics of the underlying *parent material*, because soil processes have not operated for sufficient time for the soil to be in equilibrium with its environment as with a mature soil. An immature soil is still in the process of adjustment to conditions and has poorly developed horizons, a high mineral content and a low percentage of organic material. A newly built-up surface such as a sand dune would develop an *aeolian* regosol, strongly influenced by the parent material. On steep slopes, where there may be downslope movement of the thinly weathered surface *debris*, soils such as lithosols develop. Processes are interrupted by the disturbance to the limited *soil profile*.

immigration is the movement of people into a country as distinct from *emigration* which is the movement of people out of a country.

impermeable describes a rock that neither absorbs water nor allows it to pass through. Granite is a good example. Such surfaces produce more *runoff* and therefore a greater number of streams.

import controls are *tariffs* or *quotas* designed to limit the number of overseas goods entering the domestic market. In extreme cases, some goods are prevented from entering at all, which is called an *embargo*. Other forms of import control are collectively called non-tariff barriers, which discriminate against imported goods in a more subtle way. One example of such a non-tariff barrier would be requiring excessive paperwork for imports, resulting in delays at frontiers.

import penetration: a measurement of the share of the home market taken by importers.

import substitution: this is where a country tries to establish its own industry which will produce goods that were once imported. The government then protects the developing industry from foreign *competition* by imposing *tariff* barriers.

incised meanders are formed when streams or rivers have cut deeply into the landscape. The nature of the landforms produced by incision depends to some extent on the rate of vertical *erosion*. When incision is slow, and lateral erosion is also taking place, an ingrown meander is produced. The valley becomes asymmetrical with steep cliffs on the outer bend and more gentle slip-off slopes on the inner bends. With rapid incision, where downcutting dominates, the valley is more symmetrical, with steep sides and a *gorge*-like appearance. These are described as intrenched meanders. Where incision is due to a relative fall in sea level, giving *rejuvenation* along the river course, *knickpoints* and *river terraces* will also be formed.

income at the personal level can be seen as wages and salaries. At the level of an economic organisation it is the revenue received when goods or services are sold to a customer, which may be an individual or another firm.

independent variable: when investigating the degree of *correlation* between two sets of data, changes in the independent variable are not related to the changes in the other variable (the *dependent variable*). For example, in a study of rainfall total and altitude, altitude is the independent variable: changes in altitude do not depend on rainfall total. Rainfall is the dependent variable: it is assumed that rainfall will vary with altitude. When plotting a *scatter graph* to investigate the relationship between two variables, the independent variable is plotted on the horizontal x-axis.

index numbers are convenient ways of showing change in a set of data over time. For example, let us take a car plant that produced 250,000 cars in 1995, 300,000 in 2000 and 350,000 in 2005. If we make 1995 the ***base year*** and let the value for that year be equal to 100, then the other values will be set relative to this. Therefore, in 2000 the index would be 120 [$= 100 \times (30{,}000 \div 25{,}000)$] and in 2005 it would be 140. The index tells us quickly that in 2000 output had risen 20% compared with 1995 and in 2005 by 40% compared with 1995. Index numbers are useful for displaying trends in that easy comparisons can be made between different time periods and between countries with very different absolute values for the variable. However, in using index numbers absolute changes are hidden, or ignored; they only display relative or percentage change from the base point.

index of dissimilarity is a statistical measure of the degree of difference between any two sets of paired percentages, for example, comparing the origin of imported goods into the UK at two different time periods. The index is calculated using the formula:

FORMULA: $\text{Index} = \Sigma(a - b)$

where a and b are the pairs of percentages to be compared. So, imports for the first date would be listed as a, and imports for the later date as b for each area of origin. Calculate $(a - b)$ for each pair of percentages, i.e. the percentages for each region of origin and sum these ignoring the + or − answer. Providing both sets of percentage data add up to 100, it does not matter whether a is greater than b or b is greater than a – the same result occurs. The index ranges from 0, perfect similarity (a − b for each set is 0), to 100, where the percentages are as dissimilar as is possible. The calculation of a single numerical index allows a more objective comparison of data.

index of similarity is a statistical measure of the degree of relative spatial concentration or dispersal of a particular activity or element of the population. It compares two data sets in terms of their spatial distribution. For example, the percentage of an urban area's retired population living in each ward of the city could be compared with the percentage of the total population living in each area. This could indicate the relative concentration of retired people, perhaps highlighting areas of need for particular services. The degree of similarity between the two sets of data can be calculated using the formula:

FORMULA: $\text{Index} = \dfrac{\Sigma d}{100}$

where d is the difference between the two percentages for each of the city wards. In adding up Σd, use only the + or the − values, as they should be equal. The index can range from 0, where both sets of percentages are identical (each d value = 0, therefore $\Sigma d = 0$) to 1. A low index indicates that the distribution of retired population is similar to the distribution of the whole population, retired people are neither over- or under-represented in each ward. A high index indicates that the two data sets are less similar; there are areas which are over-represented in terms of the percentage of retired popula-tion. It is not simply that there are higher percentages of retired population in some wards, they are at a greater concentration in relation to the distribution of the total population; they are over-represented in some areas.

indigenous means belonging to that area or region such as resources found within a national territory, plants that are native to a particular environment or ***ecosystem*** or the native population.

industrial estates are areas of modern, often *light industry*, that have a planned layout with purpose-built road networks. Many were established by local and national government incentives through assisted area schemes and the development of enterprise zones. They have been developed to provide employment, or to diversify employment, in areas of high unemployment. They have been built on relatively cheap reclaimed *derelict land* in the areas around the edge of the *central business district* as well as on the margins of urban areas. The more central locations often house small manufacturing and distribution firms as well as specialised industrial service functions, for example, the local security firm or companies preparing paint sprays for larger firms. Estates on the *urban fringe* are often occupied by the single storey, large space users such as the 'assembly' companies (vehicles to electrical goods) or *food processing* firms. These peripheral sites are not only cheaper, they also offer excellent access to the national market via the road network and they are near to the suburban housing areas which provide a potential workforce. Many of these outer industrial estates contain firms which have relocated from more central areas in response to inner area rebuilding and the shortage of land for expansion in central areas.

industrial heritage refers to the *conservation* and preservation of elements of the historical industrial landscape. These may be preserved in their existing form, protected from further degeneration, or they may be enhanced or improved by rebuilding in the style of the period in which they operated. Examples of such heritage schemes include Beamish (County Durham) where a street has been 'rebuilt', bringing in authentic shops, public house, garage and terraced houses from surrounding areas. The Ironbridge Gorge Museum includes six different sites in the area including original features such as Darby's iron forge, but also providing visitors with the opportunity to look at pottery and tile making, a history of the industry along the gorge, and the Blists Hill site where part of a settlement has been rebuilt. Other impressive, but smaller scale heritage sites include those managed by the *National Trust* such as the textile mill at Styal in Cheshire. The aim of these sites is to protect and preserve examples of the nation's industrial history while providing an educational resource for future generations. They are increasingly becoming tourist attractions and they generate employment; many of them are living museums, and employ people to portray the way of life and activities of the period in which they are set. Such developments are important in retaining industrial buildings as they generate some income, but it is not a very reliable source of funding for the preservation of our heritage.

industrial inertia: occurs when an industry remains concentrated in an area even though the original factors which caused it to be located there no longer continue to apply. Fixed investment in factory buildings and machinery, together with the general advantages of *infrastructure* in the area (housing, roads, railways etc.) have retained the industry. This will continue to be so for as long as the industry remains profitable.

industrial location theory attempts to explain the operation of geographical factors and the decision taken on the location of individual firms and industries. Weber's theory, published in 1909, emphasises the role of *raw materials*, markets and transport costs in determining the best or optimal location. Because of the assumptions made in the model relating to the decision-making process the optimal point was the *least-cost location*. This was achieved where the combined costs of transporting the raw materials to the plant and the product to the market were at their lowest. Weight- or bulk-gaining processes generally located near to the market while bulk-reducing industries located near to, or at, the location of the raw material which lost the greatest proportion of its weight during manufacture. Weber devised the material index to indicate the industry's preference for a market or raw material location. Because Weber had viewed the market as a single point in the landscape and emphasised costs as a factor, Losch explored the idea of maximising profits rather than minimising costs and he focused on revenues from a market area. Revenue would fall with distance from the point of production because of the cost of transporting the goods to the customer. In this way the edges of the market area could be defined where revenue was zero. Both Weber and Losch make underlying assumptions about the behaviour of decision-makers (economic man) which limit the application of the theories. The market-oriented approach seems to be more relevant to weight-gaining industries such as brewing, bakeries and assembly activities such as furniture, while least cost approaches are suited to the more traditional heavy industry such as iron and steel. Smith's theory of **spatial margins to profitability** takes into account both revenue and costs, and defines areas in which a firm can make a profit. Individual firms may locate in a sub-optimal location where they make a profit, but not the optimum profit. The exact location may reflect other factors: availability of knowledge and ability to use this information, and the aims of the decision-makers. Instead of assuming that all decision-makers aim to gain maximum return in any situation, it accepts that they may be be satisfied with a less than optimum profit in return for other benefits such as leisure time, or a pleasant environment in which to enjoy family life. This approach was summarised by Pred's work on the behavioural matrix. Because of the element of uncertainty regarding the outcome of such decisions, this is referred to as a probabilistic theory; earlier ideas by Weber and Losch were deterministic, they always gave a fixed outcome reflecting the particular factors.

industrial revolution (new) refers to the growth of new industries and businesses which are often based on high technology and which can be developed in a wide variety of locations. There has been a change from the traditional, single item, mass-produced output from a largely unionised workforce (**Fordist** industry) to a **flexible manufacturing system** with a more varied range of products within a company which can respond quickly to market needs.

industrial types: a classification of the main sectors of industry:
- *primary sector*
- *secondary sector*
- *tertiary sector*
- *quaternary sector*.

inequality refers to the uneven distribution of economic resources and wealth/income. This may be applied to individuals and groups in a society but can also refer to inequality among nations as shown by the **development gap**. Kuznets suggested that levels of economic inequality varied as countries moved through the **development continuum**; the **Kuznets curve** shows that countries with low levels of development have relatively equal distributions of wealth. However, as the country develops, **economic growth** tends to be concentrated in certain regions and wealth is generated by certain individuals. Greater inequality is then evident. As development continues, redistribution of wealth may occur through social welfare mechanisms and wealth spreads to other regions and a wider range of social groups. This reduces the level of inequality. These changes in inequality can be related to processes such as **cumulative causation** (**Myrdal**) and the concept of **core-periphery** (**Friedmann**). Kuznets used **Gini coefficients** to measure the degree of inequality between different groups in society. Geographical inequality can be graphically displayed using a **Lorenz curve**.

infant mortality is the number of deaths of children under the age of 1 year expressed per thousand live births per year.

infectious diseases are diseases that can be passed from one person to another (*transmittable diseases*). They are a major source of *morbidity* and *mortality* in both economically more developed and less developed regions. Diseases are caused by microbial viruses, *bacteria* and parasites. Transmission of an infectious disease may occur in a variety of ways:
- airborne spread of spores (e.g. influenza)
- direct contact, such as:
  - person-to-person (e.g. impetigo and sexually transmitted diseases)
  - inoculation, such as transfusions (e.g. hepatitis BC or HIV) or by contaminated needles
  - insect bites, such as mosquitoes (*malaria*) and ticks (Lyme disease)
  - entry through the skin (e.g. larvae that survive in soil or water (hookworm))
  - from objects such as bed linen and books
  - spread by contaminated food and water (e.g. swimming pools).

infield-outfield: a system of farming in which the small, enclosed fields near to the farmhouse (the infield) were heavily manured and cropped continuously while the areas further away (the outfield) were extensively grazed, cultivated at lower intensity with crops that required less attention or left fallow for long periods. Although at one time this was common in peasant agricultural areas in Europe it has been largely superseded, except perhaps for examples such as crofting in Scotland. Some present-day systems of subsistence cash cropping in West Africa, Bangladesh and southeast Asia have a similar arrangement around the village, which acts as a source of *labour* and the market for the crop, with roughly circular patterns of *land use* decreasing in intensity as distance increases.

infiltration is the passage of water into a soil. Water is drawn into the soil by gravity and capillary attraction. Infiltration takes place at a higher rate at the start of a rainstorm, but as the soil becomes more saturated then the infiltration rate falls steadily. Infiltration rates are influenced by:
- how much water is in the soil before a storm, which in turn is a reflection of the rainfall preceding a storm. A very wet soil has a lower infiltration rate than a slightly damp soil. However, a very dry, bare soil tends to have a low infiltration rate since small particles block the main water passageways, and water often rests on the surface
- the texture of the soil – sandy soils have a greater ability to allow water to pass through them, whereas clay soils have much lower infiltration capacities
- vegetation – plants promote a thicker soil and freer draining texture. They also have roots which enable water to pass into and through a soil more easily.

informal sector refers to the form of employment commonly found in the cities of *economically less developed countries* where people have had to find work for themselves, often illegally, on the streets and in small workshops. It is not part of organised industry, but does employ many thousands of people in unskilled and semi-skilled jobs. Some characteristics of the informal sector are:
- self-employed, small family enterprises based on little capital investment
- hours are irregular and wages are uncertain as bartering is the norm
- many jobs involve selling goods or providing services on the streets or at small roadside stalls

- *raw materials* are cheap, and often recycled waste, and so the quality of the goods is usually of a low standard
- a significant proportion of the workforce are children attempting to supplement the family income.

information and communication technology (ICT) is the means by which information can be stored, processed and communicated almost instantaneously to any part of the world. Advances in computers, telecommunications and micro-electronics have allowed this to occur and the pace at which they have taken place has been remarkable. The knowledge of and access to 'information' is now of crucial importance in a whole range of activities and it has had a number of geographical effects:

- *Homeworking* is becoming more widespread. People can earn an income from work undertaken at home and can communicate with employers via electronic systems. This has caused some *decentralisation* of employment.
- In-home shopping has already established itself in the USA and may soon spread to Western European countries. Telephone banking and other financial services are already available.
- Video-conferencing has reduced the need for people to travel within and between countries. It also reduces the expense of maintaining city centre office locations if there is less need for business people to meet face to face.

The adoption of the new developments in information and communication technology has largely taken place in pre-existing *core* regions and also in *transnationals*, thus reinforcing their economic superiority. Consequently, although technology offers the chance to decentralise and disperse economic activity, it seems to be concentrating it even more.

infra-red is that part of the sun's *radiation* (*insolation*) which has a wavelength greater than the visible red rays. The wavelengths are those greater than 0.75 micron (0.75×10^{-6} m). Infra-red rays penetrate hazy skies and darkness. Hence landscapes obscured by *cloud* can be photographed using infra-red-sensitive photographic plates.

infrastructure is the name given to the road, rail and air links, sewage and telephone systems and other basic utilities which provide a network that benefits business and the community. One of the main advantages which industrialised countries have over less developed ones is the existence of an efficient infrastructure. Building up such a system is very expensive, requiring a great deal of *capital* to be set aside, capital which poorer countries find difficult to afford. Successive British governments have been criticised for allowing our transport infrastructure to fall behind that of our main European competitors.

inheritance laws have an effect on farming in some societies, if, on the death of the farmer the land has to be divided equally among the sons. This often leads to the breakdown of the farm into numerous scattered small fields, a process known as *fragmentation*. In parts of Britain, the tradition of equal division was known as *gavelkind*, as distinct from the situation whereby the eldest inherits (*primogeniture*).

inner city: in most cases the inner city refers to those areas of older (mainly nineteenth century) residential and industrial development lying between the *central business district* and the suburbs of major cities and *conurbations*. In the *Concentric urban model* model it corresponds to the zone in transition. The inner city, though, is seldom a continuous area nor always central in location. They are generally areas of decline in both population and employment opportunities and have become synonymous with **deprivation**, areas of run-down housing, poor environmental conditions, high levels of unemployment and many other characteristics of poverty and social malaise.

innovation means bringing a new idea into being either within the market-place or the workplace. For new products the sources of innovation may be based on **new technology**, new design or a wholly new invention. Process innovation is also of great significance as it can lead to a major cost advantage over competitors. When the British company Pilkington invented a new way of making glass more cheaply and of a far higher quality (the float glass process), it provided not only a direct competitive advantage but also earned considerable sums in licensing fees from overseas manufacturers.

inputs are the elements that go into producing goods or services, such as **raw materials**, **capital** and the workforce (**labour**).

inselberg: an 'island mountain' or an isolated hill standing above an extensive **pediment** or plain resulting from **erosion**. They are steep-sided and often dome-like in appearance with a marked break of slope at their base. They are found in the humid tropical regions, the **savanna**, and also desert areas and there is some disagreement about their origin. Some writers argue that they are formed by the parallel retreat of scarp slopes during a long period of erosion; inselbergs would therefore be the residual mass that has not been eroded. Other authorities suggest that they are caused by differential tropical weathering in which the effectiveness of the processes may have been influenced by the density of joint spacing. This weathered **regolith** would later be removed to expose the unweathered basal rock. This would produce inselbergs as 'exhumed' features, brought to the surface by the removal of overlying layers of weathered **debris**. This latter theory is currently favoured.

insolation is the heat energy from the sun; incoming solar radiation. This consists of the visible spectrum together with ultraviolet and **infra-red** rays. Insolation varies with latitude as a result of variations in the length of the day and the angle of incidence of the sun's rays. These global variations influence the **general atmospheric circulation**. Of the total solar **radiation** that reaches the outer layer of the atmosphere, about 51% reaches the ground surface (34% directly and 17% as indirect or scattered radiation from particles in the atmosphere). About 14% is absorbed by the atmosphere and around 35% is reflected or scattered back by **clouds**, atmospheric particles and the Earth. Different sources quote different values for these various components. It is important to appreciate that these can only be presented as average figures; latitudinal variations of **albedo** due to the nature of the surface and the density of cloud cover are difficult to determine with a precise accuracy.

instability is the tendency of an **air mass** to continue to rise because it is warmer than its environmental surroundings. When air is forced to rise, for example, over a mountain barrier or along a front, it cools according to **adiabatic** laws. If the rising air cools at a slower rate than the environmental air then it will be warmer than its surroundings and will continue to rise, following the principle that warm, or less dense, air rises. When the **environmental lapse rate** is high, this tends to produce instability. A high ELR is defined as a rate which is greater than the **dry adiabatic lapse rate**, i.e. greater than 1°C per 100 m. Instability leads to greater convectional movement and overturning of the air which produces deeper cumulus and cumulonimbus *cloud* with heavy rain and more 'violent' forms of weather such as *hail*, thunder and lightning. When cold *polar* air masses move towards Britain they are warmed from below as they move into more southerly latitudes. This steepens the lapse rate and therefore encourages instability.

intake is an area of upland or moorland that is enclosed or fenced in order to enable some form of land improvement. This could involve drainage schemes to allow more productive use of the land or the use of *fertiliser* to improve upland pasture.

integrated pest management is a way of attempting to limit pest damage to crops with a much lower use of chemicals. Other ways in which pest damage can be limited in such programmes include:
- the deliberate introduction of pest predators into the system
- not using pesticide in a way that also kills the predator
- the timing of crop planting to avoid the time of the year when the pests are most active
- developing physical methods of destroying pests such as small flame burners or constructing barriers against the pest
- rotating crops to avoid build-up of crop-specific pests
- if *pesticide* has to be applied, use it at the time when it will have maximum effect, perhaps by using computer technology to assess the time of maximum usefulness.

integration: the joining together of different regions or countries in order to gain benefit. The *European Union* is an example of an integrated group of countries, with cooperation taking place over several economic, social, political and environmental policies. The future extent of their integration is the subject of considerable debate within and between each of the constituent countries. Supporters of further integration point to the greater economic and political power that would ensue. Opposition comes from those people who believe that there would be less control over internal affairs of each country.

intensive agriculture is characterised by high *inputs* of capital and/or *labour* per unit area. Such levels of input would normally produce high outputs per unit area or high yields. This system may occur with both *subsistence* and *commercial* economies. An intensive subsistence form of agriculture is the production of paddi rice in southeast Asia. Commercial examples include *market gardening*, *horticulture*, dairy farming and *battery farming*. Intensive approaches often reflect pressures such as land shortage, high population density or the production of highly perishable foods for an urban market.

interception is the process by which raindrops are prevented from falling directly on to the soil surface by the presence of a layer of vegetation. Water is intercepted by plant leaves, stems and branches particularly in the tree layer but also by species in the shrub or field layer. During prolonged or heavy rain, the capacity of the plant surfaces may be exceeded and water will drip off the leaves and branches (*throughfall*) and will run along branches and down the trunk (*stemflow*). *Evaporation* will remove some of the water held on the leaves and this is referred to as interception loss.

interdependence refers to the interrelations within a system. In economic geography, a change or a decision taken in one location can affect the operation of a part of the manu-facturing process elsewhere. A fall in price, or overseas *competition* in the form of cheap imports, can influence the success of an industry. If industry declines then there are knock-on effects on employment, social conditions, health and crime levels. In physical geography a change in one part of the *ecosystem* brings about a change in other areas. Removing trees may increase *runoff* into rivers which changes the pattern of *erosion* and *deposition* altering the system.

interglacial: the period of relatively warmer climate separating two stages of ice advance. Originally three interglacial periods were identified in Europe based on the interpretation of deposits in Alpine valley terraces. These separated the advance phases of the Gunz, Mindel, Riss and Wurm. Investigations of the isotopic ratio between the oxygen isotopes O^{16} and O^{18} in deep sea and *polar ice cores* has indicated many more, but shorter, interglacials. There may be up to 20 with lengths up to 10,000 years. Conditions in these period have been reconstructed by using pollen analysis on peat deposits preserved in *till*.

intermediate technology: the matching of technology in the *Third World* to the needs and skills of the people of the area. Too many projects, developed in the *First World* and applied to the Third World, failed because the gap between the people's knowledge and the modern technology was too great. With a lower level of technology it is possible for people to be able to be taught the technical understanding and skills that they require to become self-sufficient. In parts of the *Sahel* region in the north of Africa, to combat *drought*, the intermediate technology that has worked has been the introduction of wind pumps to bring water to the surface.

internal deformation is a form of *glacial movement*. It is the slow downhill creep of ice due to slippage both within and between the ice crystals. Ice deforms under its own weight due to gravity and the movements of tiny ice crystals. Thicker and warmer ice deforms more rapidly, although the overall movement is very slow, only around tens of metres a year. There are two main processes of internal deformation: creep, which forms fold structures, and *faulting*, which occurs when ice cannot creep fast enough and forms superficial tensional fractures. The rate of internal deformation is greatest at the base of the *glacier* where pressures are at a maximum. This type of flow can occur in both *polar* and temperate glaciers.

internationalism: the global economy has experienced internationalism as each country's economy has become less self-contained and more part of the global process of change. The UK economy of the nineteenth century was very strong on the export of manufactured goods. By contrast in an economy that is internationalised there would be a strong two-way flow in all trade categories. This is much more characteristic of the UK today.

International Monetary Fund (IMF): the banker to the world's central banks. In other words, if a country requires to borrow money, it can apply for a loan from the IMF. It is then likely that the IMF would send a team of inspectors to the country, who would advise on the conditions to be tied to the loan such as cuts in government spending.

interquartile range: a statistical method showing the dispersion of a set of values around the *median* calculated by finding the difference between the upper and lower quartiles. When calculated, 50% of all the values fall within it. The smaller the interquartile range, the more the values are grouped around the median. When divided by two, the interquartile range gives the quartile deviation.

When data is placed in rank order (highest to lowest):

FORMULA:

$$\text{Upper quartile} \left(\frac{n+1}{4}\right) \text{from top of rank order}$$

$$\text{Lower quartile } 3\left(\frac{n+1}{4}\right) \text{from top of rank order}$$

interstadial: a brief time period during a glacial phase when the climate improves sufficiently for limited *glacier* or *ice sheet* retreat. It is shorter than an *interglacial* period but may be long enough for some aspects of vegetation development as in the Allerod inter-stadial during the last ice phase in Europe.

inter-tropical convergence zone (ITCZ): the movement of the *trade winds* towards the equator results in *convergence*. This, combined with intense heating, leads to rising air which, in turn, is facilitated by *divergence* aloft. This zone of convergence is known as the ITCZ and forms the equatorward side of the *Hadley cell*. The ITCZ is associated with an area of *precipitation* that moves north and south of the equator through the year following the apparent movement of the overhead sun. (See also *atmospheric circulation models*.)

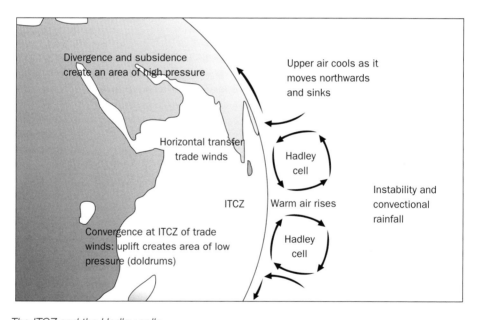

The ITCZ and the Hadley cells

intervening opportunity theory: this was put forward by Stouffer in 1940 as an attempt to explain levels of movement between places. Interaction between places can be predicted using the *gravity model* which considers the size of the two settlements and the distance be-tween them as the basis of the calculation. Stouffer suggested that the amount of movement over a given distance is proportional to the number of opportunities at the point of destination, but is inversely proportional to the number of opportunities between the point of origin and the destination. A large number of intervening opportunities reduces the number of movements to the final destination; some of the potential movement is 'intercepted' en route.

This can be applied to shopping. The presence of closer, good quality shopping outlets will tend to reduce the number of shoppers who are willing to travel to a large, but more distant shopping centre. The formula for predicting movement is:

FORMULA: $P_{ki} = \dfrac{O_i}{O_j}$

where P_{ki} is the number of people living at k who will travel to i to make a purchase, O_i is the number of opportunities to buy the item at i, i.e. the number of shops which sell the item and O_j is the number of opportunities to buy the item on the route between k and i.

interventionist: an individual who believes that government intervention can help make markets more efficient, and protect individuals from socially irresponsible business behaviour. Such a person is likely to promote an active economic policy by government, as opposed to a *laissez-faire* approach.

interventionist policies are those pursued by governments that believe it is their duty to exert a strong influence over the running of a country's economy. Such intervention might include:

- 'rescue packages' to help out large firms which have got into financial trouble
- prices and incomes policies
- laws to provide stronger protection to consumers or workers
- active support for new firms
- *regional policies*
- tougher laws and higher fines for pollution etc.
- wages councils and industrial training boards
- minimum wage policy.

They were very much in evidence in the UK from the early 1940s until 1979 when Margaret Thatcher's Conservative government came to power. She believed that the government should stand back from the detailed working of the economy because markets are the most efficient way to allocate resources. Government's role, she argued, is simply to provide an efficient background against which markets can work. (See also *Thatcherism*.)

intervention price: a guaranteed minimum price set by a government for an agricultural commodity. Within the *Common Agricultural Policy* of the *European Union* the intervention price operates when prices on the open market are falling perhaps as a result of over-production of a particular crop. When the price falls to the base level of the intervention price, the EU purchases the crop at the agreed minimum price. This is a very controversial policy as it encourages, and rewards at an agreed basic level, farmers who over produce a crop for which there is no market. This stimulus to production has led to large stores of produce which have been labelled by the press as the beef mountain, the butter mountain and the wine lake. It is difficult to sell off or give away these surpluses to underfed areas of the world because of the detrimental effect this would have on world trade. As it is, producers in other countries object to the level of protection and support the EU gives to its farmers; such policies can destabilise trade and economic stability.

intrazonal: a classification of soils which covers those that reflect the dominance of a single local factor. This includes the influence of geology and extremes of drainage. Typical *azonal* *soils* are *rendzina* and *terra rossa*, both developed on rocks that have a high calcium content such as *limestone*, and gley soils, developed where conditions are waterlogged. (See also *gleying*.)

intrusive: molten rock (*magma*) which is injected into the *crust* of the Earth from beneath is said to be intrusive. This may be exposed by later *erosion* of the overlying rocks. A typical intrusive landform produced in this way is a *dyke*.

inversion: see *temperature inversion*.

invisible export: the sale of a service to an overseas customer. As well as services such as banking, airline travel and insurance, visits by foreign tourists are counted as invisible exports, since they bring in income from overseas.

invisible import: the purchase of a service from an overseas supplier.

invisible trade consists of the import and export of services. They include financial services such as banking and insurance, as well as tourism and shipping. Invisibles account for approximately a quarter of world trade. Traditionally, the UK has enjoyed a surplus of invisibles, which has helped to pay for the long-standing deficit on *visible trade*.

inward investment is the *capital* attracted to a region or a country from beyond its boundaries. An example of such investment is a Japanese car manufacturer opening an assembly plant in the UK.

ionic exchange: see *cation exchange capacity*.

iron pan: a thin layer of mineral grains cemented together by a high concentration of rede-posited iron and humus which is located in the lower horizons of a soil. An extensive iron pan is impervious to water, which causes waterlogging of the soil directly above it. An iron pan is a common feature of the *podsol* type of soil.

irrigation: the artificial watering of the land by humans during dry periods of weather. For many hundreds of years, techniques such as the use of tanks, inundation canals and the Archimedes screw have been slow, primitive and inefficient. Newer developments, involving *barrages*, large reservoirs, pumping stations and booms, have allowed vast areas of land to be cultivated. Such schemes have not been without their problems including the silting up of reservoirs and *salinisation*. In recent years smaller scale tube wells have become more widespread in the less developed world.

island arc: a chain of volcanic islands located on the continental side of an *ocean trench*. It is associated with a *subduction* zone where an oceanic plate is being pushed below a continental plate. Subduction causes some of the oceanic plate to remelt, causing silica-rich lavas (andesites) to rise to the surface and create *volcanoes*. This situation arises in the western Pacific Ocean where the islands which make up the Aleutians, Japan and the Philippines constitute island arcs.

isopleth map: one in which lines have been drawn to show the variations in the values represented. A single isopleth connects the points on the map at which the variable repre-sented has an equal value.

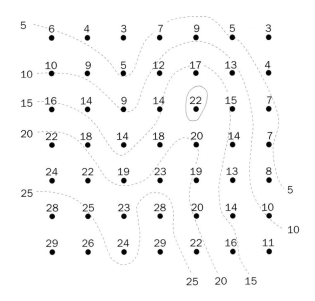

Isopleth map

Examples of isopleth maps include:
- contour maps connecting points of equal altitude
- isobar maps connecting points of equal atmospheric pressure
- isotherms connecting points of equal temperature.

isostatic readjustment concerns a localised change to sea level. During the major glaciations, the weight of the ice depressed the Earth's *crust* immediately beneath the ice. This caused a local relative rise in sea level. However, since the *ice sheets* have melted, the Earth's crust which had been beneath the ice sheets has begun to rise. This has caused a local relative fall in sea level.

isotim: a line connecting points of equal transport costs. They are generally associated with the Weber model of industrial location where an isotim is a line connecting the equal transport costs for either a raw material or a finished product.

isotope analysis: an examination of the varying forms of an element which have differing atomic weights. In geography, the main use is to identify concentrations of a certain isotope which exist under different climatic conditions. For example, warmer conditions encourage greater accumulations of the oxygen isotope 0–18 in water. If this is frozen into ice in an *ice sheet*, then it will be preserved as a climatic record of such conditions. Other isotopes can be used as a means of dating. The best known is carbon-14, which decays radioactively at a known rate, and this can be compared with carbon-12 which does not decay. (See also *carbon dating*.)

isotropic: a description of a state of uniformity. It is frequently used to describe a flat land surface where soil *fertility* and climate do not vary. There are no physical barriers to movement and so transport is equally easy in all directions.

ITCZ: see *inter-tropical convergence zone*.

jet stream: a band of very strong winds, up to 250 km per hour, which occurs in certain locations in the upper atmosphere. A jet stream may be hundreds of kilometres in width, with a vertical thickness of 1000 to 2000 m. On average they can be found at altitudes of 10,000 m. They are the product of a large temperature gradient between two **air masses** which have markedly different temperatures. There are two main locations of jet streams:

- the polar front jet – a westerly band associated with the meeting of polar and tropical air above the Atlantic Ocean at approximately latitude 60°N and 60°S. The precise location of this jet stream varies on a daily basis, but aeroplane pilots may seek to 'ride' in it when going from west to east, and to avoid it when flying from east to west
- the subtropical jet – also generally westerly and associated with the poleward ends of the **Hadley cells** at approximately 25°N and 25°S. However, in summer above West Africa and southern India this jet may become easterly. This is due to temperatures over the land in these areas being higher than over the more southerly sea areas.

JIC: see *'just in case' production*.

JIT: see *'just in time' production*.

Johnson, D. W.: formulated a simple classification of coasts in 1919. This was a genetic classification, attempting to take account of the origin of coastlines. He divided coasts into four main types:

- shorelines of submergence, where there was a relative rise in sea level due to subsidence of land or a rise in sea level, or both
- shorelines of emergence, where there was a relative fall in sea level
- neutral shorelines – this includes coasts where the exact form is due to neither emergence nor submergence but to a new form of construction or tectonic process: **deltas**, **volcanoes**, fault systems
- compound shorelines, including coasts which have an origin combining at least two of the above classes.

In practice, it is difficult to assign coasts using this classification; most of the world's shorelines show signs of both emergence and submergence due to oscillations of sea level during the **Pleistocene** period. Coasts can only be classified according to the most strongly marked characteristic. Most early attempts at classification present difficulties because they are either inflexible or descriptive. Valentin, in 1952, proposed a more dynamic approach based on advancing and retreating coastlines.

joint: a natural vertical crack in a rock. Some rocks are more jointed than others, for example, **Carboniferous limestone**. Joints are places which **weathering** and **erosion**

processes can take advantage of, and thus may be regarded as lines of weakness in that rock. Well-jointed rocks tend to weather and break up more easily.

'just in case' (JIC) production: a manufacturing system that involves the stockpiling of *raw materials* and components at the site of the production process. This was the traditional way of organising production in many Western countries and it has proven to be both inefficient and wasteful in many industries. It is being superseded in many cases by the *'just in time' production* system introduced by Japanese industrialists.

'just in time' (JIT) production: a manufacturing system designed to minimise the costs of holding stocks of *raw materials* and components by carefully planned scheduling and flow of resources through the production process. It requires a very efficient ordering system and reliability of delivery. It has been introduced to Britain by Japanese car manufacturers such as Nissan. In this case there is hourly delivery of some parts and many component manufacturers have been forced to locate near to the assembly plant. Another requirement of a JIT system is that there must be 'zero defect' or 'total quality'. Nissan have very strong links with their suppliers which are monitored rigorously.

Do you need revision help and advice?

Go to pages 325–53 for a range of revision appendices that include plenty of exam advice and tips.

kame: a *fluvioglacial landform* consisting of a mound of sand and gravel deposited by the *meltwater* from a *glacier*. The features are similar to a series of *deltas* and are formed along the front of a stationary or slowly receding glacier or *ice sheet*. At the sides of the glacier, meltwater streams often deposit material which, after the retreat of the ice, form kame terraces. Like all fluvioglacial landforms their deposits show sorting by water.

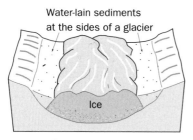
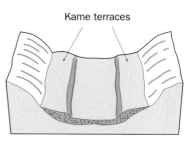

Water-lain sediments at the sides of a glacier

Kame terraces

Ice

Kame terrace formation

karst: the name given to the scenery that develops on *Carboniferous limestone* in the British Isles. Similar types of *limestones* in different parts of the world will produce the same features. The name comes from the Karst (or Carso) region of Slovenia where this type of scenery is very well developed.

katabatic refers to winds that flow down valley sides and valley floors. This typically happens at night in upland areas as the cool, denser air moves to lower levels. In the British Isles this is a very gentle feature, but on *glaciers* and in such areas as Greenland and Antarctica they can be extremely strong. (See also *anabatic*.)

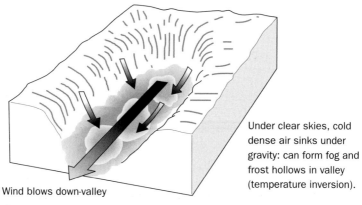

Under clear skies, cold dense air sinks under gravity: can form fog and frost hollows in valley (temperature inversion).

Wind blows down-valley

Katabatic flow

kettlehole: a depression in a glaciated lowland area that usually contains a very small lake. Their formation results from blocks of ice that become detached from the ice front as the *glacier* or *ice sheet* retreats. These blocks become covered in glacial and *fluvioglacial debris* and slowly melt. As they disappear, a hollow is produced on the site, which later fills with water.

Keynesian: a person whose economic ideas can be traced to those of J. M. Keynes. Oversimplified, these include the belief that:

- government action may be needed to push an economy out of *recession*
- allowing people to suffer, through unemployment for example, while waiting for the free market to bring the economy into balance is morally and socially unacceptable
- it is desirable for the government to follow policies that will iron out fluctuations in economic activity. (See also *business cycle*.)

kibbutz: a form of rural settlement developed in Israel and organised according to collective principles. Land is commonly owned and farming policies are jointly decided.

kinetic energy or energy due to movement, is generated, for example, by the flow of a river which converts potential energy due to height above sea level or base level into moving energy. The amount of kinetic energy is determined by the volume of flowing water and its average velocity. An increase in *discharge* and/or velocity leads to an increase in kinetic energy.

knickpoint: when a river is rejuvenated, adjustment to the new base level starts at the sea and gradually works its way up the river course, the point of change to the existing profile being known as the knickpoint. *Rejuvenation* occurs when the *base level* falls as a result of a fall in sea level or the uplift of the land. Either way, the river gains renewed cutting power, which encourages it to adjust the *long profile*. In that sense the knickpoint is the place where the old long profile joins the new. This is usually marked by *rapids* or a *waterfall*. If the process has happened more than once, there could be several knickpoints on a river.

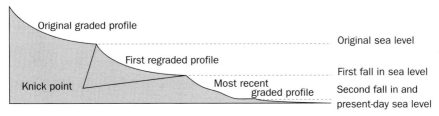

Knickpoints on the long profile of a river

knock and lochan: a landscape made up of small rocky hills (knocks) and small lakes (lochan). It is the result of *erosion* by an *ice sheet* on a surface where there are slight variations in the strength of the rock.

Kondratieff cycle: the theory that, in addition to the 5-year *business cycle*, there exists a 50-year cycle of economic upturn and downturn. This theory was put forward by the Russian economist Kondratieff in the early twentieth century. It was dismissed by many economists until the Great Depressions of the 1880s and 1930s were duly followed (50 years on) by the frequent and severe *recessions* of the period 1975–1992. The most widely accepted explanation for the cycle is that the introduction of *new technology* causes disruption, but once established it forms the basis for many new products and jobs. In the 1930s the car was displacing rail, while in the 1980s the microchip was replacing mechanical technology.

157

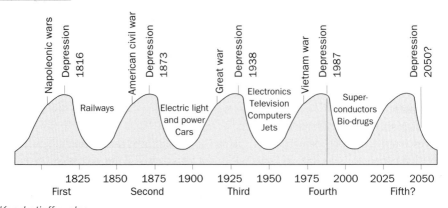

Kondratieff cycles

kurtosis is a measure of the extent to which the general shape of a *frequency distribution* graph is peaked. A distribution that has a relatively narrow but high peak is termed leptokurtic; one that is basically broad and flat is termed platykurtic; one that is between these two extremes is called mesokurtic. Distributions can also be described according to their degree of symmetry or *positive skew* and *negative skew*.

Kuznets curve is the graphical representation of Kuznets' theory that economic *inequality* increases over time while a country is developing then, after a critical average income is attained, begins to decrease. One theory as to why this happens is that in the early stages of development, when investment in physical capital is the main mechanism of *economic growth*, inequality encourages growth by allocating resources towards those who save and invest the most. However, in mature economies human capital accumulation takes the place of physical capital accumulation as the main source of growth and inequality slows growth by lowering education standards because poor people lack finance for their education in imperfect credit markets. Kuznets curves show an inverted U-curve, with inequality or the *Gini coefficient* on the vertical y-axis and economic development, time or *per capita* incomes on the horizontal x-axis. The Kuznets ratio is a measurement of the ratio of income going to the highest-earning households (usually defined by the upper 20%) and the income going to the lowest-earning households, which is commonly measured by either the lowest 20% or lowest 40% of income. Comparing 20% to 20%, perfect equality is expressed as 1; 20% to 40% changes this value to 0.5. Another context where the Kuznets curve is said to appear is in the environment. Many environmental health indicators, such as water and air pollution, show the inverted U-shaped curve. In a developing industrial economy, little weight is given to environmental concerns, raising environmental pollution by-products. After attaining a certain standard of living from the industrial production system and when environmental pollution is at its greatest, the focus changes from self-interest to social interest. These interests give greater weight to a clean environment by reducing and reversing the environmental pollution trend from industrialisation.

Kyoto Conference 1997: the environment conference that followed the 1992 *Rio Earth Summit*. In 1997, at a meeting in Kyoto, Japan, over 100 governments signed a Climate Change Protocol. This set specific targets for pollution mitigation and proposed schemes to enable governments to reach these targets in order to counteract *global warming*. Most governments agreed that by 2010 they should have reduced their atmospheric pollution levels of *greenhouse gases* to those of 1990. The Kyoto Protocol came into force in February 2005 and by 2006 had been ratified by 162 countries. However, this agreement will expire in 2012. There are three things to note about the Kyoto Protocol:

- Some countries are already polluting at levels significantly above those of 1990. The USA, for example, released 15% more **carbon dioxide** at the start of the twenty-first century than it did 10 years ago. Despite this, President Bush refused to ratify the Kyoto proposals, claiming that 'the agreement was fatally flawed' and that the emissions targets were unattainable and potentially damaging to the US economy.

- Some countries are disproportionately responsible for releasing greenhouse gases. In 1996, the USA released 21% of global carbon dioxide even though it only had 4% of the world's population. China, now one of the foremost industrial nations of the world and responsible for a considerable amount of pollution, is not required to reduce its emissions as it was not considered an industrial giant in 1997.

- Some countries, particularly in the least developed areas of the world where industrial development is limited and vehicle ownership is low, release very few greenhouse gases into the atmosphere.

Are you studying other subjects?

The *A–Z Handbooks (digital editions)* are available in 14 different subjects. Browse the range and order other handbooks at **www.philipallan.co.uk/a-zonline**.

labour is one of the *inputs* that goes into producing goods and services. There are various aspects of labour, many of which vary spatially and are therefore important elements in the location of industry. Some of these elements are:

- cost – the cost of labour to a company can be between 15 and 40% of its total costs. Differentials in global wage levels can be an influential part of the decisions taken by **transnationals** in locating plants
- quality – access to a skilled workforce has always been an important factor for some industries. Today, it is a very important consideration for many small-scale *light industries*, particularly those involved with computers and **robotics**. One of the reasons put forward for the development of the Cambridge **science park** is the supply of graduates from the university (see also **Cambridge phenomenon**)
- relations – centres of trade union activity are regarded by some employers as potential centres of militancy and therefore as a negative location factor. This has tended to become less important with the decline in trade union membership
- availability – some areas may have the required workforce for a potential employer. One of the reasons given by Nissan for locating its car plant in northeast England was the reservoir of unemployed young men, many with engineering skills. Some companies may want a large number of female employees and go to regions where they are available for employment.

labour flexibility: the ease with which a firm can change the jobs carried out by its staff. This is an important element in a firm's ability to cope with change within its market. The main factors determining *labour* flexibility are:

- workforce attitudes, including resistance to change
- traditional labour practices, which may be restrictive and entrenched, such as job demarcation
- the general skills of the labour force
- the strength and attitude of the trade unions.

labour-intensive: a work process in which *labour* costs represent a high proportion of total costs. Such a situation is most likely to exist in the service sector. This contrasts with firms that are **capital**-intensive.

labour market is the supply of *labour* (i.e. all those offering themselves for work) and the demand for labour (i.e. employers), which together determine wage rates.

labour mobility is the extent to which *labour* moves around in search of employment, called geographical mobility, or the extent to which labour moves between jobs, which is known as occupational mobility. Geographical mobility depends on such things as availability of housing, costs of moving, family ties, and the availability of information, whereas occupational mobility depends on training facilities and a willingness to learn and be retrained.

lag time is the period between the maximum *precipitation* within a *drainage basin* and the peak *discharge* of the river in that basin. The information is displayed on a *hydrograph*. The lag time varies according to conditions within the drainage basin such as soil and rock type, slope gradient, drainage density, type and amount of vegetation, extent of urbanisation, size of basin and water already in storage which will in part depend on recent weather.

lahar: a flow of wet material down the side of a *volcano*'s ash cone. Lahars occur when surface water picks up large amounts of volcanic ash and deposits it as mud over lower lying areas. In the aftermath of the eruption of Mt Pinatubo (Philippines) in 1991, *typhoons* brought heavy rainfall which caused a number of lahars to form.

laissez-faire is a political and economic philosophy which believes that governments should avoid interfering in the running of business or any other part of the economy. A laissez-faire economist places faith in the ability of the free market to maximise business efficiency and customer satisfaction. This is in direct contrast to the views of *interventionists*.

land and sea breeze: a diurnal circulation of air resulting from the differential heating and cooling between land and adjacent sea areas. By day, the land heats up more quickly than the sea. Hence the air over the land becomes warmer than the air over the sea, and low pressure forms over the land, with relatively high pressure over the sea. A wind therefore blows from the high pressure over the sea to the low pressure over the land. This is a sea breeze which can lower the temperatures of coastal areas by several degrees. By night, the land cools down more quickly than the sea. Hence the air over the land becomes cooler than the air over the sea, and relatively high pressure forms over the land, with low pressure over the sea. A wind therefore blows from the high pressure over the land to the low pressure over the sea. This is a land breeze.

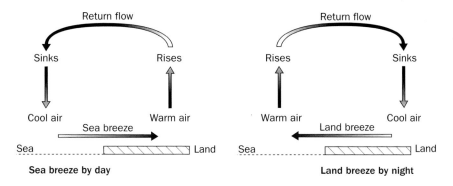

Sea breeze by day   Land breeze by night

land colonisation: areas that were not previously developed for agriculture have been exploited usually to grow commercial crops for domestic consumption or exports to bring in foreign earnings. In Brazil, the government opened up vast tracts of the Amazon Basin to provide farmland for the landless people of Brazil, particularly those in the impoverished northeast. This caused extensive damage, so the programme was stopped. In recent years, large areas of the Amazon *rainforest* have been deforested to provide pasture and land for soya bean cultivation.

landfill site: a land-based location used for the disposal of waste. The purpose is to empty waste materials into a natural or manmade hole in the ground, for example a disused quarry, and to cover it with soil when full. However, hazardous materials may infiltrate their way through the soil and enter *groundwater* supplies. In addition, the *decomposition* of the waste under the ground in *anaerobic* conditions causes *methane* to be produced. This rises to the surface and presents a potentially explosive hazard.

land reclamation includes any process which improves or recovers land usually for agricultural use, although *derelict land* in urban areas may be reclaimed for industrial, recreational or residential purposes. Drainage may be installed to improve low lying or marshy land, infertile areas can be treated with chemicals, sea areas and lakes can be drained to produce new agricultural land, moorland areas can be ditched to aid surface *runoff*.

land reform: a term used to describe the redistribution of land in an area in order to increase agricultural productivity and to raise individual standards of living. In many countries earlier this century there was an unequal distribution of land. Ownership was concentrated in a small number of large estates, while the bulk of the rural population had too little land for a reasonable living. Thus a number of wholesale changes in the ownership of land have taken place. In the communist countries the landlord had his land expropriated, and redistributed among the peasant farmers. In some of these, the smaller peasant farms were then grouped together into collectives. In some other countries the government has compulsorily purchased private land, and resold it under favourable credit terms. An increase in owner-occupancy is perceived as being an incentive to raise agricultural productivity.

LANDSAT is an American satellite orbiting the Earth at a height of 705 km. It is an unmanned remote sensing device that can monitor features on the surface.

landslide: a type of *mass movement* in which the *debris* moves downslope as a unit along a glide plane. They occur particularly where the rocks have *bedding planes* which are roughly parallel to the slope angle. They are activated by the accumulation of large amounts of soil water in the weathered surface material. This added weight increases the stress on the debris and the extra water reduces friction between the particles and lubricates the mass of material along the bedding plane.

landslip: an alternative term for *landslide*.

land use: any purpose for which land is used for human activity. Such uses can be divided into rural (agriculture, forestry, settlement, recreational space) or urban (housing, industry, commercial).

lapse rate: the rate at which temperature decreases with height. In the atmosphere the change of temperature with height is called the *environmental lapse rate*. Air which is moving through the atmosphere will cool on rising or warm up on descending, according to *adiabatic* laws. When the air is dry, i.e. it has not become saturated by cooling to its *dew point*, it changes at the *dry adiabatic lapse rate*. Once it is saturated it will cool or warm according to the *saturated adiabatic lapse rate*.

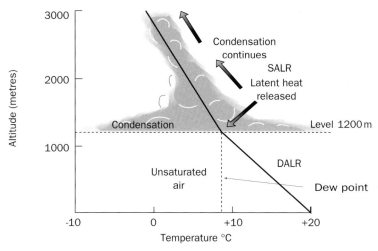

Lapse rates

latent heat is the heat energy expended when a substance changes its physical state without raising its temperature. During the process of **evaporation** when liquid water or ice changes to water vapour, heat is required to bring about the physical change from water to vapour. During **condensation**, when the process is reversed and water vapour condenses into water droplets, latent heat is released. Latent heat of evaporation causes cooling while condensation produces a heating in the atmosphere. This large-scale release of heat can accelerate rising air currents in **cyclones** and **convection** systems and it influences the **saturated adiabatic lapse rates**.

laterite is a hardened layer of ironstone formation in soils of **tropical** regions. It has resulted from the precipitation of iron and aluminium minerals either at the surface or in the subsoil. This may produce a **duricrust** layer. A lateritic horizon may be produced where the **ground-water** movements in the soil concentrate iron and aluminium oxides into a depositional zone which coincides with the depth of the **water table**. These concentrations may appear as nodules or as a slag-like layer. The term is also used in a general way to distinguish tropical soils, such as latosols, which are the result of the process of laterisation.

latifundia: a system of land holding which consists of large estates on which a large number of the local people of an area work as labourers. The system was common in the south of Italy until recent times and was often cited as one of the reasons why the region was so agriculturally backward. The Cassa per il Mezzogiorno was given the responsibility for dismantling the system in order to promote greater farm efficiency and the emergence of a new landowner class. The system still flourishes in parts of Latin America, particularly Brazil.

Laurasia is the name given to the northern supercontinent which consisted of present-day North America, Greenland, Europe and most of Northern and Central Asia. As **Pangaea** divided, Laurasia was separated from **Gondwanaland** by the Tethys Sea.

lava: when molten rock (**magma**) flows on to the surface it is known as lava. There are several forms including:
- **acid lava**, which quickly solidifies on contact with the air and produces steep-sided volcanic forms (e.g. the spine of Mt Pelée)
- basic lava, which is basaltic in character and tends to flow a long way before solidifying, producing rather flat volcanic shapes such as the Hawaiian Islands, also extensive lava

plateaux such as the Antrim Plateau in Northern Ireland (the basaltic flow can be seen at the famous tourist attraction of the Giant's Causeway).

LDC: see *least developed country*.

leaching occurs in humid environments when rainfall is greater than *evapotranspiration* and soluble bases are removed from a soil by downward percolating water. The water is naturally slightly acidic, with the most common bases removed being sodium, calcium and potassium.

least-cost location: the centrepiece of the Weber model of industrial location, as he stated that an industrialist would seek to establish a factory at that point. It is the point at which production costs, comprising the combination of raw material and transport costs, are at a minimum. Weber also stated that an industry may be located away from the least-cost location when:

- cheaper *labour* costs exist at another point that result in greater savings than the additional transport costs to that point
- cost savings from *agglomeration* also outweigh the additional transport costs involved.

least developed country (LDC): a concept first identified in 1968 by the *United Nations* Conference on Trade and Development but subsequently updated by other United Nations international conferences. The main characteristics of such a country are:

- low incomes, as measured by *gross domestic product per capita*, usually taken over a 3-year period (an annual figure of less than $800)
- human resource weaknesses, based on indicators of nutrition, health, education levels and *literacy*: specifically, the *life expectancy* at birth, per capita *calorie intake*, combined primary and secondary school enrolment and adult literacy rates
- economic *vulnerability*, shown by the low level of economic *diversification*, which itself is based on the share of manufacturing in the gross domestic product, the share of the *labour* force in manufacturing, the annual per capita energy consumption and merchandise export concentration levels. Vulnerability can also be measured by the percentage of the population displaced by natural *disasters*.

Several least developed countries are also physically isolated from the rest of the world, for example by being landlocked such as the *Sahel* countries of Africa. Other features are that most suffer from trade instability and several have political unrest. (See also *economically less developed country*.)

leisure industry: the employment and products generated by the activities people engage in for enjoyment during time that is regarded as being free from work demands or any other obligation. The leisure industry has expanded considerably in recent years in economically more developed countries as people have both more spare time and more *disposable income*. An increase in personal mobility has also allowed this to occur. The leisure industry is more than just tourism. People engage in 'at-home' activities, as well as activities outside of the home. Some pursuits are active, whereas others are passive. Since the industry has grown at such a rapid rate, concern is now being expressed at some of the environmental impacts that have been produced. (See also *honeypot sites*.)

less favoured areas are regions within the *European Union* that are given extra funding because they are difficult environments to live and work in.

lessivage: this is a particular kind of *leaching* resulting from clay particles being carried downward in suspension. This process can lead to the breakdown of the *peds* (the aggregation of different particles that give the soil its structure).

levée: that part of the bank of a river that is raised higher than the *floodplain*. Naturally, they form when the river floods as the sediment is dropped. It is usual for the coarsest to be dropped first, forming a small bank along the channel.

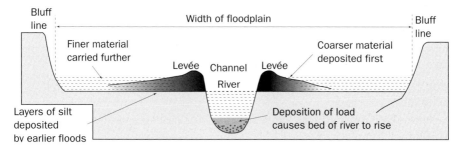

Position of levées on river floodplain

Subsequent floods will increase the size of this bank. The river, with channel sediment build-up, will now be flowing at a higher level than the floodplain. Therefore authorities sometimes strengthen the levée and also increase its height. On the Mississippi River (USA), for example, the levée strengthening began in 1699. By 1738 they had built 68 km of bank. By the 1990s the length of the levées was 3200 km.

level playing field: a phrase that sums up the need within a *market economy* for all firms to be competing on the same terms. If a national government is subsidising one industry, they will have an unfair advantage over those from a country which offers no state aid to the same industry.

lichen: a very simple plant which will be one of the *pioneer* species when bare rock areas (*lithoseres*) are colonised. Lichens are undemanding plants as they are capable of living without soil, with no supply of permanent water and in extremes of temperature. Lichens help in the break-up of the rock surfaces. As conditions improve with the creation of a thin layer of soil, other plant species begin to take over from the pioneers as a *succession* develops, although lichens may be part of that new community.

life expectancy is the average number of years from birth that a person can expect to live.

light industry is that part of *manufacturing industry* that does not fall into the classification of 'heavy'. *Heavy industry* is one in which large weights of materials are handled, therefore light industry handles small amounts of material per worker involved. Light industries are not generally sources of pollution and are often found in modern locations such as *industrial estates*.

limestone: a *sedimentary rock* composed wholly or largely of calcium carbonate and formed by either:
- the accumulation of the skeletal remains of marine creatures, for example *chalk*, or
- chemical precipitation, for example oolitic limestone.

Both types are usually accumulated in layers and compacted. They may also contain other materials such as clays and quartz. (See also *Carboniferous limestone* and *coral reef*.)

linear: a description of a feature that has long, narrow dimensions. A linear settlement is one that is strung out along a routeway such as a road or waterway, or along a confined river valley. Settlements that seek to avoid **flooding** also extend themselves along a raised terrace. Many towns have grown outwards in a linear pattern alongside a major road. This is known as ribbon development.

linkages refer to the relationships between one industry and another. Linkages may take a variety of forms:
- they may be to the consumer of the industry's product (forward linkages)
- they may be to the provider of **raw materials** and components (backward linkages).

They can also be classified as:
- vertical – where the raw material goes through several successive processes
- horizontal – where an industry relies on several other industries to provide its component parts
- diagonal – when an industry makes a component which can be used in several subsequent industries
- technological – when a product from one industry is used as a raw material by a number of subsequent industries which further reprocess it.

Linkages may also involve the subcontracting of work, maintenance links, financial links or the use of common services such as packaging and wholesaling.

liquefaction is where soils with a high water content lose their mechanical strength when violently shaken during an **earthquake** and start to behave like a fluid.

literacy is the ability of a person to read and write. The literacy rate of a country is used as a means of measuring its level of economic development. It is an indication of the quality of education in a country, which in turn is an indicator of the wealth and social development of that country. Literacy may vary between different social groups, and between different sexes in a population.

lithology: the study of rocks related to their physical, chemical and textural character. It may also be used to describe a rock in terms of these features. A rock's lithology can be described by its permeability, solubility and relative hardness. Its lithology is a major factor in determining the nature and topography of the landscape it produces.

lithosere: a primary plant **succession** which takes place on a newly exposed rock surface. The surface may be rock exposed by a retreating **glacier**, or on lava following a volcanic eruption, or on new land emerging from the sea. An example of a lithosere in a temperate part of the world similar to the British Isles would be as follows:
- the area of bare rock would be colonised initially by **bacteria** which can survive where there are few nutrients
- **lichens** and mosses would also colonise the rock and they assist in the weathering of the rock to form a thin layer of soil in which more advanced forms of plants can grow
- as these plants die, the bacteria convert their remains into humus which results in a more fertile soil
- grasses, herbs and small flowering plants then colonise the area, which in turn give way to shrubs
- these shrubs will be replaced by faster growing trees such as rowan, which then face competition from slower growing trees such as ash and oak. These trees eventually form the climatic **climax vegetation** for the area.

(See also **pioneer**.)

lithosphere: this consists of the *crust* of the Earth and the rigid upper section of the *mantle* and is approximately 80–90 km thick. It is divided into a series of plates. (See *plate tectonics.*)

little ice age: a cool period in Europe (and possibly covering a wider area) in which many Alpine *glaciers* advanced. It lasted from around 1400 until the nineteenth century.

littoral: the environment between the highest and lowest levels of the *spring tides*. Geographers sometimes use the term simply to denote the area at the edge of the sea or ocean, and to others it represents the *beach* environment.

livestock are animals that are kept or farmed for use or profit. Livestock farming may be extensive as in ranching on the Pampas grasslands of Argentina or it may be intensive as in dairying in northwest Europe. Livestock may also be kept by *nomadic* herders who generally lead a subsistence existence. Some livestock may also be kept as draught animals.

load: the material transported by a river. It may be transported by:
- *suspended load*
- *saltation*
- *solution*.

Local Agenda 21: at the *Rio Earth Summit* in 1992, it was agreed that the best starting point for the achievement of *sustainable development* is at the local level. Each *local authority* should be required to draw up its own Local Agenda 21 strategy following discussions with its citizens about what they think is important for the area. The principle of sustainable development must form the central part of that strategy. Local Agenda 21 regards sustainable development as a community issue, involving all sections of society including community groups, businesses and ethnic minorities.

local authority: the body responsible for the operation of local government and for the administration of public affairs in each locality by representatives of the local community. Although subject to certain central government controls, local authorities possess a considerable amount of self-governing responsibility and discretionary power. Local authorities have a wide range of responsibilities, providing services 'from the cradle to the grave'. These can be divided into five broad categories:
- protective services – police, fire, consumer protection, animal diseases, licensing (safety measures and regulation)
- environmental services – environmental health, highways, traffic, transport, planning, emergencies (disaster/hazard contingencies)
- personal services – education, careers service, social services, health (now under trust management), housing (replaced by *housing associations*)
- amenity services – leisure facilities, museums, galleries, parks and gardens, libraries
- trading services – these are services for which the local authority make commercial charges; some services such as day nurseries, school meals, residential care, car parks and sports facilities charge fees, but these are often below the full economic rate and may be scaled to reflect the user's ability to pay. These are not classed as trading services which aim to cover the full cost of provision and which charge the same amount irrespective of income. Markets are trading services as are smallholdings and

167

allotments, conference and exhibition centres, estate development, and beach facilities. In addition, there are some unusual trading services – civic airports, Birmingham's bank, Hull's telephone system and Doncaster's racecourse.

local government restructuring involves the reorganisation of *local authority* (government) boundaries and the reallocation of responsibilities. Major changes were suggested by the Royal Commission on Local Government in England 1966–69, which was led by Lord Redcliffe-Maud. Its aim was to establish coherent socio-geographic units with a minimum of 250,000 population, to enable efficient service provision. A maximum size of 1 million was suggested to maintain local representation. The Commission proposed that England be divided into 58 areas with a single tier of local government which would carry out all responsibilities. In the densely populated *conurbations* they proposed a two-tier structure. In such areas the metropolitan county would control many environmental services (planning, highways and transport), with the metropolitan districts taking charge of personal services such as housing, education and health.

In 1971 the Conservative government put forward its proposals which were introduced in the Local Government Act 1972. They established, in England and Wales, 48 county councils with a second tier of 333 district councils. Six metropolitan counties were created, within which there were 36 metropolitan districts. A third tier existed in England – the parish council. However, in 1985 the Local Government Act abolished the metropolitan counties and the Greater London Council was abolished in 1986.

Local government boundary commissions, one each for England, Scotland and Wales, are permanent bodies set up under the Local Government Act 1972. These bodies review local government areas and their electoral arrangements. They undertake reviews of principal areas at intervals of 10–15 years, unless otherwise directed by the minister. They make proposals to the Secretary of State, who accepts or amends proposals and presents the decisions to parliament for ratification. The English Boundary Commission, under the chairmanship of Sir John Banham, reported the findings of the latest major reviews during 1994. Following some reconsideration, the eventual pattern was a mix of a *unitary authority* in some areas, with two-tier structures remaining in others. Local preferences and needs may continue to be recognised in this way.

Other political changes have had an effect at a more local scale. Privatisation and *deregulation* have reduced the role of, and the need for, local government. Examples include the sale of council houses, the deregulation of bus services, the privatisation of environmental services such as refuse collection and grounds maintenance, and the moves towards the local management of schools (LMS). The *private sector* should be used where it is more cost-effective than local government provision, enforced by compulsory competitive tendering (CCT).

location quotient: a statistical measure of the geographical concentration of an activity in a region compared with the national average. The LQ is calculated using the formula:

FORMULA: $LQ = \dfrac{X_A/Y_A}{X/Y}$

If the LQ was applied to the concentration of industrial activity:
- X_A = employment in industry X in region A
- Y_A = total employment in all industries in region A

- X = national employment in industry X
- Y = total employment in all industries nationally

This can be simplified as:

$$LQ = \frac{\% \text{ total workforce in region A working in industry X}}{\% \text{ total national workforce working in industry X}}$$

If textiles employs 12% of the workforce in the region, but nationally 4% of the total workforce are employed in this industry, then LQ = 12/4 = 3. An LQ of 1 indicates that an industry is represented in a region in exactly the same proportion as nationally. Less than 1 suggests that the industry is under-represented; the region has less than its fair share of the activity. Greater than 1 suggests that the industry is over-represented; there is a geographical concentration, the region has more than its fair share. LQs are useful in measuring the relative concentration of an activity at a particular time, but care must be taken when comparing different time periods. Because there are four components to the equation an industry in a region could increase its work-force but produce a lower LQ if its rate of increase in the industry was lower than the national rate of change. Interpretation only relates to the degree of concentration.

lodgement till is a form of ground *moraine* deposited under a *glacier* or *ice sheet*. Towards the snout of a glacier as the forward velocity is reduced bottom melting releases rock particles from the sediment-enriched basal layer of the ice. These are laid down under pressure beneath the slowly moving ice, plastered to the bedrock. Because of the slow movement of the ice above the lodgement layer, the till stones may display some orientation with their long axis in the direction of ice movement. This contrasts with the more haphazard *deposition* of *ablation till*.

loess is a wind-blown (*aeolian*) deposit. It is regarded as a *Pleistocene* sediment de-rived from the *outwash* plains of the last ice advance. It stretches across Europe from the Paris Basin, where it is called limon, through Belgium and south Germany into Poland. It is a yellowish-brown loam that is easily weathered to produce very fertile soils. In China it has produced massive deposits up to 300 m thick. Loess is a relict deposit; this form of wind action now seems to be less important and loess is not being added to under present-day conditions.

logarithmic scale: a scale that is divided into a number of cycles each representing a tenfold increase in the range of values. If the first cycle ranges from 1 to 10, the second cycle extends from 10 to 100, the third from 100 to 1000 and so on. For very small values, for example when plotting soil grain size analysis, the cycles could extend from 0.001 to 0.01, 0.01 to 0.1, 0.1 to 1.0. Zero cannot be plotted on logarithmic graph paper, nor can positive and negative values be shown on the same axis. Logarithmic scales are useful when the rate of change is of more interest than absolute change; a steeper line reveals a faster rate of change. They also allow a wider range of data to be displayed than on a similar sized piece of arithmetic graph paper. If the rate of change being displayed is increasing at a constant proportional rate, i.e. the population doubles in each unit of time, this will appear as a straight line if plotted on semi-log paper (one axis arithmetic for time, the other logarithmic for popula-tion). When the perfect *rank–size rule* relationship is plotted on semi-log paper, it produces a straight line with a gradient of –1, that is the line is descending at an angle of 45° to the horizontal.

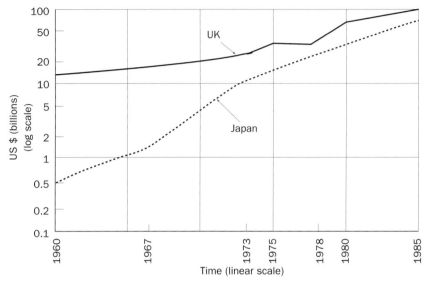

Overseas investment in the world economy made by transnational corporations based in the United Kingdom and Japan

Example of a logarithmic graph (using semi-logarithmic paper)

London Docklands is an area of urban regeneration stretching downriver from London Bridge through Wapping and Limehouse, the Isle of Dogs, the Surrey Docks and the Royal Docks. Up to 1960 London was the leading general cargo port in the UK, but the prosperity of these areas declined rapidly during the following decade. Transport technology favoured the newer container and specialised ports in the southeast. By 1981, unemployment had reached 25%, and large areas were derelict.

The government set up the first **Urban Development Corporation (UDC)**, the London Docklands Development Corporation (LDDC). With government grants and private invest-ment and the sale of reclaimed land to developers, together with improvements to the **infrastructure** such as roads and the Docklands Light Railway, the economy of the area has been transformed. There is also the London City Airport with links to major European cities. The area has developed private housing, offices, shopping, hotels and leisure facili-ties and more than 600 new firms located in the area between 1981 and 1990. These are mainly service-based activities such as financial and business services, leisure, restau-rants and catering. Some manufacturing jobs were created in printing and publishing with major newspapers relocating from Fleet Street to Wapping in the 1980s. High-technology computer industries, television and film companies have also been attracted.

Although there has been economic success, the LDDC has given less attention to the needs of the former local residents, who to some extent have been priced out of the area by the nature of the developments. A lack of housing for the less wealthy, few job oppor-tunities for the original docklands residents and poor community facilities are all criticisms which have been levelled at the LDDC.

On 31 March 1998 the LDDC ceased to exist. Some of its responsibilities for the area were handed over to the three boroughs involved, Newham, Tower Hamlets and Southwark, and to other bodies such as the British Waterways Board, which will be responsible for maintaining some of the docks. The reclamation work will be handed over to English Partnerships. By 1998 the LDDC claimed that in the 17 years of its existence the following had been achieved in the area:

- There had been £7.2 billion of **private sector** investment together with £1.86 billion of public money invested.
- 830 hectares of **derelict land** had been reclaimed.
- 145 km of new and improved roads, together with the Docklands Light Railway had been constructed.
- 2.3 million square metres of commercial and industrial floor space had been built.
- 24,000 new homes had been constructed.
- The Docklands area provided (by 1998) employment for 85,000 people and homes for 83,000.
- There were (in March 1998) 2,690 businesses now trading within the area.
- The Docklands had become a major tourist area, which by 1998 had attracted 2.1 million visitors.

longevity is the increase in **life expectancy**. With increased economic development and improved medical provision there have been significant changes in mortality in recent years. Life expectancy at birth in the UK in 1960 was 67.9 (males) and 73.7 (females). By 2005 this had risen to 76.6 (males) and 81.0 (females). With declining or steady **birth rates** this greater life expectancy produces an ageing population with a greater percentage of the population over 60. Within the EU the percentage of population aged 60 or over rose from 16% in 1960 to 20% in 1990 and is predicted to rise to about 30% by 2020. This is sometimes referred to as the 'demographic time bomb'; there is a smaller percentage of economically active population to support an increasing retired and elderly population. As increasing numbers are living beyond the age of 75, an age when many become more dependent on the state or family help, the cost of health provision and care is high. Many retired people do have private incomes through pensions and this is creating a group of consumers with considerable purchasing power and influence, the so-called 'grey market'.

long profile: illustrates the changes in the altitude of a river's course from its source along its channel to its mouth. In general a long profile is smoothly concave in shape, with the gradient being steeper in the upper course and becoming progressively less steep towards the mouth. However, irregularities frequently exist. These irregularities may be represented by **waterfalls**, **rapids** or lakes. There may also be marked changes in slope, known as **knickpoints**, which are a product of **rejuvenation**. (See also **graded profile**.)

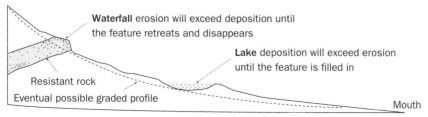

Waterfall erosion will exceed deposition until the feature retreats and disappears

Lake deposition will exceed erosion until the feature is filled in

Resistant rock

Eventual possible graded profile

Mouth

A typical long profile of a river

longshore drift is the movement of sediment along the coast by wave action. When waves approach the shore at an angle material is pushed up the beach by the *swash* of the breaking wave in the same direction as the wave approach. As the water returns down the beach the *backwash* drags material more directly down the steepest gradient which is generally at right angles to the beach line. Over a period of time sediment moves in this zig-zag fashion down the coast. Obstacles such as *groynes* and piers interfere with this drift and accumulation of sediment occurs on the windward side leading to entrapment of beach material. *Deposition* of this *debris* takes place in sheltered locations, such as at the head of a bay, and where the coastline changes direction abruptly when spits develop.

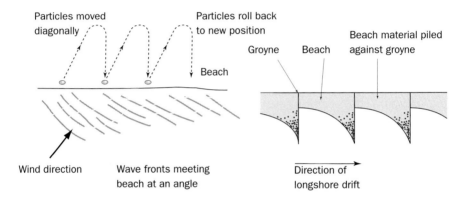

Longshore drift

long-term climatic change takes place over several hundreds, thousands or even millions of years. Evidence for such change over a long timescale comes from the presence of coal and *limestone* which must have formed in environments much different from the present. The climatic change which has taken place since the end of the *Pleistocene ice ages* is also an example of long-term climatic change. Several theories exist to explain these climatic changes. Some state that there have been variations in the *solar energy* emitted, others state that there have been cycles of celestial movements of the sun and the Earth. However, no clear consensus of opinion exists.

Lorenz curve illustrates the degree of unevenness in a geographical distribution. It is drawn on graph paper and makes use of cumulative percentage data. The vertical y-axis carries the cumulative data and points are plotted in the order of the largest, which is then added to the second largest, then to the third largest, and so on. The horizontal x-axis simply records the cumulative process. The plots are then connected by a line. If another line is drawn on to the graph to represent an even distribution, the degree of unevenness can be seen. The greater the deviation the plotted line has from the line of even distribution, the greater the degree of unevenness. A highly concave Lorenz curve represents a high level of unevenness and therefore high level of concentration.

Worked example: unevenness of population distribution

Assume a country has 10 regions, A to J:

Region	A	B	C	D	E	F	G	H	I	J
% of national population in region	10.5	1.6	12.2	1.8	35.3	6.7	2.1	25.3	3.5	1.0

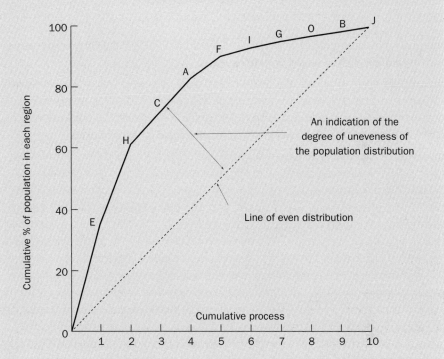

Lorenz curve: an example to show the unevenness of the population distribution of a country

magma is the name given to molten material under the surface of the Earth which has risen from the *mantle*. It is the source of *igneous rocks*.

magnitude refers to the size of a natural hazard event. A variety of magnitude scales exist for several hazards and these categorise events according to size and/or energy and enable people to understand the processes and to model the likely impacts. Examples include the Saffir-Simpson scale for *hurricanes* and the *Richter scale* for *earthquakes*.

malaria is an *infectious disease* that affects up to 40% of the world's population. It kills over 1 million annually and affects up to 500 million people. The main area affected is sub-Saharan Africa, where there are nearly 80% of all cases. Estimates suggest that up to half the population is affected. Malaria is a parasitic, mosquito-borne disease spread by the anopheles mosquito in tropical areas. It is spread by the female of this insect having a blood meal ('biting') from an infected person and 'biting' an uninfected person. Mosquitoes breed well in hot and humid conditions with areas of standing water. *Deltas*, irrigation channels and fertile farming lands are all good breeding grounds. The disease has fever and flu-like symptoms and, if untreated, can cause convulsions, coma and even death. Children may suffer learning impairment and brain damage. It has a heavy financial burden on the economy of many less developed countries. Direct costs include medical costs, preventative measures such as mosquito bed nets, loss of earnings and, on a national scale, public health spending. Indirect costs include lost productivity for employers, lack of continuity in education, reduced investment in an area (as, for example, investment in tourist facilities is reduced) and more emphasis on the growth of subsistence crops at the expense of cash crops due to the impact of disease on *labour* supply during harvest.

There are several ways in which people and countries can be helped:

- Drug treatment – originally quinine was used, and more recently chloroquinine. New drugs are being developed as the disease becomes resistant to those in common use.
- Controlling the means by which the disease is spread using insecticides. DDT was widely used in the past, but this caused health and environmental problems. Pyrethrine and other insecticides are now used instead. Insecticide-treated mosquito nets are effective, but they are often too expensive for local people.
- Killing the mosquitoes at the larval stage – draining and filling in suitable breeding sites, covering water tanks, drying out irrigation tanks and stocking water sites with fish (e.g. the grass carp), which eat the larvae.

In 1998 a multi-agency programme (with input from the *World Health Organization*, UNICEF, *United Nations* and the *World Bank*) was established for the research and control of the disease. 'Roll Back Malaria' aims to halve deaths from malaria by 2010.

malnutrition results from some form of **diet** deficiency either because the quantity of food intake, as measured in calories per day, is too low, or there is insufficient balance between proteins, energy foods, vitamins and minerals. Although proteins are needed for building body tissue, the most vital element in the diet is carbohydrate which provides energy. If this falls below minimum needs body functions divert some of the protein intake to make up for the deficiency. When the intake of protein and carbohydrate are satisfactory the minimum require-ments of protein and minerals are normally achieved. Simple measures of **calorie intake per capita** per day do not present the full picture; the proportion of food from starch and proteins is also important. Malnutrition weakens resistance and exposes people to the dangers of killer diseases.

Malthus, Rev T. R.: the author in 1798 of 'An Essay on the Principle of Population as it affects the Future Improvement of Society', which was an attempt to show the link between population and **resources**. Malthus based his theory on two principles:

- population, if unchecked, grows at a geometric or exponential rate,
  i.e. $1 \to 2 \to 4 \to 8 \to 16 \to 32$
- food supply increases at best at an arithmetic rate, i.e. $1 \to 2 \to 3 \to 4 \to 5 \to 6$.

When the population growth outstripped the capacity of the resources, Malthus sug-gested that the growth would be curbed by preventative checks, limiting population growth (postponement of marriage, moral restraint in terms of sex, vice, contraception) and positive checks that would reduce the population size such as **famine**, disease and war. The predictions of Malthus did not come true in the nineteenth century for a number of reasons including the vast improvements made in agriculture, the opening up of new agricultural lands in North America in particular and the **emigration** of millions from Europe to parts of the world that were considerably less densely populated. In the late twentieth century, however, there was a revival of his views under the heading of Neo-Malthusianism. This was as a response to the dramatic acceleration in population growth in **Third World** countries after their mortality rates began to drop. Rapid popula-tion growth was seen to impede development and to bring about a number of social and economic problems. This led to countries in those areas taking positive steps to reduce population growth assisted by a number of international agencies.

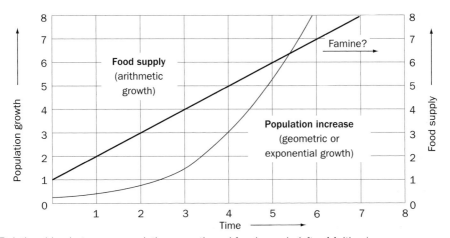

Relationships between population growth and food supply (after Malthus)

mangroves are coastal wetland forests made up of salt-adapted evergreen trees found in the inter-tidal zones of tropical and subtropical latitudes. The trees have interlacing stilt or aerial roots that trap **sediments** and sometimes lead to the creation of new land on the shoreline. They also play an important role in protecting that shoreline from **erosion**. Mangrove swamps support much wildlife in the nutrient-rich waters and are also a supply of wood for building and for fuel to the local inhabitants. Mangroves have been under threat for a number of years and for a variety of reasons:

- the clearance of the land for agriculture, particularly rice cultivation
- excessive demand for **fuelwood** from increasing populations
- conversion of the swamp areas to fish ponds (**aquaculture**)
- commercial exploitation, particularly by the Japanese, for timber to produce wood chips for cellulose and paper-making.

Manning's roughness coefficient was devised by Robert Manning, an Irish engineer, as a general measure of channel resistance. It varies from 0.03 for smooth sections of straight, natural streams, to 0.10 for those sections that are rocky or heavily vegetated.

mantle: the layer of the Earth between the **crust** and the **core**. The mantle is separated from the crust by the **Mohorovicic** discontinuity and consists of two layers, the upper or lithosphere (together with the crust) and the lower or **asthenosphere** which is in a semi-molten state. The mantle extends to a depth of 2900 km. The mantle is composed mainly of silicate rocks, rich in iron and magnesium.

manual worker: an employee who works with his/her hands, usually in a factory context. Manual workers are often divided into those that are skilled (electrician, welder), semi-skilled (van driver, production line worker) and unskilled (cleaner, road sweeper).

manufacturing industry: consists of companies that convert **raw materials** into finished goods or which assemble components made by other manufacturing companies. Manufacturing represents the **secondary sector** of employment.

maquis: a type of vegetation growing in the European **Mediterranean** areas. It grows in areas of **impermeable** rock, such as granite, as opposed to **garrigue** which is found on more permeable rocks. Maquis tends therefore to be taller and denser. Like garrigue, this is said to represent a **plagioclimax** because of human interference with the natural vegetation.

marina: a dock or basin providing a mooring for pleasure boats. With the increase in leisure time seen in the late twentieth century, there has been a great deal of pressure for the development of such facilities. Many urban waterfronts which have deteriorated and declined in the twentieth century have been revitalised and often include a marina as part of that redevelopment.

marine/maritime climate: the climate of those areas lying next to the sea which come under the influence of it. In **temperate** areas of the world this will mean a cooler summer than the interior parts of the continent and a milder winter. Coastal regions are often wetter than those further inland, particularly where the coast is backed by a mountain range such as in northwest USA. (See also **continental climate**.)

market: the demand for a good or a service. The term can also refer to a specific place which is the market for a good or service. Specific car assembly plants, for example, can be the market for one manufacturer of car components.

market economy is an economy that allows **markets** to determine the allocation of resources. The principal benefits of the market mechanism are that it is automatic and leads to greater efficiency. Markets, however, are not wholly automatic and are always modified or regulated in some way by governments. The extent to which market regulation takes place is a political decision. In the countries that were run as **command economies** (the former communist states), free markets were almost completely superseded by central planning. This system proved unsuccessful, however, not only because it was expensive and inefficient to operate, but also because it stifled individual initiative, and ultimately led to lower living standards. A compromise between a wholly planned and a market economy is called a **mixed economy**.

market gardening: the intensive production of vegetables, fruit and flowers for sale. Known as 'truck farming' in the USA. (See also **horticulture**.)

market processes operate in an environment where the ability to pay the going rate will take precedence over any local or national concerns. Objectors to a planning proposal cannot afford to outbid the developer. Consequently the latter very often succeeds and the development goes ahead with the minimum of consultation. Consultation under market processes often takes the form of an opportunity to voice objections or counterproposals, but with no right of arbitration or right of appeal.

marram grass is associated with the **stabilisation** of coastal sand **dunes**. This species is able to tolerate a dry, mobile habitat of shifting sands. Once it roots its horizontal stems or rhizomes through the sand, it binds the sand together allowing further colonisation by other species of grass.

Marxism was a view of society and economic development put forward by Karl Marx in the mid-nineteenth century. He regarded the economic means of production as the key to understanding society and its class structure and patterns of behaviour. The population was seen as a commodity or resource in that it provided **labour**, an important factor of production. As with any other resource, labour costs money to produce in that children required feeding, clothing, and educating, and workers needed housing. However, in return for this expenditure the capitalists were able to use the resource, and Marx claimed that labour was the only resource which produced a surplus, i.e. the value of the resource to the user was greater than its production cost. As labour was such a 'unique' resource, population growth could be viewed as beneficial to **economic growth**. Marx was not greatly concerned by population growth in relation to resources, unlike **Malthus**, because population provided a means of exploiting resources. Marx considered resources to be sufficient for population; but resources were unevenly distributed.

mass movement (mass wasting) is the downslope movement of weathered material under the influence of gravity. One way of classifying such movement is by its speed and nature:
- slow flows such as **creep**
- rapid flows such as **earthflows** and **mudflows**
- **slumping**, **landslips** and **landslides**
- **subsidence** and **avalanches**.

The rate of mass movement depends on the degree of cohesion of the weathered material, the steepness of the slope down which the movement takes place, and the amount of water contained in the material. A large amount of water adds weight to the mass, but more importantly lubricates the plane along which movement can take place.

mass production: the system devised by Henry Ford that turned **raw materials** into finished products in a continuous moving and highly mechanised process. Its greatest advantage is that it has a high productivity, but this relies on manufacturing very large numbers of an identical product for a mass market. Another feature is that workers on the assembly line are given single tasks for the sake of production efficiency. One of the most famous phrases of Henry Ford epitomises the spirit of mass production, when he said of his Model 'T' car, 'You can have any colour you want ... so long as it's black.' In recent years, mass production has given way to more *flexible manufacturing systems*. (See also *Fordism*.)

McDonaldisation is a term used by the sociologist George Ritzer. He describes it as the process by which an industry takes on the characteristics of a fast-food restaurant. It is a re-examination of the principles of *rationalisation*. He highlighted several components of McDonaldisation:

- efficiency – the optimal method for accomplishing a task, usually the fastest. Efficiency in McDonaldisation means that every aspect of the organisation is geared towards the minimisation of time. McDonaldisation also develops the notion that quantity equals quality and that a large amount of product delivered to the customer in a short amount of time is the same as a high-quality product
- predictable, standardised and uniform services – this means that no matter where a person goes, they will receive the same service and receive the same product every time when interacting with a McDonaldised organisation. This also applies to workers in those organisations. Their tasks are highly repetitive, highly routine, uniform and predictable.

mean: see *arithmetic average (mean)*.

meander: a sinuous bend in a river. An explanation of the formation of meanders in a river has caused some problems for geomorphologists. In low flow conditions straight channels are seen to have alternating bars of sediment on their bed and the moving water is forced to weave around them. This creates alternating shallow sections (*riffles*) and deeper sections (*pools*). The swing of the flow that has been induced by the riffles directs the maximum velocity towards one of the bends, and results in *erosion* by undercutting on that side. An outer concave bank is therefore created. *Deposition* takes place on the inside of the bend, the convex bank. Consequently, although the river does not get any wider its sinuosity increases. Once created, meanders are perpetuated by the corkscrew movement of water called *helical flow*. This is a surface movement of water across to the concave outer bank with a compensatory subsurface movement back to the convex inner bank. Eroded material from the outer bank is transported away and deposited on the inner bank. The cross-section of a meander is therefore asymmetrical. The outer bank forms a *river cliff* with a deep pool close to the bank, the inner bank is a gently sloping deposit of sands and gravels, called a *point bar*. As erosion continues on the outer bank, the whole feature begins to migrate slowly both laterally and downstream. In some parts of the world meanders regularly change their course causing difficulties for those who determine land ownership boundaries.

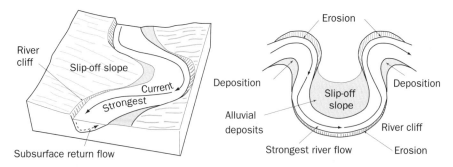

measures of central tendency are attempts to describe or summarise a set of data by calculating a single numerical figure. There are three main measures: *mean*, *median* and *mode*.

mechanical weathering: see *physical weathering*.

mechanisation is the substitution of human or animal forms of power by a mechanical system. Examples include the use of tractors and combine harvesters in farming, the use of production lines and robots in factories, and the use of motor vehicles for transport.

median: the middle value of a set of numbers arranged in order of size. In general terms, if there are 'n' numbers in a data set, then the median is the (n + 1)/2 th value. It is one of the *measures of central tendency*. Unlike the mean, it is an actual value in the distribution of numbers, except where there are an even number of values in which case the median is the average of the middle two.

Mediterranean climate: a feature of the Mediterranean area of Europe, California, central Chile, Cape Province in South Africa and parts of southern Australia. The main characteristics of the climate are:

- hot dry summers
- warm wet winters
- an annual temperature range of approximately 15°C
- an annual *precipitation* of 400–700 mm.

The summers are the result of the poleward movement of the *inter-tropical convergence zone* which causes the subtropical high pressure belt to migrate into these areas. Subsiding air and dry offshore winds bring *arid* conditions. In winter, the inter-tropical convergence zone and subtropical high move equatorwards, and the area comes under the influence of westerly winds. These blow onshore and bring rain, often associated with *depressions*. Incursions of colder air from the poles also frequently occur encouraging the further development of frontal rain systems. California and Central Chile have adjacent cold offshore *ocean currents*, and *advection fogs* are a regular event along the coastline. The movement of different *air masses* into the Mediterranean region of Europe also causes local winds to arise. In winter, cold air from the north is channelled violently into the Rhône Valley to create the Mistral. In summer, hot dry dusty winds such as the Sirocco blow from the Sahara.

M

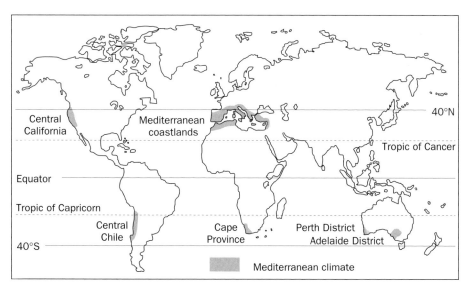

Areas with a Mediterranean climate

Mediterranean vegetation: the climatic *climax vegetation* associated with those areas of the world experiencing a *Mediterranean climate*. In all cases, it is *xerophytic* or *drought*-resistant. The original vegetation of the Mediterranean areas of Europe was mixed forest of conifers and broad-leaved evergreens. However, because of human interference little of this remains. The characteristic vegetation is now a mixture of small trees, shrubs, low scrub and grassland. Local names for this type of vegetation include *maquis* and *garrigue* in the Mediterranean and *chaparral* in California. Most plants in these areas have a range of adaptations to resist the summer drought:

- thick waxy evergreen (*sclerophyllous*) leaves to reduce water loss by *transpiration*, for example, the arbutus tree
- the ability to close their stomata, for example, the mastic tree
- very small leaves or thorns to reduce transpiration, for example, gorse
- deep root systems, for example, the almond
- bulbs and tubers which remain dormant in the soil and burst into life during the wet winters, for example, the hyacinth
- thick gnarled barks to reduce transpiration, for example, the cork oak and olive.

Many trees are also pyrophytes or resistant to fire. The occurrence of many such trees supports the view that the Mediterranean vegetation is essentially a *plagioclimax* – the outcome of the activities of humans.

megacity: a city with more than 10 million people, of which there were 20 in the world in 2007.

meltwater: the water that issues from a *glacier* or *ice sheet*. Meltwater may be found on the ice, in the ice or below the ice, but most often beyond the ice margin. Meltwater gives rise to *fluvioglacial landforms*.

Mercalli scale: this measures the intensity of an *earthquake* event and its impact. It is a 12-point scale running from Level I (detected by seismometers but felt by very few people – approximately 2 on the *Richter scale*) to Level XII (total destruction with the ground seen to shake – approximately 8.5 on the Richter scale).

mesas occur in desert areas where the strata is formed of sedimentary rock with horizontal *bedding planes*, and water erosion has removed much of the rock leaving behind mesas and *buttes*. Mesas are plateau-like features that are flat on top ('mesa' is the Spanish word for a table) and have steep edges often falling away to a *wadi* or *canyon*.

metamorphic rocks were either *igneous rocks* or *sedimentary rocks* in origin, but they have been subjected to alteration, becoming changed in both character and appearance. The original rocks may have been changed by increased temperatures or increased pressures or a combination of both. The intrusion of *magma* into a region may cause such a change, the rocks around the intrusion forming a metamorphic auriole. Metamorphic rocks are frequently crystalline, with the crystal aligned in one direction or divided into light and dark bands, and they rarely contain fossils. Examples of metamorphic rocks are slate, gneiss, schist and marble.

methane is natural gas sometimes referred to as 'dry gas'. It is an oil-associated gas which can be produced from gas-rich oil fields or from separate gas fields. It is a mixture of hydrocarbons in a gaseous state at normal (ambient) temperature and pressure. This gas can be liquified by refrigeration to $-161.4°C$ when it takes up 1/600 the volume of the gas. This liquified natural gas (LNG) is easier to transport – an important factor since there is considerable trade between the main producing areas and major markets although the USA is both the world's largest producer and consumer. Algeria, Saudi Arabia and the Persian Gulf states are major exporters; other significant producers are the UK, Norway, the Netherlands, Australia and New Zealand.

microclimatology: the study of climatic conditions and variations over a small area. This could include the changes in temperature, *precipitation*, *humidity*, wind speed and *evaporation* within a local climate such as a forest or on the shores of a lake. *Urban climates* contrast with those of their rural surroundings and different forms of vegetation or standing crop produce varied temperature and wind speed profiles at different heights above the ground.

mid-oceanic ridge: a range of submarine mountains formed at a *constructive plate margins* (divergent margins). They occur mainly in the mid-Atlantic, the East Pacific and the mid-Indian Ocean. They are formed largely of basaltic rocks and are associated with shallow foci *earthquakes*. The ridges can rise 3000 m above the ocean floor and extend up to 4000 km wide; some locally rise above sea level to produce islands, e.g. Ascension Island. The form of the ridge crest is controlled by the rate of separation of the plates. New *crust* forms as the plates diverge as basaltic magma moves towards the surface. Where there is a slow rate of separation, 10–50 mm per year, as in the mid-Atlantic, there is a marked *rift valley* along the axis of the crest. This is 30–50 km wide and up to 3000 m deep. This gives inward facing scarps with volcanic activity along the rift floor. Where there is more rapid spreading, 50–90 mm per year, as in the Galapagos ridge, the rift is less marked and the ridge appears smoother. With rapid separation, over 90 mm per year, as in the East Pacific Rise, the crest is smooth and unbroken. On either side of these ridges the ocean floor reveals alternate bands of reversed magnetism set into the rock as it formed. This form of *palaeomagnetism* provides evidence of *sea floor spreading* and plate movement. The age of rocks increases with distance from the mid-oceanic ridge.

migration is a term used for any kind of movement and in the case of population movement it usually refers to a permanent change of residence. The *United Nations* defines 'permanent' as a change of residence lasting more than 1 year. Recently the term has been applied more widely to include movements which are seasonal or even daily but such temporary moves are perhaps more accurately described as circulatory movements. Migration can be classified as either voluntary or forced and a distinction can be made between internal migration, within a country, and external or international migration. Voluntary moves are more selective, i.e. only certain individuals or groups of people may be affected by a particular set of factors; forced migration is less selective, generally the whole population is forced to move, it is not the result of individual choice.

Millennium Development Goals (MDGs) were originally developed by the *Organisation for Economic Co-operation and Development* and were highlighted out of the eight chapters of the *United Nations* Millennium Declaration, signed in September 2000. The United Nations Millennium Declaration established 2015 as the target date for achieving most of the MDGs, with 1990 generally used as a baseline. The eight goals act on the global causes of *poverty*. They are to:

1 Eradicate extreme poverty and hunger:
  • reduce by half the proportion of people living on less than one US dollar a day
  • reduce by half the proportion of people who suffer from hunger

2 Achieve universal primary education:
  - ensure that all boys and girls complete a full course of primary schooling
  - increased enrolment must be accompanied by efforts to ensure that all children remain in school and receive a high-quality education
3 Promote **gender** equality and empower women:
  - eliminate gender disparity in primary and secondary education preferably by 2005, and at all levels by 2015
4 Reduce child mortality:
  - reduce the mortality rate among children under five by two-thirds
5 Improve maternal health:
  - reduce by three-quarters the maternal mortality ratio
6 Combat HIV/**AIDS**, *malaria*, and other diseases:
  - halt and begin to reverse the spread of HIV/AIDS
  - halt and begin to reverse the incidence of malaria and other major diseases
7 Ensure environmental sustainability:
  - integrate the principles of sustainable development into country policies and programmes; reverse loss of environmental resources
  - reduce by half the proportion of people without sustainable access to safe drinking water
  - achieve significant improvement in lives of at least 100 million slum dwellers, by 2020
8 Develop a global partnership for development:
  - develop further an open trading and financial system that is rule-based, predictable and non-discriminatory. Includes a commitment to good governance, development and poverty reduction – nationally and internationally
  - address the least developed countries' special needs. This includes **tariff**- and quota-free access for their exports; enhanced debt relief for heavily indebted poor countries; cancellation of official bilateral debt; and more generous official development assistance for countries committed to poverty reduction
  - address the special needs of landlocked and small island developing states
  - deal comprehensively with developing countries' debt problems through national and international measures to make debt sustainable in the long term
  - in cooperation with the developing countries, develop decent and productive work for youth
  - in cooperation with pharmaceutical companies, provide access to affordable essential drugs in developing countries
  - in cooperation with the **private sector**, make available the benefits of new technologies – especially information and communications technologies.

In 2007 the **United Nations** Secretary-General Ban Ki-moon launched the Millennium Development Goals Africa Steering Group with major development partners (African Union, **European Union**, African Development Bank, Islamic Development Bank, **International Monetary Fund** and **World Bank**), aimed at specifically targeting the goals at Africa and to boost Africa's failing efforts to meet the goals the world had set itself to cut poverty, hunger, maternal and *infant mortality*, and other social ills by 2015.

millionaire city: a city with more than 1 million people. India and China have the most millionaire cities in the world.

mineral deposit formation: the type of structure in which minerals are found. A mineral ore is an accumulation in sufficient concentration to warrant commercial exploitation and the formation or structure in which it is located determines the relative ease, and cost, of extraction. There are three main types of formation: igneous ore bodies, sedimentary ore bodies and alluvial deposits.

As *magma* was intruded into the crustal rocks it cooled and different mineral constituents separated to form magmatic ore deposits such as magnetite (iron ore), copper and nickel, chromite and platinum. Liquids and gases were often forced upwards through *fissures* to form veins and lodes in which different minerals solidified at different temperatures and accumulated at varying depths below the surface. In Cornwall copper replaces tin at depth; often these associated or secondary minerals influence the decision to exploit a deposit. These formations are generally difficult and expensive to mine; they follow irregular courses, their composition varies and they often end abruptly.

Sedimentary ores are found in more uniform, dipping or horizontally bedded layers, resulting from *deposition* on the floor of the sea or a lake. These include iron ores such as limonite, manganese and *phosphates* as well as evaporites such as gypsum, potash and rock salt. Some deposits result from the downward *percolation* of mineralising solutions; bauxite and some nickel deposits are formed in this way. These formations are much easier to exploit, particularly if the 'overburden' (the geological deposit covering the sedimentary ore) is poorly consolidated material such as glacial drift. These deposits can be mined by opencast methods.

Where veins, lodes and sedimentary formations have been subjected to *erosion*, minerals may occur as deposits in sands, silts or clays along valley floors or on the continental margins in shallow oceans. These are termed placer deposits and include tin, gold and diamonds. They can be worked by pumping or *dredging*, a relatively inexpensive form of *mining*, because the minerals have already been removed from their natural setting by erosion.

mining is the extraction and preparation of minerals for industrial use, and the commercial exploitation of a reserve is influenced by a number of factors. Geological considerations determine the ease of accessibility to the mineral body. This reflects the *mineral deposit formation*. Other factors such as the depth of the deposit will influence the choice of extraction method. Shafts and galleries, associated with lodes and veins will be more expensive than opencast and *dredging* techniques. The quality, or metal content, of the ore reserve is also relevant to the decision to exploit a mineral. High content can offset the high costs of exploitation, as in Kiruna and Gallivare (north Sweden) where the ore is 55–70% iron. The Lorraine ores are of 24–35% content; these need to be mined near to their market or where they are easily removed from the sedimentary structures. The size of the reserve is also important as this influences the potential life span of the activity; a large reserve can adequately return the investment required to develop the mine.

Economic factors such as the availability of capital for investment, a local *labour* force (or funds to attract an outside workforce) and the accessibility of the mining area in relation to developed transport *infrastructure* are also influential. Political considerations such as the strategic value of reliable mineral supplies and the political stability of the mining region may also influence potential investors.

mist is an atmospheric condition where visibility is reduced but is greater than 1 km. It is a suspension of water droplets in the air, formed in a similar way to *fog*.

mixed economy: a combination of a *market economy* and *centrally planned* enterprises. *Market*-led economies have certain perceived disadvantages, the most important of which is the need for governments to regulate the workings of the market either through laws or by running parts of the economy through state enterprises. This is known as a mixed economy. In the UK in the past, industries such as coal, steel, water and electricity and services such as British Rail, British Airways and the Post Office have been run indirectly by government. Since 1979, however, there has been a move towards *privatisation*, increasing the influence of the market.

Centrally planned economy	Mixed economy	Market economy
State runs and organises production	Some production run by state, e.g. utilities	Market-led production
Labour works for state	State and private sector complete for labour	Wages determined by market forces
Prices controlled	Prices determined by supply and demand	Prices determined by supply and demand

mode: the value that occurs most frequently in a set of data. It is a *measure of central tendency*.

Mohorovicic: a Yugoslavian seismologist who discovered the junction between the Earth's *crust* and the *mantle* in 1909 by studying the different speeds of *earthquake* waves as they passed from the crust to the mantle. This boundary is termed the Mohorovicic discontinuity, or Moho, and it lies at a depth of up to 40 km beneath the continental areas and between 6 and 10 km below the ocean floors.

monoculture occurs where agricultural production relies on one dominant crop. This may take the form of *extensive agriculture* as with wheat growing in parts of the Canadian prairies or may occur with intensive systems such as rice cultivation in southeast Asia. A concentration on one crop can cause problems with mineral depletion in the soil leading to infertility, and for commercial producers the economic success of this type of system is very dependent on annual yields and the effect on market price. *Overproduction* in high yielding years can lead to a low return as a result of low prices on the world market.

monopoly exists, in theory, when a single producer controls the supply of a particular product to a market. In practice such complete control rarely exists, but if one organisation controls a sufficiently large proportion of the market then it can exert control over prices by restricting output. In the case of water supply, consumers can only effectively purchase water from their local water company, thus they hold a monopoly for their region. The danger with monopolies is that they can exploit the consumer, charging higher prices or offering a poor service, or they can waste resources through inefficiency due to lack of *competition*.

monsoon refers to a seasonal reversal of wind direction. The best-known monsoon occurs in India and southeast Asia. This occurs because:
- in June the *inter-tropical convergence zone* moves north but extends itself further north over northern India because of the intense heating that takes place there. This low pressure draws in warm moist unstable air from the Indian Ocean bringing heavy rain. Rainfall totals are further increased by the uplift of the air over the foothills of the Himalayan mountain range, and by intense *convection*

- in January the inter-tropical convergence zone moves south over the equator. At the same time the centre of Asia experiences intense cooling and a large area of high pressure develops. Winds blow away from this high pressure cell bringing very dry conditions to the majority of the Indian subcontinent. The eastern coast of India and Sri Lanka may also get rain from this monsoon as the winds pass over the Bay of Bengal and may become moist.

The occurrence of the wet monsoon in June is unreliable because upper atmospheric conditions have to be right to allow the winds to come from the south. Failure may result in prolonged **drought**. Conversely, excessive rains may cause serious **flooding**.

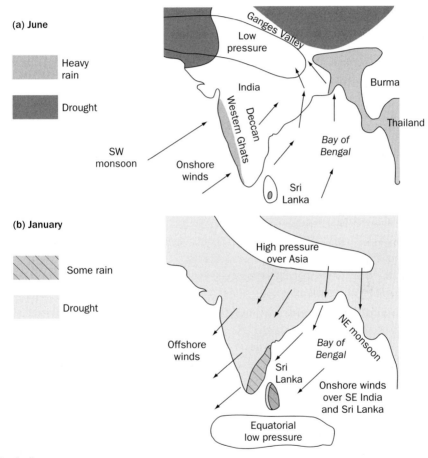

The Indian monsoon

mor is an acidic form of **humus**. It is common in wet and cold environments and is associated with **heathland** areas and **coniferous woodland**. Mor is dark brown to black in colour, poorly decomposed and lacking in nutrients. Few species of soil fauna can tolerate its acidic conditions and thus earthworms are rare.

moraine is the collective name for the **debris** which a **glacier** transports and then eventually deposits. The debris has either been eroded by the glacier along its sides or base, or has fallen on to it from rock faces above. There are several types of moraine:

- lateral moraine – material found along the edges of the glacier
- medial moraine – formed by two lateral moraines merging when a tributary glacier meets a main glacier
- end moraine – material deposited across the front of the glacier. There are a number of variations of end moraines (see *terminal moraine*)
- *supraglacial* moraine – material resting on top of the glacier
- *englacial* moraine – material within the glacier itself
- *subglacial* (ground) moraine – material being carried along the valley floor beneath the glacier.

(See also *lodgement till*.)

morbidity: is concerned with illness and the reporting of disease. In the UK, 2001 *census* respondents were asked how well they felt and whether they had a limiting long-term illness. Some diseases are so infectious that by law they must be reported; these are usually included in international surveillance programmes. Plague, *cholera* and yellow fever are the most serious, but *malaria*, influenza and typhoid are other examples.

morphology refers predominantly to the shape of a feature. The morphology of a physical landform is its relative size, shape and field location. The morphology of a settlement refers to its shape in plan as well as to the degree to which its buildings are grouped together.

mortality refers to the death of people. It is measured by a number of indices:
- the crude *death rate* – the number of deaths per 1000 people in 1 year
- *infant mortality* – the number of deaths of children under the age of 1 year expressed per 1000 live births per year. It is useful as a barometer of social and environmental conditions and is very sensitive to changes in either
- the case mortality rate – the number of people dying from a disease divided by the number of those diagnosed as having the disease
- attack rate – the number of cases diagnosed in an area divided by the total population over the period of an *epidemic*.

mountain and valley winds: see *anabatic* and *katabatic*.

mountain building: see *orogenesis*.

mountain climates are generally cooler and wetter than lowland areas of the same latitude. Temperatures fall by 1°C per 150 m rise, although the decrease in pressure, *humidity* and the amount of dust particles means that more short-wave *radiation* penetrates to ground level. This gives high daytime surface temperatures on slopes facing the sun. Poleward facing slopes, and those shaded from the sun, receive much lower *insolation*. Differential heating of air on valley sides and valley floors during the day may lead to warm air rising upslope, or up valley, as *anabatic* winds. As higher areas cool down, the cold dense air may descend into the valley floor, or move down valley as a *katabatic* wind. As moist air passes over mountains it is cooled to *dew point* producing *condensation* and relief, or orographic, rainfall. This tends to increase with altitude up to about 2000 m above sea level, but above this altitude the low moisture content of air due to low temperatures produces a decrease of *precipitation* with height. As air descends after crossing a mountain barrier it is warmed adiabatically producing the *fohn* effect.

mudflow: a form of *mass movement* in which *debris* of varying sizes may be transported in a matrix of saturated clay. These are more rapid flows which can occur on relatively low slope angles compared with *earthflows*. They occur in areas which experience torrential rain falling on ground which has limited protection from vegetation cover. This allows the weathered *regolith* to become saturated, increases the pore water pressure in the debris and reduces the frictional resistance between particles. As a result the debris moves downslope.

mull is a mild *humus* which is soft, black in colour and rich in nutrients. It is produced by the action of *bacteria* and earthworms when the soil is not too acidic and it is commonly found beneath lowland deciduous hardwoods and grasslands.

multi-cultural society: a social grouping that contains members from a wide variety of national, linguistic, religious or cultural backgrounds. It is often an emotive issue, especially when 'cultural' differences are interpreted as racial differences. Although many modern groups are thought to have descended from three broad racial types – negroid, caucasoid and mongoloid – the distinctions are now so blurred that race has little scientific relevance. Although skin colour remains as a visible distinguishing feature, people differ from one another in terms of ethnic differences, language, religion and culture. Multi-cultural societies are the result of *migration*, both forced and voluntary moves, but may also generate movement as persecuted minorities seek to escape oppression. Groups within a multi-cultural society may be integrated to a greater or lesser degree; the operation of *apartheid* in South Africa illustrated a lack of integration while Singapore represents a more tolerant attitude between the dominant Chinese (76%), Malays and Indians (22%), and the minority Europeans and Eurasians.

multi-lateralism: the development of trading agreements and negotiations between groups of countries, as distinct from bi-lateral talks which take place between two parties. Negotiations may be undertaken on behalf of a *trade bloc* such as the *European Union*.

multinational: see *transnational*.

multiple deprivation: indices of deprivation were first introduced in the UK in 2004. They are used to measure relative levels of deprivation in each administrative area of the UK. This should allow the government to rank areas according to need and to ensure that the most deprived areas receive most help. Seven indicators of deprivation are combined to give a measure of multiple deprivation. Each of these indicators is itself made up of a number of different measures. The seven basic indicators are: income; employment; health and disability; education, skills and training; barriers to housing and services; living environment including air quality and traffic accidents; crime.

multiple hazard regions are areas that are prone to a range of *hazards*. Although these areas are subject to individual hazard events, the hazards may be inter related, in that one hazard may increase the likelihood or incidence of another hazard. In the area around Los Angeles a number of physical and human factors interact to produce a cumulative hazard effect. In the *Mediterranean climate* of California, the summer *drought* and hot, dry Santa Ana wind descending from the Rockies, combine to make the *xerophytic chaparral* vegetation very dry and prone to *fires*. This hazard can produce a knock-on effect on other hazards. The increase in *atmospheric particulates* and the *temperature inversion* over the urban area operate to trap pollutants from vehicle emissions as *photochemical smog* over the city. The reduction in vegetation cover reduces *interception* and soil protection and this can make steep slopes more prone to *landslip* when saturated by the winter rainfall. The urban expansion of the city has meant that housing and road building have increased

M

187

the instability of the slopes. The area lies on a series of fault zones; the San Andreas to the east and the San Fernando and Santa Monica faults on the edge of the built-up area. The area experiences tremors and *earthquakes*, hazards in themselves, but also factors which promote further landslips. Economic activity, such as oil exploration, has affected the sub-geology, causing coastal subsidence which can increase the potential threat from tidal surges, *flooding* and *tsunamis*. *Intensive agriculture* in this environment requires *irrigation* which can lead to seepage into the subsoil and sprinkler systems installed to reduce the risk of bush fires add to ground saturation. Although these multiple hazards are not all activated at the same time, they are clearly interrelated. As an urban area, Los Angeles also experiences hazards such as high crime levels, race riots, gang warfare and the more usual traffic problems.

multiplier effect: a new or expanding economic activity in an area creates extra employment and raises the total purchasing power of the population, which in turn attracts further economic development creating more employment, services and wealth. This has been described as a case of 'success breeds success'. The multiplier effect is part of the wider process of *cumulative causation*, put forward as a model by *Myrdal*.

multi-racial society: see *multi-cultural society*.

Myrdal, G.: Swedish economist who, in the 1950s, first put forward the idea of *cumulative causation* and the *multiplier effect*.

Aiming for a grade A*?

Don't forget to log on to **www.philipallan.co.uk/a-zonline** for advice.

NAFTA: see *North American Free Trade Area*.

National Health Service (NHS) was established in 1948 as a state provider to organise healthcare for the people of the UK. It is paid for by taxes, so that patients do not have to pay for hospital treatment or for visits to and from a doctor. However, many people do pay for pre-scribed medicines. The service has undergone a number of changes in recent years, but for most people the main point of contact continues to be the family practitioner who provides primary care. Secondary care is provided by hospitals.

At the local level, some practitioners have grouped together to form partnerships and, together with other facilities such as ante-natal and family planning clinics, health cen-tres have been created. More recently, other elements of healthcare, such as low grade operations and diagnostics, have been taken to a more local level with the creation of poly-clinics. Rationalisation has also taken place at the secondary level, with the closure of many small 'cottage' hospitals. Larger district hospitals have been created providing a full range of special facilities. Similarly, accident and emergency services have often been concentrated in one hospital within each urban area.

In recent years, the NHS has come under *competition* from private health provision, encouraged by health insurance companies such as BUPA. Private hospitals can be located in leafy suburbs in converted large houses, but equally they may be in inner-city areas both for accessibility and to be close to NHS provision. Long-term care of the elderly is also being phased out of NHS provision and handed over to private organisations. Residential homes for the elderly are frequently concentrated in certain parts of a city, either in purpose-built facilities or in converted large properties.

Market processes have now been introduced into the NHS itself, with the development of health trusts. These have 'opted out' from control by regional health authorities and receive their funding directly from central government. It has created a situation where hospitals compete with each other to provide the required services, but the aim is to increase choice for patients as well as to improve the quality of the service. The distinction between public and private health provision is now blurred.

nationalisation describes the transfer of firms and industries from the *private sector* to the *public sector*. In the UK most nationalised industries were private concerns which had failed, e.g. the railways, steelmaking, coal *mining*, British Leyland (now Rover) and Rolls-Royce, but since 1979 there has been a policy of denationalisation or *privatisation* which has returned most of them to the private sector.

nationalism: loyalty and devotion to the state, such that national interests are placed above individual or global interests, i.e. the assumption that nations are the primary focus of political allegiance. It can be present in states of widely diverse types.

National Parks: areas of outstanding scenery where human activity is carefully controlled and the environment is preserved or improved for the enjoyment of people and the *conservation* of native plants and animals. There are National Parks in many parts of the world and there are slightly different criteria for their establishment and operation from country to country. The first National Park was Yellowstone in the USA, established in 1872.

In the UK, the National Parks were set up by an act of parliament (the National Parks and Access to the Countryside Act 1949). Between 1951 and 1957, ten parks were set up, nine of these in the uplands of England and Wales. They are the Lake District, Yorkshire Dales, North York Moors, Peak District, Snowdonia, Brecon Beacons, Exmoor, Dartmoor, Northumberland and the Pembrokeshire Coast. Since then, other parks have been set up in Scotland (Cairngorms and the Loch Lomond and the Trossachs Park) and the New Forest in southern England. The Broads in East Anglia has been given similar status. For a number of years, there have also been consultations over the creation of a National Park in the area of the South Downs (southern England). The 1949 act laid down two statutory duties for the park authorities:

- to protect and enhance the character of the landscape; to protect the countryside and take care of the distinctive ways of life found within them
- to provide access and facilities for relaxation and outdoor recreation.

Other recommendations point to:

- conservation of nature, archaeological remains and architecture
- encouragement of quiet outdoor activities
- higher levels of signposting of 'rights of way'
- seeking access arrangements over open land
- developing policies of tourism and traffic management.

Unlike parks in many other regions of the world, those in the UK have permanent populations and much of the land is privately owned. Therefore there is not automatic access for the public to all areas. Current conflicts within the parks include:

- The popularity of the parks is a threat in the *honeypot sites* through traffic congestion, footpath erosion and ecological damage.
- Planning applications for time-share development and leisure parks are increasing; inflated house prices often act against local housing needs.
- Economic support – these are often less favoured areas for farm support systems and schemes, which may create job opportunities that often conflict with environmental aims (quarries, power sites, extraction, destruction of woodland/farmland).
- The use of attractive land for military training (e.g. Pembrokeshire) still persists, although some areas have been released with the 1990s defence cuts.
- The planting of coniferous forest rather than *deciduous woodland* to the detriment of wildlife and scenery.

Developments in parks must conform to the criteria of the relevant park planning board; buildings must be in keeping with the surroundings and other buildings, using similar styles and building materials. There are controls on the location of tourist facilities; in some areas of the Peak District, minibus services link with a 'park-and-ride' scheme (e.g. Manifold Valley). The aim is to limit development that does not enhance the nature of the park. In 1968 the functions of the National Parks Commission were taken over by the Countryside Commission, later to become the Countryside Agency.

National Trust: an organisation in the UK founded in 1895 for the purpose of promoting the preservation of, and public access to, buildings of historic or architectural interest and land of natural beauty. The two trusts (one for England and Wales and one for Scotland) are dependent financially on donations, legacies, admission fees and the annual subscription of members. By the early twenty-first century, the trusts had acquired around 200,000 ha of land and over 350 stately homes, buildings and gardens.

NATO: see *North Atlantic Treaty Organization*.

natural increase: the increase in population within a country calculated by finding the difference between *birth rates* and *death rates*. If in 1 year, for example, the birth rate was 30 per thousand and the death rate 12 per thousand, then the natural increase rate would be 18 per thousand or 1.8%.

natural resources are those features which are needed and used by people and include such things as climate, soils and *raw materials* (iron ore, coal, crude oil etc.). Natural resources form a category distinct from human resources, this including such features as people and *capital*.

nature reserve: an area set aside by government for the purpose of preserving certain plants, animals, or both. It is distinct from a *National Park*, in that the park protects land and wildlife partly for public enjoyment, whereas a nature reserve protects animals and plants for their own sake. Endangered species, for example, are often kept in nature re-serves which protect them from hunters. Numerous nature reserves in the USA have served this purpose, particularly in the case of birds. In medieval times, landowners established game preserves for the protection of the animals that they hunted, and this was the origin of the reserve. The idea of protecting animals to keep them from dying out arose only in the nineteenth century.

neap tide: a tide with a low range; the high tide is relatively 'low' and the low tide is relatively 'high'. These occur twice a month coinciding with the first and last quarter of the moon. At this point the sun, moon and earth are at right angles, with the Earth at the apex. This reduces the tide-producing pull of the sun and moon on the waters of the Earth because the gravitational forces between the three bodies are in opposition. This position is referred to as 'in quadrature'. With *spring tides* the bodies are in line which produces a much stronger combined gravitational effect to create a high tidal range.

nearest neighbour analysis is a precise mathematical means of measuring point patterns. The technique involves using a simple formula, giving an index which shows how far the measured pattern of settlement for example is from being 'random', i.e. generated by chance. Greater or smaller indices indicate that the pattern has a tendency to be either spaced or clustered. The index value runs from 2.15 (evenly spaced), through 1.0 (random) to zero (highly clustered).

FORMULA: $Rn = 2\bar{d} \sqrt{\left(\dfrac{n}{A}\right)}$

where d̄ = mean distance between nearest neighbours

n = number of points

A = area in question

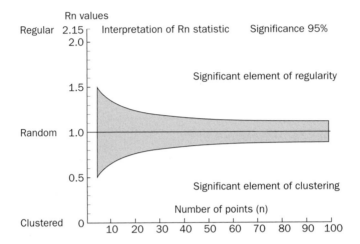

negative correlation: a *correlation* where an increase in the value of one *variable* is matched by a decrease in the value of the other variable. For example, a correlation between *birth rate* and the *GNP* of countries would be negative, as generally the higher the birth rate of a country, the lower the GNP.

negative feedback is when a *system* acts by lessening the effect of the original change and ultimately reversing it. On a river system, for example, increased *erosion* may cause so much *debris* to accumulate within the valley, that the river cannot move it, which will result in *deposition*. As a consequence, there will be less downcutting and the prevention of slope steepening. Thus, the original increase in the erosive capacity of the river has turned into a period of deposition.

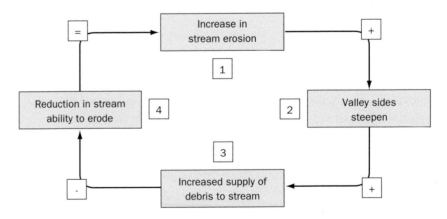

Negative feedback on a river system

negative skew is a bias within a *distribution* towards high values.

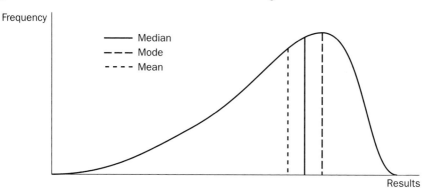

neighbourhood watch: schemes that began in the UK in the early 1980s, where the public and police combined in an attempt at crime prevention. The aim was for an alert neighbourhood to deter crimes of opportunity, such as the theft of an unlocked car, burglary of an unprotected house, and crimes against the person. In return, the police give advice on domestic security methods. This followed from the experience in the USA, where the most enthusiastic groups were the people of the *inner cities*, the main victims of crime.

neo-colonialism: the acquisition or retention of influence over other countries especially one's former colonies, often by economic or political measures.

neoliberalism is a politico-economic philosophy and set of policies that establish market expansion and privatisation, less government regulation and trade liberalisation. It was the basis of economic policies carried out in the 1980s under the governments of Margaret Thatcher in the UK and Ronald Reagan in the USA. (See also *Thatcherism*.)

net migration is the balance between *immigration* (people moving into a country) and *emigration* (people moving out).

net primary productivity (NPP) is the amount of organic matter that is available for man and other animals to harvest or consume. Gross primary productivity is the rate (in dry grammes per square metre) at which matter builds up in green plants as a result of photosynthesis. As plants respire some of this is lost and NPP is the gross PP less the amount lost due to plant respiration. World variations in NPP are due mainly to differences in moisture and temperature, and the availability of nutrients. There are considerable variations from 2200 dry gm/m²/year in tropical *rainforest*, 1250 in temperate forests, 900 in *savanna* areas to 90 dry gm/m²/year in desert and semi-desert areas. The average marine value is 152. The stage of plant succession also influences the level of NPP. On land nutrients are stored in the soil; in sea areas this lack of nutrient supply is the greatest limitation on NPP. Nutrients are lost due to sinking in the deepest oceans, these are at depths below the limit of light penetration. The most productive parts of the oceans are on the margins and the *continental shelf*, or where upwelling water brings nutrients to the surface.

newly industrialised country (NIC): a key element in the process of *globalisation* has been the emergence of newly industrialised countries, i.e. those that have undergone rapid industrialisation since the early 1960s. *Transnationals* from the developed nations were seeking new areas for their operations and Japanese companies, in particular, looked to their Asian neighbours such as South Korea, Hong Kong, Singapore and Taiwan (the

193

Asian Tigers). These countries had advantages for the development of *manufacturing industry*:

- They had a reasonably well developed level of *infrastructure* such as roads, railways and ports.
- Their populations already possessed a degree of skill as they were relatively well educated.
- Their cultural traditions revered education and achievement.
- They were in relatively good geographical locations.
- They were in possession of government support.
- Their laws (on *labour*, taxation, pollution) within which companies had to operate were less rigid than in the parent countries of the transnationals, allowing much easier, and more profitable, operations.

This was considered the first phase of NIC development. As wage levels rose in those countries, transnationals looked for cheaper locations along with emerging large companies within the Asian Tigers. This saw the growth of a second phase of NICs such as Malaysia and Thailand. In recent years, both China and India have emerged as the target for *foreign direct investment* by the transnationals of many nations. This is regarded as the third phase of NIC development. India shows a change from the previous pattern as much of the new development has been in the service sector.

new technology is used to describe the rapid changes in communication and other processes resulting from the exploitation of the silicon chip. The *innovations* resulting from the use of new technology can be split into three main types:

- by process, that is advancements in *manufacturing* technology and automation
- by product, that is new product opportunities using micro-electronic technology such as the fax machine and electronic games
- by communication links, that is *information and communication technology*.

new town: built initially in the UK as a response to the New Towns Act of 1946. This attempted to regulate the movement of people to, and to relieve overcrowding within, the major cities such as London, Liverpool and Glasgow. Slum clearance schemes moved many people out of inner-city areas, and they were attracted to new houses with all amenities in a semi-rural environment. Since then a number of 'generations' of new towns, designed to be 'self-contained and balanced', have been built mainly to take *overspill* populations from expanding cities, but also to act as *growth poles* in areas of high unemployment. The third generation created the 'new cities' of Milton Keynes and Telford. New towns have had varying degrees of success. Several of the earlier ones were built too near to London and were quickly 'swallowed up' by urban growth. Washington, in the northeast, has proved to be successful having attracted many foreign firms. However, Skelmersdale, near Liverpool, still has high unemployment, few job opportunities and poor local services. New towns are also found in other parts of the world, and serve a variety of purposes. In South Africa they were built to house workers in nearby cities, and in Israel they have been built to colonise new areas of desert and to house Jewish immigrants in recently annexed lands.

NIC: see *newly industrialised country*.

niche: the role a single *organism* or plant plays in an *ecosystem*. It may refer to the organism's place in a *food web*, or to a precise description of a plant's *habitat*. A niche *glacier* is a small upland body of ice resting on a slope or in a shallow hollow. They are common on north-facing slopes.

NIMBY: meaning 'not in my back yard', this term is applied to situations where a proposal may be wanted by a group of people for a common good, but no one is prepared to have it near to where s/he lives. Examples of such situations include the building of new routeways or the development of a potentially hazardous industry. Most people agree that these are either useful or necessary. However, strong resistance from those affected most by their construction and use is a regular feature of such schemes. Often the area with the least powerful 'voice' is chosen, and sometimes more than once, leading to a concentration of negative developments in an area.

nitrates are nutrients that are essential for plant growth. The main sources of nitrates are the soil itself, *fertiliser*, animal manures and silage liquor. A popular misconception is that the high levels of nitrates in *groundwater* and in some rivers and lakes are due to excessive applications of nitrogeneous fertiliser which find their way into the water. Research has shown that relatively small amounts of nitrates are left over from fertiliser to be leached away. However, levels of nitrates in groundwater, rivers and lakes are increasing, and in extreme cases cause **eutrophication**. Nitrates are produced naturally by microbe activity in the autumn as they break down organic material after a harvest. Methods that have been suggested to farmers to reduce the amount of nitrates being leached out by the autumn rains include:

- the non-application of fertiliser during this time
- not leaving land bare as ploughing causes a surge of microbe activity and therefore nitrogen release
- to sow winter crops early so that roots can take up the nitrates and thereby reduce leakage.

nitrogen cycle: *nitrates* that exist in soil are taken up by the roots of plants, for example grass. Herbivores, such as cattle and sheep, eat the grass and then release the nitrogen as ammonia by excretion. The excreta decays on or in the soil and the ammonia is converted by *bacteria* into nitrates which are then leached back into the soil. There are additional *inputs* of nitrogen from the atmosphere, as well as outputs via seepage into *groundwater*. (See the diagram on page 24.)

nivation: the process by which a hollow containing a snow patch becomes wider and deeper. *Freeze–thaw* action beneath the snow patch causes the underlying rocks to disintegrate. During warmer periods *meltwater* from beneath the patch flows away from the hollow and washes away the weathered material. In time the hollow may enlarge so that it contains sufficient snow to last through a summer. When this is the case, the hollow becomes an embryo *cirque*.

nomadic: a description of an existence which involves a group of people roaming from one place to another seeking pasture or water for their animals. Nomads do not have a permanent home. Nomadic herding usually exists in areas where the climate is too extreme to support permanent settled agriculture. The climate may be too *arid*, cold or hot, and the vegetation is sparse. Nomads will tend to follow seasonal rainfall, or will move away from snow-covered grazing grounds. Examples include the Tuareg of the Sahara, the Rendille of northern Kenya, and the Lapps of northern Finland. In some parts of Africa, nomadism is being actively discouraged. *Overpopulation* and *overgrazing* have caused resources to be overstretched. Nomads are being forced to move near to small towns and to become sedentary farmers. Wells and other services have been provided. However, there is a reluctance by the nomads to part with their animals, and this has led to extreme forms of overgrazing around the settlements.

non-governmental organisation (NGO): a body that has not been created by a statutory act but appears to have major responsibilities. They are created mostly by volunteers, but are later supported by individual or corporate benefactors. Many are charities and have international responsibilities such as organising relief and humanitarian operations, as in the case of the Red Cross. Others are national in their outlook and may have environmental concerns, such as the *National Trust*. *Greenpeace* is an example of an NGO with significant worldwide following and influence. Frequently, NGOs are given the responsibility of administering funds from governments, as they are seen to be impartial.

non-renewable resources are those that are *finite*, as their exploitation can lead to the exhaustion of supplies. In the energy field, coal, oil and natural gas are classified as non-renewables.

non-tariff barriers are hidden barriers to trade imposed by governments because they wish to restrict imports without being seen to do so, perhaps because it would be contrary to international regulations under the *General Agreement on Tariffs and Trade (GATT)*. Such barriers may take several forms:

- constantly changing technical regulations which make compliance difficult for importers
- forcing importers to use specified points of entry where documentation is dealt with only slowly
- regulations which favour domestic producers, e.g. packaging, and labels which conform to local language requirements.

normal curve: the bell-shaped curve of *normal distribution*.

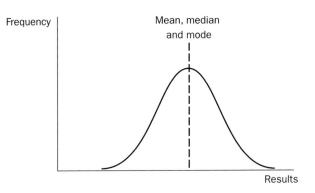

normal distribution: a *dispersion measure* that is bell-shaped so that half of the variables lie to the left of the *mean* and half to the right. Due to the symmetry of the curve, the mean, *median* and *mode* all coincide at the same value. If data is collected on an event that occurs many times over, variations in the result will tend to form a pattern that is known as normal distribution. This means that the results will tend to be clustered around the average, and will be split equally between results above and below the average (see the diagram above). Normal distributions can be standardised so that the width of the distribution is about six *standard deviations*. The areas within each standard deviation can be very useful in statistical analysis, for example in *significance testing*.

North: the term used to indicate the wealthier nations of North America, Europe, the former USSR, Japan, Australia and New Zealand. This category therefore contained the countries of the *First World* and the *centrally planned economies* of the *Second World*. The term *economically more developed countries (EMDC)* is now often used for this group. The term 'North' first appeared in the *Brandt Report* (1980). (See also the *South*.)

North American Free Trade Area (NAFTA) whose membership consists of the USA, Canada and Mexico, is an attempt to create the equivalent of the *single European market* within the North American continent.

North Atlantic Treaty Organization (NATO) is a military alliance established in 1949 set up to organise collective defence when members were attacked by an external party. From its initial numbers, NATO has expanded to include 26 members by 2008. These are Belgium, Bulgaria, Canada, Czech Republic, Denmark, Estonia, France, Germany, Greece, Hungary, Iceland, Italy, Latvia, Lithuania, Luxembourg, Netherlands, Norway, Poland, Portugal, Romania, Slovakia, Slovenia, Spain, Turkey, the UK and the USA. Albania and Croatia were the next candidates to be considered for membership.

North/South divide is the line that separates the richer countries of the *North* from the *economically less developed countries* of the *South*. The difference in economic development between the two areas is known as the *development gap*.

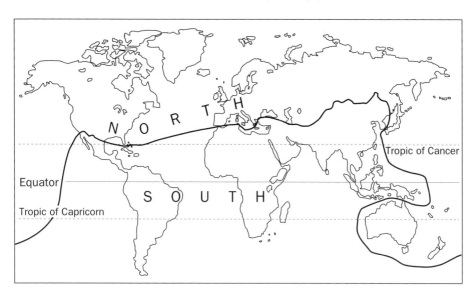

NPP: see *net primary productivity*.

nuclear energy involves the use of radioactive energy produced by nuclear reaction. The heat released by the reaction is used to produce steam which drives turbines to produce electricity. Uranium is processed, enriched and converted to uranium dioxide which is used in the reactor. This undergoes *nuclear fission* which releases large amounts of heat. Only small amounts of uranium are needed to produce a given output of heat compared with other forms of fuel. The plants are expensive to construct and decommission and there is a major concern relating to the disposal of the *waste* material. This must be seen in the context of world energy needs; it is unlikely that demand can be met from renewable sources and yet other sources such as coal, oil and gas are *finite*. There is no shortage of uranium and operating costs are very competitive. However, there are questions of safety and environmental damage, but so far nuclear plants have a good record in relation to other forms of electricity production and they produce less CO_2 and SO_2. Nevertheless there is a public fear which is used by nuclear opponents such as *Greenpeace* and Friends of the Earth. Disposal of waste is a problem because it is radioactive and requires specialised treatment and there is strong public concern that some contamination will occur in transit or at the point of disposal.

nuclear fission is a reaction in which a heavy nucleus splits into two parts which then emit neutrons, releasing energy in the process. The fission of uranium releases two or three neutrons which are able to split more atoms; thus a chain reaction is established which releases energy.

nuclear fusion: involves joining light atomic nuclei with heavier nuclei. This fusing together of different atomic masses creates energy without any loss of mass. This was achieved for the first time in 1991 at Culham Laboratories, a Research Establishment near Oxford.

nuclear waste consists of both high-level and low-level radioactive material. Power stations produce high-level waste which consists of used fuel rods and cells which have been removed from the reactor. These are transferred to a reprocessing plant, such as the Thorp plant at Sellafield, where reusable uranium and plutonium are separated out to leave radioactive waste. This is currently put into steel-clad glass containers which are stored at Windscale, in Cumbria. Low-level waste includes clothing and materials used in hospitals where exposure to radium and X-rays can present a low-level risk. These material are disposed of in controlled chemical and *radiation* dumps and may be buried in suitable geological structures.

nucleated: a term applied to a settlement in which the buildings are grouped around some focal point such as a church, a manor house, a green or a water supply. Historically this may reflect the search for a wet point site in a porous rock area, or the need for a dry point site in a wet landscape. It could be a defensive point or a central location in an area of open fields. Nucleated settlements display various forms: *linear*, T-shaped or cruciform, or they may reflect the shape of the green: circular, rectangular or lens-shaped.

nuée ardente is a glowing cloud of hot gas, steam and dust, volcanic ash and larger *pyroclastics* produced during a violent eruption. This descends the slopes of a *volcano* at high velocity as in the case of the eruption of Mt Pelée in 1902, giving rise to the term *Pelean eruption*.

null hypothesis: a negative assertion which states that there is no relationship between two variables which are being tested, e.g. 'there is no relationship between building height and distance from the centre of the *central business district*'. It assumes that there is a high probability that observed differences between two sets of data are due to chance variations. If the null hypothesis can be rejected statistically, then we can assume that any differences between the data sets are not due to chance but are the result of differences between the two *statistical populations*. We could then suggest that the *hypothesis* was acceptable.

nunatak: a rocky mountain peak which projects above an *ice sheet*. Because it is not covered by ice the peak is not planed off or rounded and *freeze–thaw* action maintains the steep sides. After deglaciation, the jagged and angular peak contrasts with rounded surfaces which were covered by ice.

nutrient cycle: the circulation of minerals around the *ecosystem*. Nutrients are taken up by the root system of plants. These are then used and released as the plants shed organic matter such as leaves, or when the plant dies. Some nutrients pass along the *food chain* as herbivores consume leaf material. These return minerals to the system through excreta or on their death. The litter which accumulates on the surface is broken down by micro-*organisms* and fungi which return nutrients to the soil store to be used in the cycle again. The three main storage areas within the system are the litter layer, the soil and the *biomass*. The relative importance of each of these stores depends on the nature of the *biome*. In *tropical rainforest* areas, a very high percentage of the nutrients are stored in the biomass which consists of several layers of plants. High temperatures and rainfall produce rapid decay of the litter layer and nutrients are quickly returned to the soil from where the many plant roots take up minerals. In *coniferous forests* the litter is the dominant store. Lower temperatures restrict breakdown and there is a slow return of nutrients to the soil. This slow replacement is insufficient to make up for the loss through *leaching*. The biomass is a relatively low proportion of total nutrients; it is made up of one main tree layer, consisting of species which have thin needle-like leaves and the acidic litter restricts the lower layers of vegetation.

nutritional requirement: the amount of food needed to sustain an individual. The daily intake of calories necessary to sustain a person averages about 2200 calories per day but this must be seen as a very general figure because an individual's need will reflect a variety of factors: climate, level of activity, type of employment, body size. It is not simply a measure of *calorie intake*; proteins, energy foods, vitamins and minerals are essential elements of a *diet*. If these needs are not met then *malnutrition* may result. Carbohydrates which provide energy are critical, and the intake of these is generally the measure used to show differences in diet, expressed as calories per person per day.

A–Z Online

Log on to A–Z Online to search the database of terms, print revision lists and much more. Go to **www.philipallan.co.uk/a-zonline** to get started.

oasis: an area in the middle of an *arid* region which is made fertile by the presence of water. Oases are usually small, but they can cover over a hundred square kilometres. The reason for the presence of water is usually the occurrence of water bearing strata (an *aquifer*) on the surface at that point.

obesity is defined as abnormal or excessive fat accumulation that may impair health. Body mass index (BMI) is a simple index of weight-for-height that is commonly used in classifying overweight and obesity in adult populations and individuals. It is defined as the weight in kilograms divided by the square of the height in metres (kg/m^2). The *World Health Organization (WHO)* defines 'overweight' as a BMI equal to or more than 25 and 'obesity' as a BMI equal to or more than 30. These cut-off points provide a benchmark for individual assessment, but there is evidence that risk of chronic disease in some populations, such as those in Asia, increases progressively from a BMI of 22. WHO's latest projections indicate that globally in 2005:

- approximately 1.6 billion adults (age 15+) were overweight
- at least 400 million adults were obese.

At least 20 million children under the age of 5 years were overweight globally in 2005. Childhood obesity is a big problem in the USA, where over 35% of children are overweight. WHO further projects that by 2015 approximately 2.3 billion adults will be overweight and more than 700 million will be obese. The fundamental cause of obesity and being overweight is an energy imbalance between calories consumed on one hand and calories expended on the other hand. Global increases in overweight and obesity are attributable to a number of factors including:

- a global shift in *diet* towards an increased intake of energy-dense foods that are high in fat and sugars but low in vitamins, minerals and other micronutrients
- a trend towards decreased physical activity due to the increasingly sedentary nature of many forms of work, changing modes of transportation, and increasing urbanisation.

obsolescence occurs when a product, service or machine has been overtaken by a new idea that provides the same function in a better or more attractive way. This often occurs as a result of *new technology*, though it may also be due to changing lifestyles, fashions, or scarcity of resources. The threat of obsolescence encourages most firms to update their products regularly and to look for new products that can replace declining ones. Some firms look to gain financial advantage from two specific types of obsolescence.

- built-in obsolescence means that the product will not be designed to last more than say, 3 or 4 years
- planned obsolescence is the creation of a feeling on the part of customers that they should replace items that are in fact usable. The car industry is a good example, where the addition of new features to existing models by manufacturers means that the customer is persuaded of the need for change.

Both these types of obsolescence are regarded by environmentalists as undesirable ways of using the Earth's resources.

occlusion: in a mid-latitude *depression* when the *cold front* catches up with the *warm front*. This then forms an occluded front. If the air behind the warm sector is colder than the air in front of it, a cold occlusion occurs and if the air is warmer, a warm occlusion.

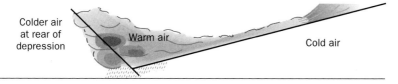

Cold occlusion

ocean basin: an area of ocean beyond the edge of the *continental shelf*.

ocean currents: large-scale movements of water within the oceans, that are part of the process of the horizontal transfer of heat polewards. Other currents carry colder water towards the tropics. In *temperate* regions, warm ocean currents generally accentuate the moderating effect of the oceans, particularly where the *prevailing winds* are *onshore*. The influence of the North Atlantic Drift on Western Europe is a good example. Ocean currents are largely set in motion by the prevailing surface winds associated with the *general atmospheric circulation*.

ocean ridge: see *mid-oceanic ridge*.

ocean thermal energy conversion is a method of exploiting the temperature difference in *tropical* waters between the warm surface of the ocean and the colder water at depth. This would be a form of *renewable energy* which it is claimed would be one of the world's most environmentally benign. A number of small experimental schemes have been established off India and Taiwan, for example.

ocean trench: a narrow, deep depression in the ocean floor which corresponds with the *subduction* zone associated with *destructive plate margins* (convergent margins). The Marianas Trench, on the western margin of the Pacific Ocean, extends to depths below 11,000 m. They are typically arc-shaped and correlate with the location of deep focus *earthquakes*, although intermediate and shallow earthquakes are also formed. These form on the margin of the trench towards the line of volcanic islands which form the *island arc*. Examples include the Chile-Peru Trench where the Nazca Plate (oceanic) subducts beneath the South American Plate (continental) and the Japan Trench where the Pacific Plate (oceanic) subducts beneath the Eurasian Plate (continental).

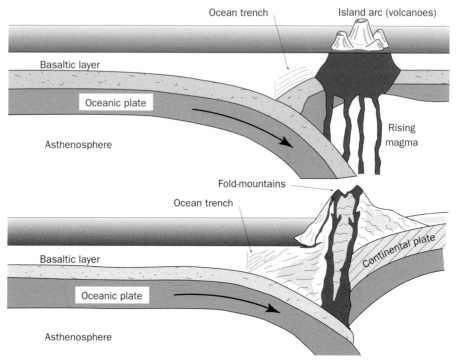

Location of ocean trenches

OECD: see *Organisation for Economic Co-operation and Development*.

offshore: this term may be used in a number of ways:
- referring to movement from the land towards the sea; an offshore wind can develop at night because the sea retains its heat while the land radiates heat rapidly causing cooling. The relatively warm air over the sea rises and is replaced by cooler air moving off the land to create a land-sea breeze
- in relation to coastal geomorphology, offshore features develop on the seaward side of the wave breakpoint, e.g. offshore bar
- in manufacturing many *TNCs* have located 'offshore', i.e. they are assembling the final product in a less economically advanced country using components produced in advanced industrial nations. Many Japanese companies have taken advantage of cheaper *labour* and land costs; the products are generally high-tech, high value/weight ratio goods which can withstand the transport cost incurred in moving components into the assembly area and in moving products back to an advanced market. Companies from the USA have also moved across the border into northern Mexico for similar reasons; location 'offshore' can also help a company to penetrate the local market.

oligopoly exists when a small number of companies dominate the market for a particular product, thus accounting for a very high percentage of output, e.g. car production. These firms tend to imitate each other's behaviour in terms of model range and development in order to preserve their share of the market. This can appear to be a competitive situation, but this *competition* is usually in the form of special offers rather than price reductions as there is often a degree of collusion which maintains price levels and profits.

onshore: a sea to land movement; during the day heating of the land causes warm air over the land surface to rise. This is replaced by cool air moving in from over the sea creating a sea-land breeze.

OPEC: see *Organization of Petroleum Exporting Countries*.

opencast mining: a method of extraction which involves extensive excavation to remove the overlying material (overburden) covering the mineral deposit. This method is used particularly for working coal or iron ore where the mineral has been laid down in relatively gently dipping or horizontally bedded geological structures as in a sedimentary *mineral deposit formation*. Large-scale earth movers and drag cranes are used to expose and extract the mineral. It is a lower cost form of *mining* which can allow even relatively low-quality deposits to be commercially exploited, particularly where the overburden is easily removed, as in the case of unconsolidated glacial material in the iron ore 'ranges' of the Lake Superior field.

optimum population is the theoretical number of people which, when working with all the available resources, will produce the highest standard of living and *quality of life*. This is a dynamic situation, because as technology improves, new resources may become available which will mean that a higher optimum population can be carried. If there are too many people for that living standard to be achieved, then the area is said to be *overpopulated*. Some areas have resources that are not fully exploited with the present population and therefore could stand a population increase. Such areas are said to be *underpopulated*.

order refers to goods and services, in that low-order items are obtained frequently and cost little per unit (bread, newspapers), whereas *high-order* items are more expensive and bought less frequently (furniture, clothing). People are prepared to travel much greater distances, therefore, to obtain higher-order goods and services. This means that low-order functions (newsagent, primary school) are spaced much closer together than high-order functions (secondary school, hospital).

ordinal data is presented in terms of the relative importance (or order of magnitude or rank) of data rather than absolute values. Data can be classified into broad groups, e.g. sediment size given as fine, medium or coarse, or settlements placed into general groups based on size.

Ordnance Datum (OD): the *mean* sea level based on observations at Newlyn, in Cornwall. Heights on Ordnance Survey maps are given in metres (or feet on earlier editions) above OD.

organic farming is a form of agriculture that does not use chemical *fertiliser*, *pesticides*, insecticides or *herbicides*, but instead favours animal and green manures and mineral fertilisers such as fish and bone meal. The farming system is intensive, but it involves crop rotation and the use of fallowing periods.

PROS:
- the increased organic matter in the soil enables it to retain more moisture during dry periods and allow better drainage during wet spells
- less likely to cause *soil erosion* or exhaustion as, apart from more organic matter, it will have more soil *fauna* (worms etc.)
- should not harm the environment as there will be no *nitrate* runoff to cause *eutrophication* in rivers
- kinder to wildlife as there is no pesticide to kill insects and birds

- self-sustaining, as it produces more energy than it consumes
- less harmful to people as fewer chemicals are used

CONS:
- yields can drop considerably in the initial period when compared to conventional farming
- weeds can increase and need to be controlled by hand *labour*
- the produce is more expensive when sold than conventional farming products.

organic matter: see *humus*.

Organisation for Economic Co-operation and Development (OECD) was set up in 1961 and aims to coordinate international aid for *economically less developed countries* as well as providing a forum for discussion on *economic growth* and trade. Its members are mainly drawn from the *developed* world, but some of its most effective work has been in the collection and publication of worldwide economic and social data, standardised so that inter-country comparisons are possible.

organism: an individual living plant or animal. As well as plants, the organisms in a soil are the *bacteria*, the fungi and the animals which live in the soil such as earthworms and wood-lice. They are more active and plentiful in warmer and more aerated soils. They are responsible for the *decomposition* of leaf litter, for the fixation of nitrogen in the air into *nitrates*, and for the binding together of individual soil particles to give a good structure. Burrowing animals also assist in the circulation of air and water through the soil.

Organization of Petroleum Exporting Countries (OPEC) is the name of the *cartel* which sets output *quotas* in order to control crude oil prices. The members of OPEC are countries in the Middle East, South America, Africa and Asia, but not the USA, USSR or the European producers such as the UK. In the 1970s, OPEC had great power when it controlled 90% of the world's supply of crude oil exports. Its influence, however, has been reduced (36% of world production in 2008), as the higher price of oil allowed more expensive fields, such as the North Sea and Alaska, to be brought into production. Its headquarters are in Vienna. Members of OPEC are Algeria, Angola, Ecuador, Iran, Iraq, Kuwait, Libya, Nigeria, Qatar, Saudi Arabia, United Arab Emirates and Venezuela. Indonesia was suspended from membership in 2008 as it became a net importer of oil, taking the membership to 12 countries.

orogenesis refers to mountain building. Active belts of mountain building are formed either by vulcanicity or by the breaking and bending of the Earth's *crust* by *tectonic processes*, or a combination of both of these. Vulcanicity involves the accumulation of volcanic rock by the extrusion of molten material from within the Earth. Tectonic processes involve either the movement of continental crust towards oceanic crust, or the convergence of two continental crusts. In both cases intervening *sediments* are compressed, uplifted and folded to create mountains. The Himalayas are thought to have been formed in this way when the Indian subcontinent moved north to collide with the Eurasian plate and uplifted the sediments of the former Sea of Tethys.

orographic: relates to processes caused by uplift over a mountain range. Orographic rainfall results when moist air is forced to rise over a mountain range. The rising air cools, the water vapour in the air condenses, *clouds* are formed, and eventually rain will fall. Such rainfall is common on the western side of Britain, and on the western coast of Canada. As the air descends on the leeward side of the mountain, it warms adiabatically, *condensation* ceases, clouds dissipate and little rain falls. This is known as a rainshadow. (See also *adiabatic*.)

out-of-town location: a site for an industry or shopping complex on the edge of an urban area. Here the land is cheaper, and there is more space for future expansion. Large car-parking facilities can be built. Such a site is frequently near to a motorway interchange, making it easily accessible to places further away. Larger out-of-town retailing centres such as Gateshead's MetroCentre are also aimed at creating a day out for the family with an emphasis on shopping and other leisure activities.

outsourcing/offshoring involves the transfer of a process, both in manufacturing and services to another company or individual, sometimes known as subcontracting. Offshoring is when the work is subcontracted to another country. Many US and UK companies, for example, have subcontracted work, such as call-centre operations, to Indian companies.

outwash: the material deposited beyond the snout of a **glacier** by the **meltwater** streams which issue from it. Outwash consists of a gently sloping plain made up of sands and gravels. These particles are graded into different sizes, the larger ones being found near to the ice front and becoming progressively finer with increasing distance from it. The finest particles, rock flour, may be transported considerable distances before being deposited. Outwash deposits tend to be more rounded than glacial deposits due to having been rolled around in moving water. Outwash plains are characterised by multi-thread river channels, called **braiding**, and by small lakes, called **kettleholes**. The deposits in an outwash plain are also graded vertically. In summer, the **discharge** of the meltwater streams is greater and they can transport larger particles across the plain. In colder periods discharge is much lower and the ability to transport is reduced. Consequently, a vertical stratification can be found with alternating layers of coarse and fine particles.

overgrazing is a major cause of **soil erosion** in areas with an **arid** or semi-arid climate. The **nomadic** tribes who inhabit these areas measure their wealth in terms of the number of animals they own. In very dry periods of weather they concentrate their large herds around the few remaining water-holes, and the amount of grass available cannot cope with their needs. The vegetation cover becomes depleted, and the soil is exposed to **erosion** by either wind or water.

overland flow is the outcome of rainfall intensity on a slope being greater than the rate at which the water can infiltrate into the soil on that slope. A thin layer of water forms on the surface and it begins to move downslope under gravity. Some of this water accumulates in small surface depressions, but these also overflow when full. Subsequent coalescence of this flow takes place and **rilling** will occur. Overland flow of this type occurs in areas with high intensities of rainfall, devoid of vegetation. It can also occur in some cultivated fields, and where the soil has been compacted by vehicles or animals. Saturated overland flow may occur in the lower parts of a slope. The ground nearest to the stream channel is saturated, and subsequent rainfall cannot seep into the soil. As the storm proceeds water flows over the surface and into the stream channel. The longer the storm continues, the larger the saturated area, and the greater the amount of **runoff** over the surface.

overpopulation exists when there are too many people in an area relative to the amount of resources, and to the level of technology locally available, to maintain a high standard of living. It implies that, with no change in the level of technology or **natural resources**, a reduction in a population would result in a rise in living standards. The absolute number or density of people need not be high if the level of technology or natural resources is low.

Overpopulation is characterised by low **per capita** incomes, high unemployment and underemployment and outward **migration**.

overproduction is exemplified by the surpluses produced under the **Common Agricultural Policy** of the **European Union**. Farmers were given guaranteed prices for their produce and hence produced as much as possible. This created excesses in a range of agricultural products such as the grain, butter, beef and sultana 'mountains', and the milk, wine and olive oil 'lakes'. These surpluses were stored by the governments of the European Union or sold cheaply to developing countries.

overspill: people who are encouraged or forced by **redevelopment** schemes to move from a town or city to a new settlement beyond the limits of the original urban area. A number of **new towns** have been built as overspill settlements, for example, Skelmersdale for Liverpool and Redditch for Birmingham.

oxbow lake: a horse-shoe shaped lake separated from an adjacent river. The water is stagnant and, in time, the lake gradually silts up becoming a crescent-shaped stretch of marsh. It is formed by the increasing sinuosity of a river **meander**. **Erosion** is greatest on the outer bank and, with **deposition** on the inner bank, the neck of the meander becomes progressively narrower. During times of higher **discharge** the river cuts through this neck and the new cut eventually becomes the main channel. The former channel is sealed off by deposition.

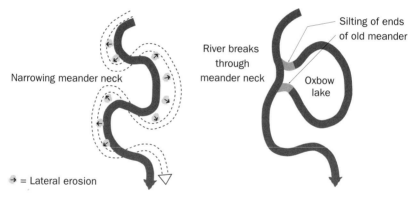

Formation of an oxbow lake

oxidation is a **weathering** process which involves the reaction between metal ions and atmospheric oxygen, or oxygenated water, to produce oxides or hydrated oxides respectively. An example is when iron II (ferrous) in mineral compounds is oxidised to form iron III (ferric). The structure of the original mineral compound is destroyed, as the oxidised mineral increases in volume. The rock containing the mineral is weakened, and becomes prone to further weathering. Oxidation can be identified on rocks by the presence of a yellow or red-brown discoloration.

ozone depletion occurs when the rate at which the **ozone layer** is formed is less than the rate at which it is destroyed. In recent years, this depletion has manifested itself by the emergence of a 'hole' in the ozone layer over Antarctica and the Arctic. Such 'holes' are also beginning to appear over other parts of the world, albeit for limited lengths of time. The damage is believed to have been caused by the use of **chlorofluorocarbons (CFCs)**, which break down ozone. In both 1987 (the Montreal Protocol) and 1989 (the 'Saving the

Ozone Layer' Conference in London) a large number of countries signed a declaration to set in motion the reduction and phasing out of the production of CFCs. The countries of the *European Union* agreed to a 100% ban on CFC usage by the end of the twentieth century. Globally, these substances are being gradually removed from the atmosphere.

ozone layer: a concentration of the gas ozone located in the *stratosphere* at an altitude of between 10 and 50 km above sea level. The ozone is formed by the interaction of solar ultraviolet *radiation* and oxygen in the atmosphere. This concentration shields the Earth from most of the harmful ultra-violet radiation from the sun. It is this radiation which causes an increase in the incidence of skin cancer and eye cataracts. Ozone is also produced on the Earth's surface by car exhaust systems, and is damaging to both humans and plants.

Do you need revision help and advice?

Go to pages 325–53 for a range of revision appendices that include plenty of exam advice and tips.

Pacific Rim: the countries on the margins of the Pacific Ocean Basin. These include Japan, Australasia, the western part of North America and the **newly industrialised countries** of southeast Asia such as Hong Kong, Singapore, South Korea and Taiwan. Whereas at the beginning of the twentieth century manufacturing was concentrated in Western Europe and North America, during the 1970s and 1980s plants were established by **transnationals** in the newly industrialised areas. The Pacific Rim now challenges the traditional manufacturing areas on the margins of the North Atlantic; this has been described as a **global shift**.

Pacific Ring of Fire: the name given to the **Pacific Rim** area where there is a **subduction** zone created by the oceanic plates sliding under those of the continent. The result is a line of volcanic activity that stretches through South America from southern Chile into Mexico and on into the Cascade Range of northeast USA. It continues through Alaska, the Aleutian Islands, Kamchatka, Japan, the Philippines, Indonesia, Papua New Guinea and on to New Zealand. This zone contains some of the world's most active and destructive **volcanoes** such as Mt St Helens (USA), Tambora, Krakatoa (Indonesia), Pinatubo (Philippines).

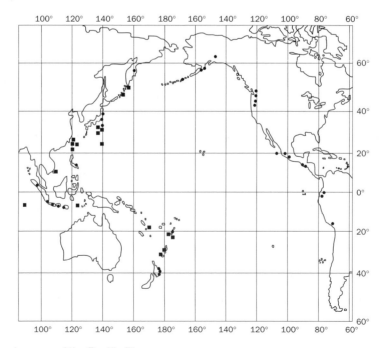

Active volcanoes of the Pacific Rim

palaeoclimate literally means fossil climate, i.e. evidence in the present landscape which indicates that the climate was different when that evidence was laid down. This can include the evidence of the landforms which resulted from that climate; features which can only have been produced by the force of *glaciers* or *ice sheets* are found in areas with very different climates today. Fossil reef *corals* which required warmer conditions for their growth, have been found in areas which do not support coral now. The evidence for reconstructing past climates are largely indirect, that is the climatic conditions can be inferred from the effect which the climate had on some other feature. These methods include *dendrochronology*, or tree ring analysis; *pollen analysis*; changes in coleoptera beetles; *ice core* and ocean core investigations of oxygen isotopes O^{16} and O^{18}. Each of these techniques works on the general principle that differences reflect variations in climatic conditions.

palaeomagnetism: evidence of the Earth's magnetic field is stored within rocks as 'fossil magnetism' in that magnetic minerals align themselves with the magnetic field which was operating at the time of their formation. When *magma* cools as it reaches the surface, for example at a *constructive plate margin*, ferromagnetic minerals will behave like a compass needle and point towards the North Pole. They also dip at an angle to the horizontal in the magnetic field. The amount of dip varies with latitude from zero at the equator to 90° at the pole. In sedimentary rocks, magnetic grains will also line up with the magnetic field. The Earth's magnetic field has reversed many times over the last 2 billion years, the time interval for these changes has varied from 20,000 to over 10 million years. These reversals are recorded in rocks on either side of the *constructive plate margin* in the mid-Atlantic and the symmetry in this magnetic striping provides evidence that the plates are moving apart. Rocks of a similar age show alignment to different patterns of magnetic field which suggests that there was more than one pole. But, as at any one period the magnetic minerals must have aligned with one North Pole, this provides evidence which indicates that it is the continents which have drifted.

pandemic refers to a disease that has affected people over a wide area of the world.

Pangaea: the large supercontinent proposed by *Wegener* as the original land mass of 200 million years ago which later split up and drifted to form the present continental areas. It consisted of *Laurasia* to the north and *Gondwanaland* to the south.

parent company: the original business that another company has developed from and whose directors still make decisions that affect the organisation as a whole. For example, the Walt Disney Company owns several television networks, animation studios and theme parks.

parent material is the rock or weathered *regolith* from which a soil has been formed. The mineral constituents and fragments in the soil are largely derived from the parent material which exerts a particularly strong influence on poorly developed or *immature soils*. As time progresses, a soil generally becomes more independent of the parent material, and is more in balance with the climate and vegetation in the *biome*, as it becomes more mature. Parent material is an important influence on *soil texture* which reflects the mineral composition of the rock and the way it is weathered.

Park model: this provides a theoretical basis for the various responses to *hazards* over time after their occurrence:
1 The pre-disaster phase.
2 The hazard occurs.

3 Immediate search, rescue and care, which may last for several hours to several days. During this time the *quality of life* in the area affected deteriorates significantly.

4 Relief and rehabilitation, which may involve outside help and aid. This will take place over several days or weeks. During this time the quality of life for the general population slowly improves.

5 Reconstruction, including strategies to modify *vulnerability* to prevent further *disasters*. This may take weeks to years, and eventually the quality of life for the population will improve greatly and may become better than before the disaster.

particle size analysis: a technique for measuring the percentage of each size of particle in a deposit. A sample of the deposit, for example a *till*, *outwash* material or river *alluvium*, is collected in the field. A known weight of the sample is passed through a series of sieves with different mesh sizes. By weighing the contents of each sieve, and comparing this with the total weight, the percentage of the total deposit which lies within each grain class size can be calculated. The particle size distribution can be plotted on a *cumulative frequency* graph.

The main grain sizes are normally defined as follows:

Clays	< 0.002 mm	(< 2 µm)
Silts	0.002–0.06 mm	(2–60 µm)
Sands	0.06–2 mm	(60–2000 µm)
Gravel	2–64 mm	
Cobbles	64–256 mm	
Boulders	> 256 mm	

Measurement in microns is often given in preference to millimetres, especially for the smaller grain sizes up to 2 mm. 1 micron is one thousandth of a millimetre, 1 µm = 0.001 mm; the equivalent grain sizes in microns are shown above.

particulate pollution refers to the solid matter in the urban atmosphere, which derives mainly from power stations and vehicle exhausts (particularly from burning diesel fuel). Such particles are usually less than 25 µm in diameter and are responsible for *fog/smog*, respiratory problems, soiling of buildings and may contain carcinogens. Other particulates in the atmosphere include cement dust, tobacco smoke, ash, coal dust and pollen. Coastal cities also have vast numbers of sea salt particulates. Particulates are sometimes referred to as *PM10s* as the bulk of particles have a diameter of less than 10 µm.

partnerships represent the cooperation between various tiers of government and other bodies, public and private, with the overall aim of achieving economic and social development or regeneration of an area. The main aims are in terms of job creation, land reclamation and environmental improvement. Areas with high unemployment or with high concentrations of vacant, derelict and contaminated land are the main targets. The redevelopment of the Salford Quays in Manchester is an example of a successful partnership. There have been a variety of urban partnerships in the UK in recent years. Early examples included the City Challenge partnerships in the 1990s, such as in Hulme Manchester. More recently, moves towards sustainable communities have been more common with the following general aims:

- urban economic sustainability, which should allow the individuals and communities who live in cities to have access to a home, a job and a reliable income

- urban social sustainability, which should provide a reasonable *quality of life* and opportunities to maximise personal potential through education and health provision, and through participation in local democracies.

pastoral: farming that involves the deliberate rearing of *livestock* for meat or other product such as milk or wool. This can be extensive in operation as in the *commercial* ranching of sheep in Australia or beef cattle in East Texas; or it can be an intensive system such as dairying in the dense market areas of Western Europe. The activity ranges from the high technology of scientific breeding and capital investment associated with commercial systems to the low technology of *nomadic* herding groups who are very dependent on the environmental conditions within their marginal production areas.

paternoster lakes are produced by glacial scouring along a valley floor which results in a series of lakes separated by depositional material or rock bars. As these lakes are linked together by streams, from above they resemble a string of beads; hence the term paternoster, a bead in a rosary.

patterned ground is a term which relates to the coarse and fine *debris* which accumulates on the surface, or just below, in *periglacial* areas. This includes features such as *stone circles*, garlands and stone stripes as well as *ice wedge* features. These are characterised by a degree of lateral sorting of debris. Daily fluctuations of temperature through freezing point and the annual *freeze–thaw* action cycle results in the formation and decay of ground ice in the weathered surface layer. This produces *frost heave*, which brings debris to the surface forming a small mound. Ice pushes stones to the surface where they are sorted by gravity down the low angle slopes of the frost mounds. On steeper slopes the debris is elongated downslope to produce garlands and stripes.

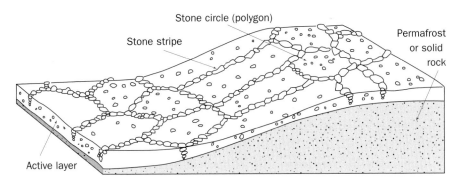

Patterned ground

peak flow is the maximum level of *discharge* in a storm *hydrograph* largely resulting from the arrival of fast *overland flow* into the channel system.

peak land value point: the point within the urban area where land values are highest. It usually occurs at a major intersection within the *central business district* where the *high-order* retail outlets tend to concentrate in this area of high pedestrian flow.

peatland: an area which has a waterlogged, *anaerobic* environment favourable for the formation of peat. This could occur on flat uplands with *impermeable* sub soil or on valley floors which have a high *water table*. With waterlogging there is a lack of oxygen in the soil, breakdown of organic matter is very slow and material accumulates into thick layers of peat which may produce extensive blanket *bogs*.

211

pedestrianisation: the creation of pedestrian-only, traffic-free areas within the central retail zones of towns and cities. This may only cover one main shopping street, the 'high street' in smaller market centres, but it may extend to several streets, or a compact block of streets in a larger regional centre. This development separates shoppers from the danger of vehicles, which can usually only gain access to the pedestrian area for deliveries out of shopping hours. Although the safety aspect is a major advantage, the exclusion of cars has presented some problems to less mobile members of the public with car parks located on the periphery of the **central business district**. The restriction on through traffic and street parking has caused some concern to retailers who fear a loss of trade. In some towns 'shopmobility' has been introduced providing small electric carts for use by disabled shoppers direct from the parking areas and there is some central parking in the pedestrian zone for disabled drivers.

pediment: a low angle, rock cut surface at the base of a steep slope in a semi-*arid* or arid area. They are broadly concave in profile and may be covered by a thin layer of **alluvium** which is being transported from the steep slope above the pediment to a more gentle slope below the feature. The presence of alluvium suggests that water has been involved in the formation. However, their origin has been the focus of considerable discussion: some researchers view the pediment as a basal slope which is the remnant of a mountain front undergoing parallel retreat, its slope transporting weathered sediment away from the steep front to the flat area beyond the pediment. Other workers suggest that the feature is the result of undercutting by running water in the form of sheet floods causing it to recede. The effects of sheet floods on semi-arid landforms is controversial; some workers argue that sheet floods do not produce this feature, they only develop as sheet floods because the feature already exists. It may be that both theories play a part in the formation.

pedogenesis: the formation of soils. The factors which influence soil development can be expressed by the formula:

FORMULA: s = f (Cl, O, R, P, T)

where s = soil characteristics and f is a symbol for 'function of'.

Cl is climate – this affects the rate of weathering of the **parent material** and **regolith** and the balance between **precipitation** and **evaporation** determines the soil moisture budget which influences the extent of downward **leaching** or upward **capillary movement** of minerals.

O is organic material and **organisms** – plants, **bacteria** and animals interact in the **nutrient cycle** returning minerals to the soil store to be used again. Organic matter provides material for **humus** within the soil which contributes to the clay-humus complex.

R is relief – as altitude increases rainfall total tends to increase, and temperature and the length of the **growing season** decrease, influencing the nature of the soil. Angle of slope affects drainage, **runoff** and soil depth. On steep slopes, faster **throughflow** and surface **runoff** increases downslope movement of soil particles which will retard soil development. Upper slopes are shedding sites, i.e. water is draining away which may promote leaching; lower angle zones at the base of slopes are receiving sites which may experience waterlogging and **anaerobic** conditions.

P is parent material – this influences the mineral composition of the weathered material which forms the basis of the soil and the degree of permeability of the sub soil which will affect soil drainage. Depth and *soil texture* will be influenced by the ease of weathering and the mineral composition of the parent material.

T is time – young, poorly developed soils which are *immature* tend to be strongly influenced by their parent material, but mature soils are in balance with the *ecosystem* and tend to reflect the climate and vegetation more closely, they have become independent of the parent material.

peds: soil units which are produced by the grouping together of sand, silt or clay particles. Peds are bound together by the gums and mucilages formed during bacterial breakdown to create *soil structure*. Peds are the 'building blocks' of the soil and the spaces between peds house microfauna which are important to soil formation. Soil ped arrangements (structure) fall into five categories: structureless, platy, crumb, blocky and prismatic.

Pelean eruption: a violent form of volcanic eruption which is accompanied by an explosion of gas, ash, and pyroclasts in the form of an incandescent cloud, *nuée ardente*, which travels at great speed down the flanks of the *volcano*. It is named after Mt Pelée on the island of Martinique in the Caribbean which erupted violently in 1902.

Peltier diagram: shows the relationship between weathering (type and rate) and climate (mean annual temperature and rainfall).

per capita means per head of population and is often used when making comparisons between regions or countries which have different base populations, e.g. when quoting an individual measure such as income, *GNP*, or *calorie intake* per capita.

perception is the way in which an individual or a group views a particular environment or situation. For example, different groups will view and interpret the attractions and hazards of a landscape in different ways according to their perception. Perception is the process of evaluating and storing information received. This information may be part of the cultural background or tradition in society, or it may be provided by the media in the form of television documentaries, newspapers, news bulletins, films. Information may be transferred verbally from one person to another. Some or all of these sources could influence the image that an individual has of a particular place or environment.

percolation: the downward vertical movement of water within a soil. The water then enters the *groundwater* store. The rate of percolation depends on the size of the pores through which the water travels. Sandy soils have high rates of percolation because of the large voids between each particles, whereas, clay-enriched soils have low rates due to the very small voids between clay particles.

perennial: lasting through all seasons of a year or existing for several years. Perennial may be applied to:

- plants that live for several seasons or years
- irrigation schemes where the water is stored behind a dam and released at regular intervals to maintain a constant water supply throughout the year.

periglacial literally means on the fringe of, or near to, an *ice sheet* or *glacier*. It is also used to describe any area which has, or has had, a very cold climate. This includes high mountainous areas, the *tundra* areas of northern Europe and northern Canada, as well as most of southern England during the *Pleistocene ice ages*.

periglacial landform refers to a feature created largely through the actions of *periglacial processes*. (See also *active layer*, *blockfield*, *ice lens*, *ice wedge*, *permafrost*, *patterned ground*, *pingo*, *scree*, *stone circles* and *thermokarst*.)

periglacial processes are caused by the ground freezing hard and often for considerable periods of time, yet also thawing for short periods of time. They take place in *periglacial* environments.

The processes include:
- *solifluction* (see also *active layer* and *permafrost*)
- *frost heave* (see also *ice lens*, *stone circles* and *patterned ground*)
- thermal contraction (see also *ice wedge*)
- *frost* weathering (see also *blockfield* and *scree*)
- thermal subsidence (see also *thermokarst*).

periphery refers to an area of low or declining economic development, associated with the models of *Myrdal*, *Hirschman* and *Friedmann*. It is usually applied to a region within a country which is suffering from high unemployment, high rates of selective out-migration, low personal incomes, low living standards and high rates of crime. The periphery is often in an unfavourable geographical location, with a poor and/or deteriorating resource base. Periphery can also refer to those parts of a country which do not constitute the *core* of that country.

permafrost is perennially frozen ground. It lies beneath approximately one-fifth of the world's land surface, with extensive areas in Russia, northern Canada and Alaska. It also exists beneath the sea in the Beaufort Sea, and at high altitudes in the Rockies and Central Asia. In northern Russia, the permafrost extends to depths of over 1000 m. In some areas, there exists above the permafrost an *active layer* – soil that thaws out seasonally.

There are several types of permafrost:
- 'wet' permafrost – where all of the pore spaces and voids are filled with ice
- 'dry' permafrost – where the soil and rock are unsaturated
- continuous permafrost – where it is present in all localities
- discontinuous permafrost – where there are some small scattered unfrozen areas
- sporadic permafrost – where there are only a few patches of permafrost, for example on a poleward-facing slope, in an otherwise unfrozen area.

The development of permafrost is due to:
- very cold temperatures all year round, the mean annual temperature being below $-5°C$
- low *precipitation* totals, so that snow does not insulate the ground from the cold
- a limited vegetation cover.

permeable: a description of a substance that allows water to pass through it. Permeability can be divided into two types:
- *porosity*
- perviousness – this occurs when rocks have *joints* or *fissures* along which water can flow. Examples include *Carboniferous limestone*, which has its joints and *bedding planes* widened by solution, and granite, which has joints because of the manner in which it has been formed. (See also *impermeable*.)

personal investigation: an enquiry based on first-hand and/or second-hand evidence completed by a student. A range of techniques and skills should be used by a student in carrying out such an enquiry:

- the identification of the aim of the enquiry, usually in the form of testing a *hypothesis*
- the collection of evidence, including measuring, mapping, observations, questionnaires, interviews and sampling techniques where applicable
- the organisation and presentation of the evidence using cartographic, graphic and tabular forms, possibly making appropriate use of information technology
- the analysis and evaluation of the evidence, noting any limitations of the evidence
- the drawing together of the evidence (synthesis) and formulating a conclusion to the hypothesis.

pesticide: a chemical applied to crops to control pests and disease. Some people have expressed concern at the rate at which the use of pesticides is increasing. Instructions for usage are not always followed correctly, safety regulations are not always conformed to, and equipment is not always properly maintained. There have been several claims that pesticides have damaged human health in some farming areas, and may be the cause of birth abnormalities. Pesticides have affected 'innocent' wildlife in many areas, killing birds, bees and other fauna.

pH indicates acidity by measuring the concentration of hydrogen ions. This is represented on a *logarithmic scale* from 0 to 14 (pH3 is 10 times more acidic than pH4 and 100 times more acidic than pH5). This is particularly used when describing soils, most soils in the UK being slightly acidic (pH5 to 7). Normal rainfall has a pH of between 5 and 6.

phosphates are deposits which are mined for use as *fertiliser*. As rock phosphate is in limited supply, large amount of fish can be converted to phosphorus-rich fertilisers. Another source is where there are masses of fish-eating birds; these birds excrete matter rich in phosphorus which is then mined if the deposits have built up to the extent that they are commercially exploitable. Such deposits are known as guano.

photochemical smog: a form of *smog* that occurs in large cities and can be dangerous to health. When exhaust fumes and factory emissions are trapped by *inversions* of temperature, they can combine with sunlight to produce ozone. Such smogs can cause breathing difficulties, vomiting, eye irritation and a general lethargy. In some photochemical smogs the presence of substances containing nitrogen could increase the risk of cancer.

photosynthesis: the green pigment chlorophyll, present in green plants, is able to capture light energy and convert it into food energy by manufacturing carbohydrates (energy storing chemicals) using elements such as oxygen, carbon and hydrogen. Photosynthesis, therefore, allows the plant to grow and increase its *biomass*.

$$\text{FORMULA:} \quad 6CO_2 + 6H_2O + \text{solar energy} \rightarrow C_6H_{12}O_6 + 3O_2$$

| carbon dioxide | water | | glucose | oxygen (released) |

photovoltaic cells contain semi-conductor crystals of silicon (or gallium arsenide) which, when exposed to the sun, generate electricity, which can be stored in batteries. They are used in the *Third World* for small-scale local electricity production, but a wider application has run into problems of scale, cost and maintenance.

physical weathering (mechanical weathering) involves the breakdown of rocks into smaller fragments through mechanical processes such as expansion and contraction mainly due to temperature change. Two such types of physical breakdown are *freeze–thaw* weathering and insolation weathering or thermal fracturing. In addition salt weathering, caused by the growth of salt crystals, and **pressure release** or dilatation can also bring about weathering without temperature changes.

The effect of these weathering processes is to widen joints and cracks which allow deeper penetration of weathering action and the surface may become littered with fragments which become detached from a rock outcrop; surfaces may be buried under a layer of weathered **debris**, the **regolith**. The products of weathering vary according to the constituent minerals in the rock. Breakdown may produce large blocks, block disintegration, or smaller fragments which reflect granular disintegration. Some researchers classify biological, or biotic, weathering as a separate form of breakdown, but others regard some of the biological processes as physical in operation; tree and plant roots penetrate joints and prise the rock apart in a process which is similar to joint widening by freeze–thaw action.

phytoplankton are very small plants which float near the surface of seas and oceans. Like land-based plants, they produce food by **photosynthesis**, obtaining their carbon from atmospheric **carbon dioxide** dissolved in ocean water and fuelled by sunlight which only penetrates the surface layers of the oceans. Phytoplankton form part of the base of the marine **food chain**. In places, they often form dense blooms, which may have a density of 200,000 individual plants per cubic metre of sea water.

pie chart (graph): a representational technique consisting of a form of **proportional circle** in which the area of the circle is divided into segments. These segments represent a percentage of the whole figure.

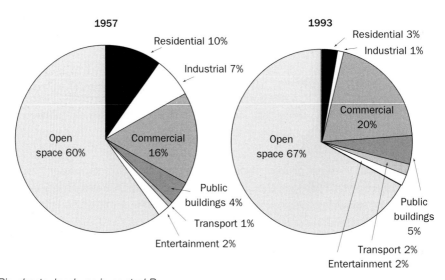

Pie charts: land use in central Bury

piedmont: the area at the foot of the mountains, as used in the term piedmont **glaciers**. These are formed when valley glaciers flowing from the mountains extend on to lowland areas, spread out and merge. A well-known example of a piedmont glacier is the Malaspina Glacier in Alaska (USA).

pingo: an isolated, dome-shaped hill on a flat **tundra** plain. They usually consist of a high proportion of sands, but with an **ice core**. The top of the hill may have a crater-like depression. Pingos vary in size from a few metres to over 60 m in height, with a diameter of up to 300 m.

Two types have been recognised. The open-system pingo occurs in areas of thin or discontinuous **permafrost**, where surface water has penetrated the ground and circulated in the unfrozen **sediments** beneath the permafrost. When this water rises towards the surface through hydrostatic pressure, it freezes to form localised bodies of ice which force the overlying sediments upwards. Closed-system pingos are found in areas where there is more continuous permafrost. Any unfrozen parts, say beneath a lake, become increasingly saturated. As the permafrost advances, it causes an ice core to form, which expands and lifts the surface.

pioneer: a plant or a community of plants that are the first to colonise an area. These are generally simple and hardy plants which will ultimately improve the conditions allowing other plants to grow. Under different conditions, different pioneers will colonise the area. On bare rock (**lithosere**), for example, the colonisers will be **lichens** and mosses, but on a sandy area (**psammosere**) the pioneers will include sea couch, lyme and **marram grass**.

plagioclimax: when human interference has permanently arrested and altered the natural vegetation so that the climatic **climax vegetation** has not occurred, the resulting community is known as plagioclimax. Much of the vegetation of the British Isles falls into this category.

plagiosere is a plant **succession** that has been shaped by human action.

planning blight is where uncertainty and delay in the development of a particular urban plan can cause an area to deteriorate as people refuse to invest in property that might be demolished. If the plan is delayed for some time, residents wishing to move suffer, as they cannot find buyers even for their devalued property.

planning permission (planning consent) is required from a **local authority** before any new construction can be started. This applies to shop signs, house extensions and change of use, as much as to house, factory and office construction. It is designed to protect the environment from haphazard development.

planning processes attempt to provide a means by which planners can:
- listen to the local community
- listen to the organisation responsible for a proposal for change
- have overall development control.

Any refusal that a planner may give to a proposal may lead to an appeal, or may result in the developer going to a higher body, for example, to regional or national government. Planning processes are therefore costly, but are perceived as being more democratic and open. Planners may:
- require modifications to be made to the original proposals
- negotiate modifications in an attempt to offset objections from opposers to the proposal
- request additional provision by the developers in order to allow the development to proceed.

Planning processes therefore attempt to weigh up short-term advantages and disadvantages of a proposal for change, as well as consider wider and more long-term benefits and problems which may arise.

plantations are farms with large-scale agricultural activity aimed at producing cash crops of high value. They were originally developed in tropical areas by European and North American merchants. Large areas were cleared and a single crop of either a bush or tree was planted. They are an example of an extreme form of *monoculture*, and the crops include sugar cane, coffee, rubber, tea and bananas. Plantations have a high capital input. The land needs to be cleared and planted; estate roads, housing, and schools need to be built, together with some local processing facilities. They also have a high *labour* input – manual labour is needed for cultivation and harvesting. They are frequently managed by European/North American/*Pacific Rim transnational* companies, with the labour being supplied by indigenous people who work for low wages. Most of the produce is sent out of the country of production and much of the profit of the enterprise leaves the host country. The produce of plantations is subject to fluctuations in world prices and demands.

plate tectonics is the collective name for a group of concepts in which the structural complexities of the Earth's *crust* are ascribed to the interactions of moving crustal plates. The Earth's crust is divided into a series of blocks or plates which float like rafts on the underlying *mantle*. Their movement gives rise to a series of landforms, submarine features and hazardous events. (See also *conservative plate margin*, *constructive plate margin*, *destructive plate margin*, *earthquake*, *folding*, *island arc*, *mid-oceanic ridge*, *ocean trench*, *sial*, *sima* and *volcano*.)

playa: a shallow, *ephemeral* and saline lake found in desert or semi-desert areas of the world. It is formed after rare rainstorms, after which the water evaporates rapidly leaving a very flat area of clay, silt or salt. Some larger playas, for example the Dead Sea and the Great Salt Lake, Utah, have not dried out completely. However, the latter is only a vestige of what was once the much larger Lake Bonneville, the former bed of which is used for attempts at the world land speed record.

Pleistocene: a geological time period stretching from 2 million years *BP* to 10,000 years BP. It was characterised by a series of alternating cold phases (glacials) and warm phases (*interglacials*) know collectively as the *ice ages*.

plucking: a process of glacial *erosion* by which a *glacier* freezes around a rock on a valley side or bottom and subsequent movement of the ice causes the rock to be pulled away with it. It is believed that this can only be effective when the bedrock has been well-weathered, or is well-jointed, prior to the glacier passing over it. As the process goes on, some of the underlying joints in the rock may also open up as the weight of the overburdening rock is removed. This process of *pressure release* enables plucking to continue to operate.

plume: an upwelling of *magma* caused by radioactive elements inside the *mantle* below the Earth's *crust*. It may cause a *hot spot* to develop where lava breaks through to the surface, causing active *volcanoes* to occur. The term is also used in the context of *urban climates* to indicate their effects downwind of an urban area into a rural area.

pluralism is used across a wide range of topics to denote a diversity of views as against one single approach or method.

pluvial: a period during the *Pleistocene* when wetter conditions existed in those areas of the world which are currently *arid*. There is much evidence of more water being available in desert areas in the past:

- the beds of huge former lakes which can be found in southwest USA, such as Lake Bonneville
- expanses of fossil soils of a humid type, including *laterites*
- vast river systems in the central Sahara which are now dry and inactive

- plant and pollen remains (oak and cedar) and animal fossils (antelope and rhino) in the Sahara
- evidence of former human occupation of Neolithic culture.

PLVP: see *peak land value point*.

PM10s: an alternative name for the causes of *particulate pollution*.

podsol: a type of soil associated with the northwest European and Siberian type climates and *coniferous woodland* or *heathland* vegetation. In these types of environment, *evapotranspiration* rates are less than *precipitation*, causing water to move downwards through the soil. Due to the cool and wet climate the decay of the needle-shaped leaves is very slow, and a thick acidic form of peaty humus (*mor*) is created on the surface.

The downward *percolation* of acid water through the soil, particularly after snowmelt, causes the rapid *leaching* of soluble bases and the movement of clays (*lessivage*), and the *eluviation* of iron and aluminium *sesquioxides* (*podsolisation*). The result is that the upper layers of the soil become depleted of these substances and bleached of colour, leaving a sandy ash-grey horizon. However, they are redeposited lower down the profile. At first, a very thin layer of redeposited iron, called an *iron pan*, may be found in some podsols. Beneath this, a darker horizon consisting of redeposited humus, and a clay-enriched orange-brown horizon are found.

The cold climate also discourages *organisms*, and there are few earthworms to mix the soil horizons. Horizon boundaries are therefore sharply defined. Podsols are not naturally fertile because of their acidity. They can be improved by deep ploughing to mix the upper and lower horizons, and by the frequent application of lime and fertiliser.

podsolisation is an intense form of *leaching* which operates in cool climates where the rate of *precipitation* is significantly greater than the rate of *evapotranspiration*. The vegetation of such areas is *coniferous woodland* and *heathland*. These produce a highly acidic form of humus – *mor*. Rainwater passing through this becomes highly acidic, so much so that it is capable of breaking down clay particles and mobilising the *sesquioxides* or iron and aluminium. These are removed from the upper part of the soil leaving a bleached ash-grey horizon. They are redeposited lower down the soil, in some cases forming a thin *iron pan*. The resultant soil of this process is a *podsol*.

point bar: a gently sloping bank on the inside of a *meander* composed of sands and gravels. The water in the river is moving more slowly here and this results in *deposition*. Initially coarser particles are deposited at a point just downstream of the steepest part of the inside bend.

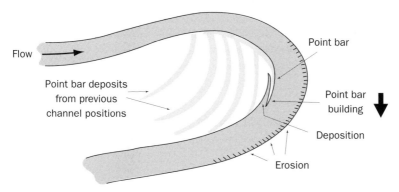

Point bar

As this grows, finer *sediments* are deposited in the shallower and calmer water between it and the inside bend. Vegetation may encroach, trapping more sediment.

polar: used to describe natural phenomena that owe their origin to being at or near to either of the poles of the Earth. Some examples include:

- polar *air masses* – associated in the northern hemisphere with northern Canada, northern Russia and the Arctic Ocean
- the polar *front* – the zone where tropical and polar air masses collide in the high and mid-latitudes of each hemisphere. In the northern hemisphere this is in the North Atlantic ocean at approximately 50–60°N (see also *fronts*)
- the polar cell – a vertical circulation of air in the high latitudes. This involves the upward movement of warm air at the polar front, the horizontal movement of this air at high altitude towards the poles, the subsidence of the air at the poles, followed by a horizontal movement along the surface of the Earth back to the polar front
- polar climates – the equatorward limit of which is generally taken as the line where the mean temperatures of the warmest month is no more than 10°C.

polarisation is the uneven development that often takes place in *Third World* countries as a result of the emergence of a *core*. The core area attracts investment particularly in *infrastructure*, often to service *transnationals*. This is often at the expense of the *periphery*, which loses resources, investment and young people to the core. The term can also be used to express the accentuation of a difference between two things or groups. Alternatively, the process of division into two groups representing the extremes of opinion, wealth etc.

political system: the means by which government and/or administration can take place in a country. A wide range of systems exist around the world, and have existed in the past. Types of systems, which are not always mutually exclusive, include:

- totalitarianism – ruled by one governing party or individual. No rival loyalties or parties are permitted
- democracy – government by the people, either directly or through elected representatives
- *command economies*
- *capitalism*.

pollen analysis: a method of identifying *climatic change*. Each plant species has a distinctive pollen grain. If a grain lands in an area where conditions are *anaerobic*, such as a peat bog, it will resist decay. Several grains trapped in this way are therefore representative of the vegetation growing at that time, which in turn is a reflection of the climate of that time. If differing layers of peat in which varying combinations of pollen grains are found can be dated by other means, then it is possible to ascertain the climatic conditions that prevailed at different times in the past.

polluter pays: the increasingly held view that governments should force the originators of *pollution* to pay the costs of removing the contamination and making good the damage that has been caused.

pollution means contamination of some kind. It is usually referred to in the context of the environment, where the pollution can take many forms – air, water, soil, noise, visual and many others.

pollution reduction policies: there are a number of ways in which governments and other organisations have tried to reduce atmospheric pollution in cities:

- *Clean Air Acts* – after the London 'pea-souper' of 1952, the UK government decided legislation was needed to prevent so much smoke entering the atmosphere. The act of 1956 introduced smoke-free zones into the UK's urban areas and this policy slowly began to clean up the air. The 1956 act was reinforced by later legislation. In the 1990s, for example, very tough regulations were imposed on levels of airborne pollution, particularly on the levels of *PM10s* in the atmosphere. Local councils in the UK are now required to monitor pollution in their areas and to establish air quality management areas where levels are likely to be exceeded. Some have planted more vegetation to capture particulates on leaves.
- Vehicle control in inner urban areas – a number of cities have looked at ways of controlling pollution by trying to reduce the number of vehicles that come into central urban areas. Many British towns and cities have **pedestrianised** their **central business districts** and have promoted 'park-and-ride' schemes. In London, attempts to control vehicle numbers have included the introduction of a congestion charge, which means that vehicle owners have to pay if they wish to drive into the centre. In Mexico City, the city council passed driving restriction legislation known as the Hoy no Circula ('don't drive today'). This bans all vehicles from being driven in the city on one day a week, the vehicle's registration number determining the day.
- More public transport – attempts have been made to persuade people to use public transport instead of cars. Such schemes have included Manchester's development of a tram system (Metrolink), the development of bus-only lanes into city centres and the encouragement of car-sharing schemes.
- Zoning of industry – industry has been placed downwind in cities if at all possible and planning legislation has forced companies to build higher factory chimneys to emit pollutants above the inversion layer.
- Vehicle emissions legislation – motor vehicle manufacturers have been made to develop more efficient fuel burning engines and to introduce catalytic converters, which remove some of the polluting gases from exhaust fumes. Catalytic converters, however, actually increase emissions of **carbon dioxide** and they only work once the car has warmed up, on longer journeys. The switch to lead-free petrol has reduced pollution in the UK.

pools and riffles: on the *meanders* of a river, the pools are the areas of deeper water, whereas the riffles are the shallower parts. The pool represents the area where the energy of a river builds up due to a reduction in friction, but across the riffle energy is dissipated. Here, a higher proportion of the river's energy is required to overcome friction so that *deposition* occurs. Pools and riffles tend to be regularly spaced along the course of a meandering river.

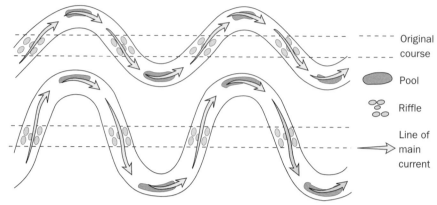

Pools and riffles

population: apart from the straightforward meaning of the term in reference to people, population is also a statistical term meaning the whole from which a *sample* will be chosen for a research exercise. This could be people, but it could equally be pebbles on a beach, farms in an area or houses in a town.

population change: the population of an area increases or decreases according to the *birth* and *death rates* as well as the movement of people in and out of the region (*migration*).

population density: this is the number of people per unit area. The density of population is obtained by dividing the total population of a country (or any region) by the total area of that country (or region). The density of population for a country can be misleading as it does not show variations between densely populated regions and those areas that are almost unpopu-lated. Population densities are often shown on *choropleth maps*.

population distribution describes the way that people are located within an area. When plotted on a map, this can reveal those areas that have a high population and those where few people live, which maps of *population density* fail to show. One of the ways of plotting this information is by means of a dot map.

population policies: since the 1970s, many *Third World* countries have begun to formu-late policies with regard to the growth of population. Most countries now have policies with regard to:

- the growth of the population
- the level of *fertility*
- levels of mortality
- the spatial distribution of population within the country
- international *migration*.

In most of the countries the policies are directed towards attempting to control the high population growth rate. Programmes include *birth control programmes*, abortion, education, sterilisation and social and economic incentives. Among the best-known of these policies is that of the one-child policy in China. The *United Nations Population Fund* has taken responsibility for raising awareness of population-related matters and assembles the support and resources needed to attain the *Millennium Development Goals (MDGs)*.

population pyramid: the *population structure* of a country is best shown by a population or age–sex pyramid. The diagram is known as a pyramid, although the structure of the population of many countries does not take on this shape.

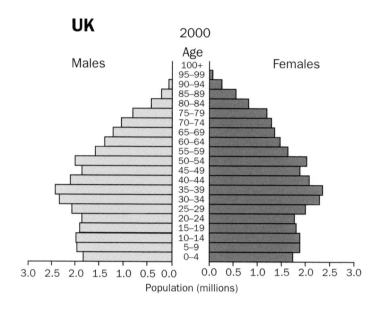

UK
2000

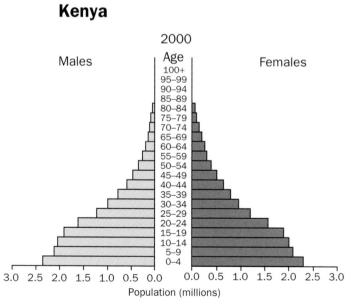

Kenya
2000

Examples of population pyramids

population structure is the make-up of the population of a country (or area). The most studied form of structure is that of age and sex and is represented by a *population pyramid*. The age, sex and *life expectancy* of a population has implications for the country's future economic and social development. Other structures that can be studied include race, language, religion, family size etc.

porosity refers to the amount of area between the particles of a rock. Such areas are called pore spaces and the size and alignment of them determines how much water can be stored or can pass through the rock. Porosity is sometimes expressed as the percentage of the total rock which is taken up by pore spaces, e.g. sandstones range from 5 to 15%, loose sand and gravels can reach 45% and clays up to 50%. Saturation occurs when all the pore spaces are full of water.

positive correlation: a *correlation* where an increase in the value of one *variable* is matched by an increase in the value of the other. For example, the number of cars per head of population will show a positive correlation with *GNP*.

positive feedback: this occurs within a *system* where a change causes a snowball effect, continuing or even accelerating the original change. For example, intense heating in the tropics causes *convective* uplift over the oceans. This draws in moist air and the system is continually fed through the release of *latent heat* from the massive *condensation* and *precipitation* that occurs within the rising air. A small disturbance can then rapidly grow into a *tropical* storm and perhaps even into a *hurricane*.

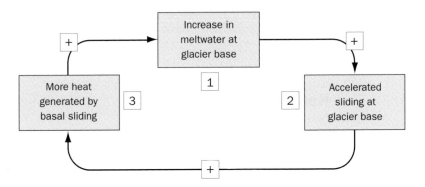

Positive feedback system at a glacier base

positive skew is a bias within a *distribution* towards low values.

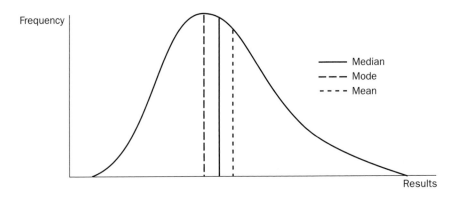

post-glacial: the period since the retreat of the last *Pleistocene* ice advance. In Britain the last *ice sheets* retreated about 10,000 years *BP* and there followed a series of climatic changes with a rise in temperature to the climatic optimum during the Atlantic period which ended about 5000 years BP. Since then the climate has become cooler. ***Pollen analysis*** and studies of oxygen isotope ratios in ***ice cores*** have enabled the post-glacial to be divided into a number of climatic periods: pre-***Boreal*** (10,000–9000 BP); Boreal (9000–7500 BP); Atlantic (7500–5000 BP); sub-Boreal (5000–2500 BP); sub-Atlantic (2500 BP to present).

These changes are closely linked to the general ***vegetation succession*** which developed after the ***ice age***. In the pre-Boreal the climate was cold and dry and species such as the birch and pine were common. Temperatures modified slightly in the Boreal, which was dry with cool winters but warmer summers. The amount of pine and birch gradually decreased and species like the elm, hazel and oak, trees which prefer warmer conditions, became more significant. The Atlantic period was warm and moist, perhaps 3°C warmer than today, and this represents the climatic climax period for the mixed ***deciduous woodland*** in Southern England with oak, elm and hazel being joined by more temperature sensitive species such as lime, and alder which tends to appear in the later stages of forest development. The sub-Boreal was cooler and drier; there was some reappearance of birch, but generally a stable vegetation. Human activity begins to have a greater impact during this period (the Bronze Age) and pollen from grasses, weeds and plantains appear in the pollen spectrum which suggests some small-scale clearance and cultivation. The end of this period sees the end of the lime in any number, an indication of the cooler conditions. The sub-Atlantic is a warmer and wetter climate; vegetation changes became increasingly affected by human activity.

post-industrial: the changes in an economy when employment in traditional industries such as ***mining*** and manufacturing is replaced by employment focused on services and technology.

potential evapotranspiration: the amount of water that could be evaporated or tran-spired from an area given sufficient water available. Potential evapotranspiration is directly related to prevailing temperature conditions. The relationship between ***precipitation*** and potential evapotranspiration for an area is represented by a ***soil moisture graph***.

potholes are cylindrical holes drilled into the bed of a river by turbulent high velocity water loaded with pebbles. Vertical eddies in the water may be strong enough for the pebbles to grind a hole into the rock. Potholes vary in size from a few centimetres to several metres in width. The name potholes is also applied to the passageways found beneath ***Carbonifer-ous limestone*** areas. These are created by the combination of sink holes, caverns, widened ***bedding planes*** and joints and springs.

poverty: an abstract notion of extreme deficiency or inferiority which is extremely difficult to quantify. The latest estimates on poverty draw on over 500 household surveys from 100 developing countries, representing 93% of the population of the developing world. The international poverty line is based on a level of consumption representative of the poverty lines found in low income countries. Since 2000, the international poverty line has been set at $1.08 a day, measured in terms of 1993 ***purchasing power parity***. Prior to this date, it was measured by the number of people living on less than $1 a day.

In assessing the progress made in reducing global poverty, the Human Development Report (HDR) published by the ***United Nations*** notes that:
- in the past 50 years, poverty has fallen more than in the previous 500 years
- poverty has been reduced in some respects in almost all countries
- death rates of children in the less developed world have been cut by more than half since 1960

- *malnutrition* has declined by almost one-third since 1960
- since 1960, the proportion of children not in primary education has fallen from more than a half to less than a quarter.

On the negative side, the HDR details that there are still substantial problems, including:

- One-fifth of all people living in the less developed world still live in poverty, with nearly 1 billion living below the international poverty line.
- Nearly 1 billion people are illiterate; one child in five does not complete primary school. In sub-Saharan Africa, a child has only a one in three chance of completing primary school. One in four school-aged children in south Asia is not being educated.
- Some 840 million people go hungry or face *food insecurity.*
- More than 1.5 billion people lack access to safe drinking water.
- Women are disproportionately poor, with half a million women in the less developed world dying in childbirth each year – one every minute of every day. A woman in sub-Saharan Africa is 100 times more likely to die during pregnancy or in childbirth than is a woman in Western Europe.
- In much of the developing world, the HIV/*AIDS* pandemic continues to spread un-checked. More than 15 million children lost one or both parents to the disease in 2005 and the number of AIDS orphans is expected to double by 2010.

The *United Nations Millennium Development Goals (MDGs)* are seeking to address many of these issues.

PPP: see *purchasing power parity*.

prairiefication: the widespread removal of hedgerows in *arable* farming areas to create vast open fields. It is thought that, since 1945, some 50% of East Anglia's hedgerows have been cleared. The main reason for the removal of the hedgerows is to allow machinery, par-ticularly the combine harvester, to be used most effectively. Hedgerows also harbour pests and weeds, and use up valuable land. Opponents to prairiefication state that hedgerows form an important part of the English landscape. Hedgerows act as windbreaks to prevent *wind erosion*, and they form the habitat of a variety of wild animals, birds and insects (including butterflies).

precipitation: water in any form which falls from the atmosphere to the surface of the Earth. It includes *rainfall*, *snow*, *sleet* and *hail*.

precipitation intensity refers to the rate at which *rainfall* falls within a given period of time. Low intensities of rainfall, sometimes called drizzle, tend to have longer periods of duration. High intensity rainfall periods tend to have a shorter duration, as in the case of a *thunderstorm*. Intensity can also vary spatially. The centre of a storm tends to have high intensity rainfall, whereas the edges of the storm have lower intensities.

precipitation processes: the mechanisms by which tiny *cloud* droplets become signifi-cantly larger rain droplets, and then fall to the ground surface. Research has identified two main processes:

- the *Bergeron–Findeisen theory*
- the coalescence theory. Unlike the above, this accounts for the occurrence of rain from clouds within which freezing does not take place, as in the tropical areas of the world. It assumes air turbulence within the cloud which causes masses of tiny cloud droplets to be swept up and down within the cloud. These rising and falling air movements cause the

droplets to collide with each other, and to grow in size. Eventually their weight becomes so great that they can no longer be supported by the rising air currents, and they fall as rain.

prediction: a form of hazard management whereby researchers try to forecast the occurrence of a hazardous event. Successful prediction enables people to be forewarned about an event, and therefore to make preparations to reduce the impact of it.

preservation: the maintenance of a landscape or building such that its current state is as close as possible to its original condition. Such features may be preserved as sources of tourist income and/or to provide an educational resource.

pressure (atmospheric): the weight exerted by the gases which constitute the atmosphere on any surface exposed to it. The atmosphere consists of a mixture of gases, which are confined by the ground or sea from below, and by the force of gravity from above, thus preventing them from escaping. The layers of the atmosphere which are closest to the ground surface have the greatest weight acting on them, and so pressure is greatest here. Consequently, pressure decreases with height. Atmospheric pressure also varies horizontally, being a direct function of temperature. When temperatures rise, air expands and rises by *convection* and pressure decreases. Conversely, when temperatures fall, air contracts and becomes more dense, causing an increase in pressure. Atmospheric pressure is measured in millibars, and the average sea level value is 1013 millibars. Points of equal atmospheric pressure are shown on a weather map by lines called isobars.

pressure gradient is the rate at which barometric pressure changes spatially between high and low pressure areas. It is indicated by the spacing of the isobars on a synoptic chart, the closer the spacing the steeper the gradient. The speed of winds is influenced by the steepness of the pressure gradient. A simple analogy can be made with the link between slope gradient and the spacing of contour lines.

pressure melting point is the temperature at which ice under pressure will melt. Normal melting point is 0°C at the surface of a *glacier* or *ice sheet*, but at depth the increased pressure marginally lowers the melting point. When ice moving down a valley meets an obstacle, the basal ice will be subjected to additional pressure. If it is close to melting point, as it is likely to be in a temperate or warm glacier, then the additional pressure applied to the ice will cause melting. Temperate glaciers are close to pressure melting point throughout their thickness, but *polar* glaciers are normally well below 0°C and the pressure melting point.

pressure release or dilatation occurs when rocks, such as granite which formed within the Earth's *crust*, are gradually exposed by denudation and the removal of overlying strata. These rocks, which were subjected to great pressure when buried under other strata, expand as the pressure reduces. This is known as sheeting, where layers of rock separate and form a system of jointing generally parallel to the surface. Pressure release, or 'unloading', can also result from removal of rock by glacial action as the density of ice is lower than the density of the underlying rock.

prevailing wind: the most frequently occurring wind direction at a given location, for example, for most of Britain the prevailing winds are southwesterly. These are not necessarily the strongest, or dominant, winds which affect that location.

primacy occurs when the largest city dominates the city size distribution of a country. If a country followed the prediction of the *rank–size rule*, the second city would be half the population of the first city, i.e. n would equal 2. If the population of the first city is greater than twice the size of the second city then this can be described as a primate distribution. With primacy, the largest city is usually the capital and leading political, economic and cultural centre in the country. It has a disproportionate share of activity, and this extreme centralisation may attract more population through *migration*. It is difficult to generalise about the factors which encourage primacy; both economically advanced and less developed countries display primacy; former colonies, such as Sri Lanka, may display primacy as development was focused on one point of entry, although others such as Brazil display a binary pattern with two dominant cities of roughly equal size.

primary data is information which is collected through a *personal field investigation*, or material derived from other sources that has not been processed. This would include such material as that contained in a *census*, telephone directories, trade directories etc.

primary sector: economic activity directly concerned with the extraction of *natural resources*; this includes *mining* and *quarrying*, agriculture, fishing, forestry and hunting. These are sometimes referred to as extractive industries.

primary succession: a vegetation succession that takes place on a surface where no soil or vegetation has formerly existed. Examples of 'new' surfaces include natural surfaces such as sand *dunes*, lava flows, tidal marshes, *landslips*, outwash plains, land exposed by the retreat of ice as well as surfaces created by human activity such as abandoned quarries, *derelict land* resulting from urban clearance, mine waste and spoil heaps. Primary successions can be divided into two broad types: *xeroseres*, which develop in dry conditions, and *hydroseres*, which are initiated in standing water. Xeroseres can be either *lithoseres*, which develop on bare rock surfaces, which are classed as dry because initially there is no soil to retain moisture, or *psammoseres* (sand dune systems), which are dry because of the extreme permeability of the surface. Hydroseres are subdivided into hydrosere successions, where they develop in fresh water environments such as a lake or pond, and *haloseres*, where the succession occurs in salt water such as a tidal marsh or estuarine mudflats. Although each succession follows its own stages of development, they all terminate in the climatic climax community for that *climax vegetation* type; in the case of southern Britain, this is the *deciduous woodland* community.

primeur crop: agricultural produce which is marketed out of season or early in the season. It particularly applies to fruits, vegetables and flowers which have a 'year-round market' in advanced urban areas. This demand is met by bringing in crops from other areas where warmer conditions in winter or early spring allow crops to be harvested and marketed earlier than in the market area. Florida, with its mild winters, markets 'early' produce in the urban area of the Atlantic coast from Washington through to Boston. The Channel Islands and the Scilly Isles produce early crops for the UK market.

primogeniture is a form of land inheritance in which the estate is passed on without sub-division to the eldest son (male primogeniture) or to the eldest surviving child. The advantage of this form of inheritance over *gavelkind* is that the lands of the estate do not become frag-mented, although in Britain family holdings may have become smaller as land was sold off to pay death duties.

prisere: a succession of plant community stages, or seres, which make up a *primary succession*.

pristine area: see *wilderness*.

private sector is that part of the economy operated by firms that are owned by shareholders or private individuals. In economies such as that of the UK it is the dominant sector; the remainder is called the **public sector**.

privatisation (denationalisation) occurs when firms or even whole industries are returned to the **private sector** after having been state-run. A policy of privatisation has been pursued by various governments since 1979. At its simplest, privatisation can involve contracting out services within a **local authority** (refuse collection, parks and gardens) or the **deregulation** of bus services. At its most extreme level, it involves the change of large state-run industries into public companies. Some of the major privatisations have been those of British Petroleum, British Aerospace, National Freight, the Forestry Commission, Sealink, British Telecom, British Gas, British Airways, British Airports Authority, British Steel, electricity supply and generation, British Coal and British Rail.

Arguments in favour of privatisation:
- state control is not the best way to run industries as efficiency can only come about through the market mechanism
- many state industries were **monopolies** that acted against the public interest
- selling shares widened the share ownership base within the country
- privatisation provides government revenue, which can keep taxes down
- there should be better industrial relations and **productivity** within the industry.

Critics have claimed that:
- in some cases the state **monopoly** has given way to privatised monopoly as there was really only room for one producer in the market
- this policy is akin to 'selling off the family silver'.

production chain: the entire sequence of activities required to turn **raw materials** into a finished product for the consumer. The chain will include **primary**, **secondary** and **tertiary sector** activities with the latter involved at every stage.

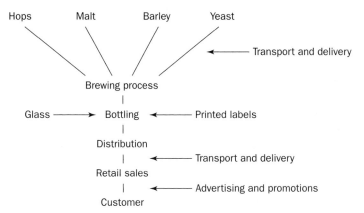

A highly simplified example of a production chain for beer production

production line: the arrangement of a **flow production** system so that parts move systematically from one stage to the next.

productivity is a measurement of the efficiency with which a firm turns production **inputs** into outputs. The most common measure is **labour** productivity, i.e. output per worker. This is important because output per worker has a direct effect on labour costs per unit of production.

The higher the productivity, the lower the labour costs per unit of production, as in the following example.

Worked example: productivity and costs for widget manufacture

	Weekly wage	Productivity (output per worker)	Labour cost per unit
Best UK firm	£250	25.0	£10
Worst UK firm	£250	12.5	£20
Average Korean firm	£280	40.0	£7

Differing levels of productivity (efficiency) are the main single explanation for variations in industrial performance and levels of national wealth. Rising productivity can cause job losses, if demand does not rise as fast as productivity gains. Yet despite this threat, high productivity remains vital to the competitiveness of every company and country.

product life cycle: the theory that all products follow a similar life course (conception, birth, growth, maturity and decline) although products pass through these stages at different speeds. Factors that affect the length of a product's life cycle include durability, fashion and technological change. An important implication of the theory is that as every product will eventually decline and die, it is necessary for firms to carry out a continuous new product development programme.

property-led regeneration: the name for many of the policies that were put forward in order to *regenerate* the UK's *inner cities*. The emphasis was on changing the image of the place, improving the environment, attracting private investment and improving confidence for further investment. Many of these schemes had '*flagship*' projects. A classic example of such a property-led emphasis is the *Urban Development Corporations (UDCs)*. (See also *London Docklands*.)

proportional symbols: representational techniques, where a symbol is drawn proportional to the value that it represents. This can be in the form of circles, spheres, bars, squares and cubes. Perhaps the most widely used is the proportional circle, where it is often found as a divided diagram, the divisions representing the components that go to make up the whole. Proportional divided circles are perhaps better known as pie graphs or *pie charts*.

protectionism describes policies of erecting barriers to trade such as *quotas*, *tariffs* and non-tariff barriers. Because of the benefits that stem from trade, protectionism is banned under *General Agreement on Tariffs and Trade (GATT)* agreements. The worst period of protectionism followed the Great Depression of 1929, as many countries put up such barriers in the hope of maintaining levels of employment.

psammosere is a *succession* of plants that develops on sand, best seen on coastal sand *dunes*. In the UK, the *pioneer* plants in such a succession are likely to be hardy grasses such as lyme, sea couch and marram. This should eventually lead to the oak tree *climax vegetation*.

public corporation: the technical term for a *nationalised industry*, i.e. an enterprise that is owned by the state, but offers a product for sale to *public* and *private sector* customers.

public enquiry: a way of judging a planning permission case that is thought to be of particular public interest. A typical example might be the building of a motorway through a site of outstanding natural beauty, or the construction of a nuclear power station. The enquiry has the power to turn down the planning application.

public sector: the organisations and activities that are owned and/or funded by national or local government. These include *public corporations (nationalised industries)*, public services (such as the *National Health Service*) and the municipal services (those run by local councils).

purchasing power parity (PPP): a measure of average wealth that takes into account the cost of a typical 'basket of goods' in a country. In low income countries, goods often cost less, meaning that wages go further than might be expected. For example, China's *GDP per capita* of $2000 per year generates a PPP closer to $7000.

push and pull factors encourage people to migrate, being the observations that are negative about the area in which the person is presently living (push) against the perceived better conditions in the place to which the migrant wishes to go (pull). A good example would be the factors that encourage rural-to-urban *migration* in the *Third World*:

PUSH:
- imbalance between resources and population, giving *overpopulation, under employment* and unemployment
- rural poverty and low wages
- physical environment often unfavourable for agriculture
- modernisation of farming leading to unemployment
- lack of social services such as education, health and housing
- break up of traditional communities through selective out-*migration*
- *hazards*, including *flooding*, *earthquakes* and *volcanoes*

PULL:
- *perception* of job opportunities
- higher wages in urban areas
- better provision of housing, education and health facilities
- the attraction of the cultural and recreational activities of a large urban area.

PYO stands for 'pick your own', a method of marketing which cuts out the middlemen in the *agricultural chain*, such as *wholesalers*. Farmers are therefore able to increase their profits, and customers benefit from very fresh produce at prices below the market level. PYO schemes reflect the growth in freezer and car ownership from the late 1970s. In the UK, PYO schemes have a wide geographical distribution. High concentrations occur in the West Midlands and the southeast, reflecting the large urban markets there and the traditional horticulture of the regions.

pyramidal peak is formed when three or more adjacent *cirques* (corries) develop on the side of a mountain leaving a very sharp mountain peak with steep sides and *arêtes* radiating from the central peak. The best example in Europe is the Matterhorn in the Alps (for diagram, see *arêtes*).

pyroclastics are materials that have been blown into the atmosphere by volcanic activity. These include cinders, ash, lapilli (small stones), pumice and volcanic bombs. Also included is the incandescent cloud of gas, known as a *nuée ardente*, that sometimes accompanies eruptions. The capital of the island of Martinique, Saint-Pierre, was destroyed by a nuée ardente from the *volcano* Mt Pelée in 1902. Nuée ardente form a deposited material known as an ignimbrite.

pyrophytes are plants which have adaptations that enable them to withstand fire. This usually consists of bark that is fire-resistant. Examples of pyrophytes are the baobab tree and the acacia, both of which are typical of *savanna* regions. For some plants fire is required before they can regenerate. In Australia, for example, plants such as banksia need the fire for their woody fruit to open and thus regenerate.

Are you studying other subjects?

The *A–Z Handbooks (digital editions)* are available in 14 different subjects. Browse the range and order other handbooks at **www.philipallan.co.uk/a-zonline**.

quadrat: a frame, usually a square, enclosing an area of known size, and used mainly for *sample* surveys of vegetation and surface deposits. A grid can be inserted by using wire or string, and this will provide sampling sites, particularly for a **systematic sample**.

qualitative data is information that can be captured but is not numerical in nature. Sources include brochures and leaflets, photographs, websites and interviews.

quality of life: this is measured on an index devised in the 1980s and known as the physical quality of life index (PQLI). The PQLI is the average of three characteristics: **literacy**, **life expectancy** and **infant mortality**. Each feature is scaled from 0 to 100 and the lowest country in a category is given 0 and the best, 100. The three 'scores' for each country are added and the average is found. This figure is the PQLI for that particular country. Almost all the countries which are normally considered to be in the **First World** have a PLQI of 90 or over.

quantitative data is information that can be counted or expressed numerically. This type of data is often collected through observation and is then processed and statistically analysed. It can then be represented visually in charts, graphs etc.

quarrying: the removal of rocks for commercial purposes from large and open surface workings. Quarrying is part of the **primary sector** of economic activity, and although it can be an important in some areas, quarry workings are not without their critics:

PROS:
- important employer in many rural regions (where there may be few alternatives)
- brings revenue into rural regions
- could improve local **infrastructure**
- may be a factor in reducing **rural depopulation**
- provides important **raw materials** for other industries such as cement, steel and chemicals
- can reduce foreign spending on imported raw materials

CONS:
- visual **pollution** of quarry scars and waste tips etc.
- noise from blasting and excavation
- very dusty in some places, both from quarrying activities and from vehicles
- increased vehicular activity on narrow rural roads.

In the UK, the quarrying of **limestone** has generally caused the greatest public concern. This is because so many limestone quarries are to be found within **National Parks**, e.g. the Peak District.

Quaternary period: the latest period of geological time spanning the last 2 million years. It covers both the **Pleistocene** (which includes the most recent **ice age**) and the **Holocene** (**post-glacial** period).

233

quaternary sector is that sector of economic activity which follows the tertiary (some believe it is part of the *tertiary sector*). This covers activities such as training and *research and development*. Industries involved include high technology (*high-tech*) and information services.

quartile: when values are ranked in order to calculate the *interquartile range*, the first necessary step is to find the quartiles. Once the *median* has been found, the quartiles represent the median of the values above the median (upper quartile) and the median of the values below the median (lower quartile). For example, in a ranked list of 19 values, the upper quartile will be the 5th figure and the lower quartile, the 15th.

questionnaire surveys involve both questions and answers and are used for obtaining opinions, views, ideas and information about the way that people behave. Questionnaires can be completed in the street, door-to-door, by post or on the telephone. For geographical surveys, the best questionnaires involve a few short and uncomplicated questions that produce clear and precise information. Good questions ask people for the pattern of their behaviour, not how they think they behave. For example, it is better to ask a shopper how many times s/he has been to the *central business district* that week, not 'how often do you shop in the CBD?'

quota: the introduction of a set figure of production or the amount of import commodities that a country will allow in from certain sources. Quotas, for example, have been imposed in the dairy industry in order to reduce the output of milk.

Aiming for a grade A*?

Don't forget to log on to **www.philipallan.co.uk/a-zonline** for advice.

radial drainage is represented by a pattern of streams radiating out from a central point. It results when land is uplifted to produce a dome structure with streams developing from the highest point. Radial drainage patterns can be seen at a variety of scales, for example, on a slag heap next to a coalmine, or on a volcanic cone, or radiating from the central dome structure of the Lake District.

radiation: the emission of electromagnetic waves from a body such as the sun or the Earth. An important feature of these waves is their wavelength, as this gives rise to different forms of radiation:

- short-wave radiation – emitted by the sun with its very high surface temperature as *insolation*
- long-wave (terrestial) radiation – emitted by the surface of the Earth. This is caused by the Earth's surface absorbing insolation and converting it into heat. The ground, warmed in this way, re-radiates the energy back to space as long-wave radiation. However, much of this radiation is absorbed by the atmosphere, thus warming it from below. (See also *insolation* and *energy budget (the Earth)*.)

radiation fog results when a body of moist air, in contact with the ground surface, is cooled to its *dew point*. This commonly occurs at night under cloudless anticyclonic conditions with only a light breeze blowing. Because the sky is clear, the ground surface cools rapidly by *radiation*, and in turn cools the layer of air immediately above it. Once the dew point has been reached, *condensation* occurs. The cooled lower layer of air is stirred by the light wind so that it cools the air above it to its dew point, and so the *fog* grows deeper. Radiation fogs commonly occur under *temperature inversions*, which prevent the air from rising. They may persist for several days in winter if the sun is too weak to disperse them. In industrial districts, an extreme form of radiation fog called a *smog* may occur.

raindrop impact: the erosional effect of raindrops falling in an intense storm. These droplets can have a very large *kinetic energy*, and on contact with a loose soil surface may cause a splash of particles both vertically and horizontally for some distance. On a flat surface, the particles merely change position, but on a slope the net transfer is downslope. This is because the particles splashed down the slope have longer trajectories than those splashed up the slope.

Raindrop impact can have two significant effects:

- larger particles of sand are left on the slope, whereas finer particles are moved down the slope
- finer particles may be removed from around a larger pebble or stone. This produces a column of finer material capped by a protective stone, called an earth pillar.

rainfall: droplets of water large enough to fall to the ground. Their diameter is usually between 0.5 and 5 mm, but drizzle is made up of smaller droplets. Rainfall is produced by *precipitation processes*.

rainforest: the name given to a vegetation type which can be found in the Amazon Basin of South America, the Zaire Basin and the Guinea coast of Africa, and parts of southeast Asia, Indonesia and northern Australia. The climatic type of these areas is known as the *Equatorial climate*.

The main characteristics of the rainforest vegetation are:

- tall evergreen trees (30–50 m) grow close together forming a high *canopy* of leaves, called emergents. The trees have very few branches except at the top
- continuous growth occurs with the constant high temperatures and rainfall. As old leaves die and fall, new ones grow giving an evergreen appearance
- smaller trees and shrubs form additional leaf canopies at lower heights
- many different species of trees are found, with hardwoods such as rosewood, mahogany and ebony being common
- leaves are thick and leathery so as to be able to withstand the strong sunlight, and they often have drip tips to enable them to shed excess water
- the tall trees have buttress roots to give them support
- the dense leaf canopy shades the ground so that there is little undergrowth except near rivers and clearings where light can reach the ground
- lianas grow from the shaded ground and reach the light in the leaf canopies by climbing the larger trees
- *epiphytes* grow on the taller trees and occasionally have roots hanging in the air
- *saprophytes* are found in large numbers on the dark forest floor
- parasitic plants live on the trees of the forests and feed from them.

Rainforests form the habitat for a huge number of birds, insects and other animals. They are, however, very sensitive and fragile *ecosystems*. Their existence relies on the rapid recycling of nutrients, but once the cycle is broken by human interference, then the forest will have difficulty re-establishing itself.

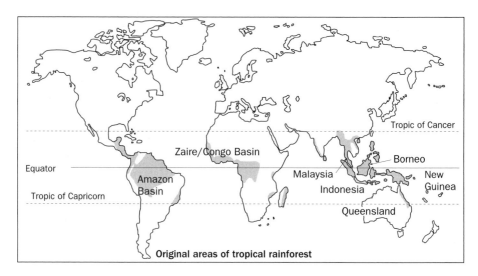

Original areas of tropical rainforest

rainshadow: a dry region of land that is leeward of a mountain range or other physical features, with respect to the *prevailing wind*. The mountains block the passage of some rain-producing weather systems. Any moist air rising on the windward side leads to *condensation* and *precipitation* there. As the air descends on the lee side, it warms up, with a consequent drop in *relative humidity*. One example is where the Andes (South America) block the prevailing southeast *trade winds*, causing a dry area on the western side of the mountains known as the Atacama Desert.

raised beach: a *coastal landform* produced by the land rising relative to the sea. As the land rose former beaches were raised above the influence of the waves, and they are frequently accompanied by a raised *wave-cut platform*, backed by relict cliffs. The existence of all of these features is evidence that the *isostatic readjustment* of land was not constant. Sea level had to be unchanging for a period of time to allow their creation. Sea shells found on the raised beaches can be dated to indicate the age of the beach.

ranching: a system of agriculture based on the rearing of *livestock*, usually cattle or sheep, which earns the lowest net profit per hectare of any type of *commercial farming*. It is practised in remote or difficult areas of the world on an extensive basis. Examples include sheep farming in central Australia and Patagonia, and cattle ranching in the Argentinian Pampas, the Great Plains of the USA, and northern Australia. Large areas of land are needed to supply the animals with sufficient grass to feed on, and consequently it is found in areas of low population density. Each farm is a very large unit, often over 100 km^2, but little additional capital is used. The output per farm worker is high, with the demand for *labour* being seasonal, particularly for sheep farming. Ranching is dependent on good *communication systems* for exporting the produce, as well as the food preservation methods of refrigeration, freezing and canning. Many ranches are now owned by business interests and run by managers on the site.

R and D: see *research and development*.

random means due to or of chance, therefore no pattern should be detectable in any situation. In *sampling*, for example, a random sample is one that shows no bias and in which every member of the population has an equal chance of being interviewed or used. Random samples are usually obtained by using *random numbers*. In describing the distribution of settlement in an area using *nearest neighbour analysis*, a result that indicated a random distribution of settlement is one where there is no pattern to be seen in that distribution.

random numbers are numbers that have no pattern to them if used systematically. They could be drawn from a hat, but more probably, be generated by computers and compiled in random number tables.

range: the term has two meanings:
- in statistics, this is a very simple method of determining the degree of *dispersion* of the values about the *mean*. It involves calculating the difference between the highest and lowest values, thus only emphasising the extremes, but indicating nothing about the remainder of the values
- the maximum distance that people are willing to travel to obtain a good or service. This will depend on the type of good or service, its value, the availability and type of transport, and the frequency of need for the good or service.

rank–size rule: an attempt to find a relationship between the population size of settlements in a country or region. The rule, promoted by **Zipf**, states that the size of settlements is inversely proportional to their rank. In other words, the second largest settlement should be half the size of the largest city, and the third largest settlement, one-third the size, and so on. It is usual to plot the relationship on **logarithmic scales** which, if the relationship is perfect, will give a straight line relationship on the graph. If the largest city is very much larger than the second, this is known as **urban primacy**, e.g. Argentina. Some countries have two cities that are of almost equal size and this is known as a binary distribution, e.g. Australia.

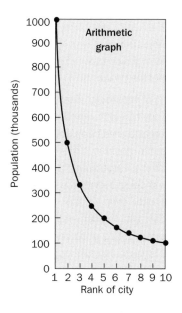 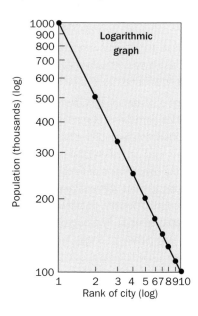

rapids are sections of fast-flowing water that develop where there is a local steepening of the gradient along the course of a river. This change in gradient may be caused by a band of harder, more resistant, rock, which slows down **erosion** and the development of the graded *long profile*.

rapid transit systems are methods of moving people around large urban areas and so avoiding the use of the car. Such systems involve light railways, underground systems and trams. In the UK there has been a growth of such systems in recent years, including the Tyneside Metro system, the **London Docklands** Light Railway and the Metrolink tram system of Greater Manchester. In 1995 Sheffield also opened a city tram network.

rationalisation means reorganising to increase efficiency. The term is mainly used when cutbacks in costs are needed in order to reduce an organisation's break-even point. This may be achieved by:

- closing one of the company's factories and reallocating the production to the remaining sites
- closing an administrative department and delegating its tasks to the firm's operating divisions
- removing a layer of management.

The term is often used by companies as a euphemism for making people redundant.

Ravenstein, E. G.: presented a paper in 1885 entitled 'the Laws of *Migration*'. His main points were that:

- most migrants travel only short distances and with increasing distance, their numbers decrease (*distance decay*)
- most migrations produce a compensatory counter-movement
- migration tends to occur in stages and with a wave-like motion
- urban dwellers are less likely to move than those in rural areas
- females tend to migrate more than males within their country of birth, but males are much more likely to be involved in international migration
- the major direction of migration is from rural to urban areas
- long-distance migration is more likely to be to large centres of industry and commerce.

raw materials: the *inputs* needed by a firm to make a product. These can be totally unprocessed materials such as iron ore, wheat, raw cotton, crude oil etc., or they can be products of an initial process which have come from another industry. Car manufacture, for example, uses a range of materials that have come from other sources.

rebranding means changing the image or perception about a product, service or a place; replacing an existing idea with a new one. This can be seen in product/service marketing or in refocusing the image of a place. Rebranding can mean three things: changing the name of the service or product, changing the brand appearance or identity or repositioning in the market, i.e. trying to reach a wider market.

There a number of reasons why businesses might rebrand themselves:

- when they open new businesses that are very different from the market perception of what they traditionally produce, e.g. 3M (originally Minnesota Mining Manufacturing)
- when company ownership has changed through a merger or a split or through privatisation of an industry, e.g. GlaxoSmithKline (merger); Accenture (management consultants split from Andersen); BT and Powergen (when nationalised industries were privatised)
- as part of standardisation on a global scale, e.g. Snickers bar (from Marathon)
- to modernise the brand image, e.g. Go (BA), First Direct (Midland Bank/HSBC), Egg (Prudential).

recently industrialised country (RIC): these are countries that have shown rapid growth in the last 20 years. Much of this growth has been based on *globalisation* and *transnationals* seeking ever cheaper locations for production, often attracted by nearness to *raw materials* and cheaper land and *labour* costs as well as the potential of a rapidly growing domestic market. RICs include Brazil, China, India, Russia, Mexico, Turkey, Indonesia and South Africa. Within this group special attention is given to the *BRIC* group (Brazil, Russia, India and China). Between them, these four countries have very large reserves of *natural resources* and massive, growing populations. Economic analysts suggest that within the BRIC economies China and India will become the world's dominant suppliers of manufactured goods and services, respectively, while Brazil and Russia will become dominant as suppliers of raw materials. These countries are not a political or trading alliance (like the *European Union*), but they have the potential to form a powerful economic bloc.

recession is that part of the *trade cycle* which is characterised by falling levels of demand, very little investment, low business confidence and rising levels of unemployment. The official definition of a recession is two successive declines in quarterly *gross domestic product*.

reclaimed land is an artificially created coastal environment that ranges from ditched *salt marsh* areas, where there is very little extra human interference, to intensively managed areas used for agriculture. Reclamation normally occurs in one of two ways:

- simple enclosure of former inter-tidal areas
- enclosure, followed by infill with sea bed material and finally pumping the area dry.

In recent years, coastal reclaimed land has come under threat from an increase in agricultural intensiveness, such as the use of *fertilisers* and *pesticides*. It is also under much greater pressure from increases in human activity such as caravaning and birdwatching.

recovery is that part of the *trade cycle* which is characterised by slowly rising levels of demand, some investment, improving business confidence and falling levels of unemployment.

recreational activity includes any activity which is undertaken voluntarily for personal enjoyment in leisure time. This could include home-based hobbies and pastimes such as gardening, listening to music or watching television, although generally the term relates to activities undertaken away from the home. Geographical research into leisure activity usually distinguishes between recreation, defined as pursuits which involve absence from home for less than one day, and tourism, which involves a longer absence. Clearly, there is only a fine distinction between recreation and tourism as activities.

recreational forest: an area of woodland, either privately owned or managed by the *Forestry Commission*, which is open to the public for leisure activities. This could include nature/woodland trails, cycle routes, hides for birdwatching and visitors' centres with onsite wardens. In some forests, provision for visitors may extend to camping facilities or chalets/log cabins for hire. It provides an opportunity to manage the area for both economic and controlled leisure activity.

recurrence interval is a statistical estimate of the average time between two events of the same *magnitude*. It is used in *risk assessment/analysis* of hazard events such as *flooding* or *earthquake* activity. A flood event with a 50-year recurrence interval would be of a magnitude that on average only occurred once every 50 years. As this is based on statistical analysis of past data, the background conditions in the drainage catchment may have changed and this would influence the likelihood of a particular event. As it is based on average data, a 50-year recurrence interval does not mean that a 50-year event could not occur 2 years in succession.

recyclable: any item that, after some form of processing, has the potential to be used again.

recycling is the reuse of materials that would normally be discarded after use. It has been encouraged by the threat of exhaustion of *finite*, *non-renewable resources*. Scrap metals such as steel and aluminium have been recycled because reprocessing consumes less energy, and there has been an increase in the recycling of waste paper and glass recently. Pressure groups have made the public more aware of the problem and councils have established bottle banks and recycling centres. In some cases the burning of rubbish has been linked to community heating schemes. Although desirable, currently it is not always economically viable because of technical and collection costs. A much larger scale of public awareness and participation is needed to develop an effective system.

redevelopment is the renewal and renovation of urban areas which have become run-down or derelict. This may be achieved by improving existing buildings through refurbishment or by widening roads to increase accessibility. Alternatively, buildings may be demolished and new structures built on these **brownfield sites**. **Property-led regeneration** schemes and **partnership** schemes between local and national government have led to the **regeneration** of inner-city areas, not only for residential developments but also for leisure and social facilities and environmental improvements.

redlining is the demarcation of areas of a city in which financial organisations are unwilling to lend money for house purchase. These areas are perceived to be in decline and they are regarded as areas in which lending money to potential purchasers would be a high risk. Building societies, banks, estate agents and mortgage brokers effectively starve these areas of investment; this discriminates against house owners in these areas and contributes to a general downward spiral in these districts. Redlining requires higher down payments on mortgages and imposes higher interest rates, effectively denying house purchase to low income families. As housing deteriorates the neighbourhood declines; home improvement loans are denied and property insurance becomes very expensive, even if it can be obtained. Businesses and services begin to fail and house prices decline rapidly. Redlining may help to inflate house prices in those areas where institutions are willing to invest. Although it is a legitimate practice, it does perpetuate decline and has important consequences for the social geography of the urban area.

reduction is a chemical process in which oxygen is removed, for example ferric irons are changed to a ferrous state. Under waterlogged conditions, the pore spaces remain filled with water and the soil becomes deoxygenated. The iron in the soil becomes reduced and the soil has a greenish or bluish tinge, which is characteristic of **gleying** which develops under **anaerobic** conditions. Reduction is the reverse process to **oxidation**.

refraction is the process by which waves change direction as they approach a **headland**; waves are turned so that their crestline becomes more parallel to the shore. This concentrates wave energy and erosive power on the projecting headland where cliffs, arches and **stacks** may develop and reduces wave energy in the bays where **deposition** occurs to produce a bay head beach.

refugee: someone who, owing to fear of persecution, is outside the country of his/her nationality and is unable, or as a result of fear, is unwilling, to return to that country. This fear of persecution may be due to race, religion, nationality or membership of a particular social or political group. This excludes people who may leave their homes but remain in their own country (**displaced persons**) and migrants who enter another country without appropriate documentation (**illegal migrants**). In recent years, terms such as economic, ecological and environmental refugees have appeared in the media, but as these people are simply seeking better economic opportunities they are not technically refugees and governments tend to regard them as ordinary immigrants who do not qualify for any special consideration with regard to political asylum.

refurbishment: improvement to existing urban areas through investment in better road access or modifications to both the interior fittings and the external appearance of buildings as part of urban **redevelopment**.

regelation: the refreezing of water into ice on the downglacier side of an obstacle. When the underside of a **glacier** moves over an obstacle, **pressure melting** takes place on the upglacier side of that obstacle. This water acts as a lubricant enabling the glacier to move over the obstacle. However, on the downglacier side, the pressure is reduced, and the water reverts back to ice.

regeneration is the investment of capital and ideas into an area to revitalise and renew its economic, social and/or environmental condition. In recent years, the most common type of area in Britain to be regenerated has been the *inner city*.

Regeneration schemes have taken place, such as:
- the City Action teams, *Urban Development Corporations (UDCs)* and the Task Forces of the Conservative governments in the 1980s and 1990s
- the English Partnerships in the late 1990s.

Key elements of these schemes were slum clearance and housing renewal, new industrial growth and development, improvements to transport systems and environmental improvements. A central theme for many of the later schemes was to encourage the development of *partnerships* with *private sector* investment in addition to that from local or central government sources.

region: an abstract concept of an area of a country, or across a group of countries, which has quite distinct boundaries, and within which there are generally similar characteristics. The notion of a 'region' has been of interest to geographers for many years, but no simple definition has emerged.

A region may be defined by its:
- physical characteristics – a similar relief (e.g. the Paris Basin) or a similar climate (e.g. the *monsoon* region)
- economic characteristics – for example, the 'old' industrial regions of the north of England
- functional interrelationships – for example, the city region, say around London, within which people travel, work, and obtain services.

The concept of the 'region' has been hijacked for administrative purposes. In the UK, for example, there are economic standard regions, health regions and regional crime squads for the police.

regional contrasts arise from the inherent differences between varying parts of a country. No country has an *isotropic* surface – they all have variations in physical characteristics and in the distribution of resources. In addition, historical and political factors have resulted in the use and misuse of individual parts of a country. Some regions develop at a greater rate than others, and consequently other regions become exploited, neglected, and more backward. Some economic theories have referred to regional contrasts by the concepts of the *core* and *periphery*. (See also *Myrdal* and *cumulative causation*.)

Regional contrasts may be measured by a wide range of economic and social indicators. These include unemployment rates, *per capita* incomes, rates of out-*migration*, education standards, and health provision. An example of a regional contrast is the so-called 'north/south' divide in the UK.

regional policy is the attempt by a government to redress *inequality* and the unevenness of economic development and social/environmental conditions within a country by stimulating investment in the less prosperous areas. Many countries have established regional policies in an attempt to encourage development in poorer regions of the country.

Regional contrasts are regarded by some to be undesirable because:
- reducing unemployment in poorer areas brings social and economic benefits
- selective out-*migration* of the better educated, more motivated, population further depresses the poorer region
- economically, national output is not maximised if there are both depressed '*periphery*' areas and overcrowded '*core*' areas (see *Friedmann*).
- politically, votes may be lost in the less prosperous regions.

In the UK, various governments, since the 1930s have initiated schemes as part of their regional policy. Regional policies are not viewed as being successful by all. Some people argue that they have had little long-term effect – the regional contrasts tend to remain. Within the *European Union*, all member states bid for EU structural funds as a source of capital for development in poorer regions. This is controlled centrally to prevent individual governments giving an unfair advantage to particular regions. Regional development agencies establish priorities within areas.

In the UK, the main form of state aid is through discretionary grant schemes:
- Grant for Business Investment (GBI), which helps fund new investment projects that lead to long-term improvements in productivity, skills and employment
- Regional Selective Assistance (RSA), operated by RSA Scotland (part of the Scottish Executive) and RSA Cymru Wales (part of the Welsh Assembly), aimed at promoting new projects, strengthening existing employment and creating new job opportunities
- EU Structural Funding including:
  - Convergence Objective (formerly Objective 1) – this covers regions whose *gross domestic product per capita* is below 75% of the EU average
  - Regional Competitiveness and Employment Objective (formerly Objective 2) – this covers all regions of the EU territory, except those already covered by the Convergence Objective. It aims at reinforcing competitiveness, employment and attractiveness of these regions
  - Territorial Cooperation Objective (formerly Objective 3) – this aims is to promote cooperation between European regions, as well as the development of common solutions for issues such as urban, rural and coastal development, economic development and environment management.

regolith: the collective name for all of the material produced by *weathering* extending down from the ground surface to the unaltered bedrock.

regression: a more accurate method of showing the relationship between two variables than the *best-fit line* on a *scatter graph*. This involves the use of semi-averages to locate a regression line to summarise the trend on the graph. The procedure is as follows:
- Calculate the *mean* values for the variable on the independent axis ($\bar{x}$) and the variable on the dependent axis ($\bar{y}$). Use the values as the coordinates to plot point A on the scattergraph.
- Using only the values of x which are higher than the mean of x ($\bar{x}$), calculate the upper semi-average. Repeat this for values of y which are higher than the mean of y ($\bar{y}$). Use these values as the coordinates to plot point B on the graph, the upper semi-average.
- Calculate the lower semi-average by using only the values of x and y which are below the mean of x ($\bar{x}$) and y ($\bar{y}$). Use these values as the coordinates to plot point C on the graph, the lower semi-average.
- Draw a straight regression line guided by the position of A, B and C. These points may not form a perfect straight line, but can be used to obtain a more accurate placement for the trend line.

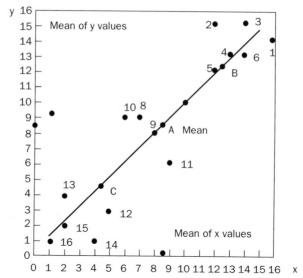

The construction of a regression line

Reilly, W. J. produced a model of retail gravitation similar in nature to that of Newton's law of gravity. Reilly stated that two centres attract trade from intermediate places in direct proportion to the size of the centres and in inverse proportion to the distance between them. In other words, larger centres attract more trade than smaller centres, and the amount of trade with a smaller place increases the nearer that smaller place is to the larger centre. This model can also be used to define the sphere of influence between competing centres by using the break-point formula. If there are two towns A and B, and B is smaller than A, then the distance of the break-point from B is:

$$\frac{\text{distance between towns A and B}}{1 + \sqrt{\dfrac{\text{population of A}}{\text{population of B}}}}$$

Both of these aspects of Reilly's model assume that the larger the town the stronger its attraction, and that people always go to their nearest centre. These assumptions may not be true as:
● there may be difficulties in reaching the larger town, and expensive forms of car parking
● the smaller centre may be easier to reach, safer and less congested
● perceptions, positive or negative, of a place may override logical considerations.

rejuvenation: a fall in sea level relative to the level of the land, or a rise of the land relative to the level of the sea enables a river to revive its erosional activity. The river will adjust to the new base level, at first in its lowest reaches, and then progressively inland. In doing so a number of landforms may be created: *knickpoints*, *river terraces* and *incised meanders*.

relative humidity: the amount of water vapour present in a body of air relative to the maximum amount of water vapour it could contain at that temperature and pressure. It is usually expressed as a percentage – a relative humidity of 100% means that the air is saturated. Since the amount of water vapour that a body of air can contain increases as its temperature rises, relative humidity is inversely proportionate to temperature. As the temperature of a body of air rises, its relative humidity decreases, and vice-versa.

relict landscapes contain features or elements which are no longer being put to active use for their original purpose. Disused canals, old mine workings/quarries, abandoned factories, old dock areas, fortifications, mounds and strip lynchets are all legacies of the past which can be studied in historical geography in order to gain some understanding of the connection between past and present human environments. Some of these relict features have become significant again as part of the growing heritage industry.

remembrement: the process of consolidation of agricultural holdings which have become excessively *fragmented*. Such a policy was promoted by the government of France, from where the term originates.

remote sensing is a general term which covers a range of techniques that can be used to study features on or near the Earth's surface from a distance above the surface. Landforms, vegetation, *land use* and weather systems can be studied through aerial survey and the use of airborne electronic devices. Since the early 1970s satellites have been used to monitor changes on the surface and in the atmosphere. *LANDSAT* provides complete coverage of the Earth as the orbit changes slightly on each revolution. Some satellites, such as Meteosat follow a fixed orbit and therefore remain above the same path; these are termed *geostationary satellites*.

rendzina: an *intrazonal* soil that develops where *limestone* or *chalk* is the *parent material* and under a grassland vegetation. The upper horizon of the soil, is rich in *humus* and given the nature of the parent material, the *pH* is between 7 and 8. Limestone and chalk when weathered leave little insoluble residue, therefore such soils tend to be thin.

renewable resources are *resources* that are either a flow of nature or living things, and can thus be used repeatedly. Living things such as forests are renewable as long as they are not used faster than they are replaced. In energy terms, renewable resources include *wind*, *tides*, *waves*, water power (*hydroelectric power*), *solar energy*, *geothermal*, *biomass* and *ocean thermal energy*.

report: this is the conventional method of presenting precise information. A report may be used to convey an assessment of a situation or the results from qualitative and/or data analysis. It should have clearly stated aims and be tightly focused on the subject under investigation. To be successful, it should be easy to read.

research and development (R and D) means scientific research and technical development. This is directed at improving the product, rather than finding out what the customer wants and thinks (known as market research). With products having shorter lives before they are replaced by other types of products, R and D has taken on a greater significance, as companies have a need for further research in order to keep ahead. Companies also need to have some high yielding products which will generate capital, without much present investment, in order to finance R and D. Such products are known as 'cash cows'.

residential segregation: the grouping of people with similar characteristics into separate residential areas. (See also *social segregation*.)

residential type refers to the type of housing tenure in an urban area; traditionally this has taken one of three forms: owner-occupied (owned outright or purchased on a mortgage or other loan); privately rented (from a private landlord); and council-rented property. The growth of **housing associations** since the mid-1970s, which have, to some extent, replaced **local authorities** as providers of housing for those who do not wish or cannot afford to buy, has added a further category.

residual: a point on a **scatter graph** that lies some distance away from the **best-fit line**; this may be described as an anomalous value. A positive residual lies above the trend line, i.e. the **dependent variable** has a higher value than would be expected; a negative residual, below the line, indicates a lower value than would be expected.

resource: any feature of the environment which can be used to meet human needs. Traditional approaches consider that it is the act of exploitation which converts the commodity into a resource, i.e. they emphasise the use of the commodity. More radical approaches stress the exchange value of a resource. On the commodity markets profits can be made without doing anything to a resource, owners buy at one price and sell on at another without perhaps ever seeing the resource or even taking ownership of it in the sense of storing it. Such speculative behaviour, which also applies to purchase of land in anticipation of its future potential value, is viewing resources in a different way. Resources may be classed as **renewable**, flow resources, or **non-renewable**, stock resources.

resource depletion: the consumption of **non-renewable**, **finite**, resources which will eventually lead to their exhaustion. This has led to greater awareness of the need for **conservation** and **resource management**.

resource management involves the control of resource exploitation and use in relation to economic and environmental costs. Sensible management of a resource includes some initial survey to determine the extent or quality of the resource, for example, soil **fertility** or land capability. This is followed by the development of strategies which might be used in the planning and exploitation stage, assessing possible conflicts which could arise if the resource use impinges on the interests of other groups. For example, the extension of the M3 through Twyford Down caused conflict with local farmers, environmentalists and villagers. Plans to extend or develop a quarry in a part of a National Park would arouse conflict. It would create employment for some, but would be an eyesore and a deterrent to tourists who provide income to others. Resource management has to try and balance these conflicting interests and perceptions, considering both economic aspects and **conservation** issues.

retailing: the sale of goods and services to the public. The term usually includes high- and low-order shopper goods, such as food, clothing and consumer durables (fridges, washing machines etc.) and is often extended to include consumer, professional and financial services such as hairdressers, solicitors and banking. Fast food outlets, restaurants and entertainment may also be included.

retail park: a shopping development usually located on the edge of an urban area, an **out-of-town location**. It has a similar background and location factors as a **hypermarket**, and it is smaller than a regional shopping centre. Whereas a hypermarket tends to have all the different outlets under one roof, a retail park consists of a number of independent companies in individual buildings served by one large car park. The Fosse Park shopping centre on the southern outskirts of Leicester near to the intersection of the M1 and M69 houses companies such as Asda, Marks & Spencer, Next, Habitat, Toys R Us, WHSmiths, Currys and Comet. There is a common car park, a petrol station and restaurants. There is a considerable amount of shopper movement from one store to another, many of the goods on offer are complementary and customers may be purchasing a variety of goods. Such developments can be seen on the edge of most large urban areas which provide a large potential market.

retreat the line is one of the strategies considered in *coastal management* involving the managed realignment of the coastline. Defences are realigned inland at a pre-determined, more sustainable, position (or natural landscape features are exploited) and the sea is allowed to encroach on the existing shoreline. Shoreline management plans (SMPs) in north Norfolk have designated several areas for managed retreat over the longer term, e.g. the section from Sheringham to West Runton. There is local opposition to this type of policy because of the threat to residential property, together with business and leisure interests.

reurbanisation is a process by which large towns and cities in *economically more developed countries* that have been experiencing a loss of population through *counterurbanisation* are beginning to slow this decline or even reverse it and begin to grow again. It represents a movement of population towards the central city area with a rise in the number of housing developments and employment opportunities. In the UK it is linked with the *redevelopment* of *brownfield sites* and the growth of city-centre apartments and waterside developments and increased employment in the financial and service sectors.

ria: an irregular coastal inlet that has formed as a result of a relative rise in sea level causing submergence of a former valley system. Valley systems in southwest England were flooded by the *post-glacial* rise in sea level. A ria is deepest at its mouth and becomes progressively more shallow inland; in cross-section it resembles the former valley profile with some alluviation along the bed of the original channel due to the reduced energy along the river following the sea level rise. The heads of the tributary inlets may also be undergoing infilling by alluviation and the development of tidal mudflats.

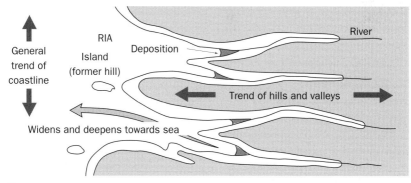

Ria coastline

ribbon lake: an elongated lake occupying the floor of a glaciated valley or trough. They occupy rock basins which have been produced by selective overdeepening; *post-glacial* melting has resulted in a lake which may also be dammed by depositional material.

RIC: see *recently industrialised country*.

Richter scale: a system devised in 1935 to measure the *magnitude*, or total energy release, of an *earthquake*. It is a *logarithmic scale* which extends up to 10; an earthquake at Richter scale 5 has a magnitude ten times greater than one at scale 4. Ground tremors record a magnitude of 2 and damage to structures such as buildings occurs at scale 6 and over. The largest shock recorded so far had a magnitude of 8.9.

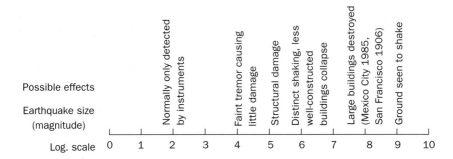

Richter scale

riffle: see *pools and riffles*.

rift valley: a structural landform which results from the downthrow of rock strata between parallel faults. The best-known example is the East African Rift Valley which extends from Mozambique in the south through East Africa extending northwards into Jordan, a distance of 5500 km. In some areas the inward-facing fault scarps are 600 m above the floor of the rift valley, and there are step faults on the valley side which indicate that a series of parallel fault lines have moved to cause a lowering of the floor section. The relief is further complicated by the escape of volcanic lava within the rift area; the valley is thought to be the result of tension in the underlying crustal rocks and it may indicate that the continental *crust* of the African Plate is separating to produce a *constructive plate margin* (divergent margin).

rilling develops when a series of roughly parallel small channels form on a slope. They are closely spaced and their maximum size is usually up to 2 m wide and 50 cm deep. They only discharge water during and immediately after heavy rainfall, but the rapid *runoff* can cause *erosion* and rills may produce *gullying*.

Rio Earth Summit: met in June 1992 to find ways of coping with the permanent environmental damage that has accompanied recent economic developments. Specifically, this meant devising strategies for coping with the effects of *greenhouse gases*, the accelerating loss of *biodiversity* and concerns over the environmental consequences of rapid population and industrial growth in the *Third World*. The Summit produced Agenda 21, which was a blueprint aimed at a clean-up of the global environment and encouraging environmentally sound developments. Particular areas that were targeted included:

- poverty alleviation and environmental health in the Third World
- family planning and expanding job and educational opportunities for women
- reduction of *soil erosion* and the promotion of environmentally sensitive agricultural programmes
- protection of natural *habitats* and biodiversity
- research and development on non-carbon energy alternatives in order to reduce greenhouse gases and to avoid *climatic change*.

risk assessment/analysis: a way of judging the degree of damage that an area may experience as a result of a hazard event such as an *earthquake*, *flooding* or *volcano*. Risk assessment relates the potential effects of the hazard to the *vulnerability* of the location.

river capture is where the headwaters of one river system capture those of another river system. A river with a steeper gradient, flowing over less resistant rock strata or with greater rainfall within the basin may extend itself by headward *erosion* to capture part of another stream. This is sometimes referred to as stream piracy. The main features that result from this action and indicate that it could have taken place are:

- elbow of capture – the point at which the headwaters of the river that has been captured sharply change direction
- wind gap – a valley where the captured stream used to flow
- misfit stream/river – the river that has been captured has a much reduced flow in the old lower course so that it now occupies a valley that is much larger than the present stream could have cut.

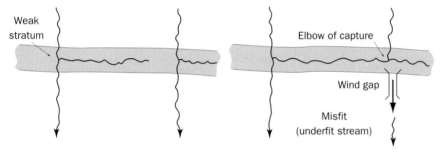

An example of river capture

river cliff: the steep slope that characterises the outer bend of a river *meander*.

river deposition: when a river loses the ability to transport its *load* (*competence*), deposition will occur. This includes the following circumstances:

- when the river slows down, e.g. entering the sea, at the inside of a *meander*, when it overflows its banks or there is a sudden change in gradient
- reduction in volume during times of *drought*
- when the load is suddenly increased, e.g. *landslide*, entry of large tributary.

Rivers generally deposit the heaviest part of their load first, grading over time and space to the finest.

river erosion: a collection of processes that contribute to the wearing away of the banks and bed of the river and which thus contribute to the *load* for transportation (see also *Hjulström curve*). The main processes are:

- *abrasion* – the use of the load to wear away the banks and bed
- *attrition* – the wearing down of the load
- *hydraulic action* – the sheer force of the water itself. Also included here is *cavitation*, where the collapse of bubbles of air within the water sends out shock waves that weaken the banks
- *solution* – related to the chemical composition of the water and the geological make-up of the river valley.

river regime is the term used to describe the annual variation in *discharge*. Rivers in differing climates experience very different regimes.

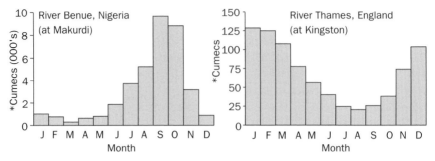

Examples of river regimes

river terrace: a remnant of a former *floodplain*, which after *rejuvenation* of the river has been left at a higher level. The River Thames has created terraces in its lower course, which are now occupied by parts of London.

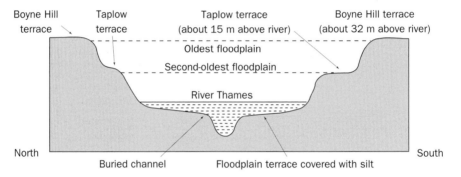

River terraces on the River Thames

robotics is the science of using robots in production processes to replace people, especially where such processes are monotonous or hazardous.

roche moutonnée: a protruding knob of bedrock found along the floor of a glaciated valley. One side of the rock is gently sloped and smooth (the stoss side), whereas the other (the lee side) is steep and uneven. It was formed by the action of a *glacier* which on encountering a more resistant protusion of rock beneath it, scratched and polished the stoss side as it slid over it, and then plucked out rocks loosened by *freeze–thaw* action and *pressure release* from the lee side. (See also *plucking* and *regelation*.)

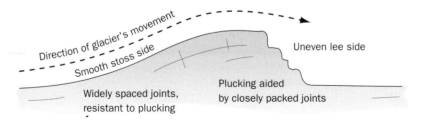

Roche moutonnée

250

rock type: the classification of rocks into three main groups:

• *igneous rocks*
• *sedimentary rocks*
• *metamorphic rocks*.

Rossby waves: the pattern of flow of the winds (known as the Upper Westerlies) which blow in the higher parts of the atmosphere. It is known that in the upper atmosphere winds blow around the planet in a westerly direction, but follow a wave pattern. The waves stretch from *polar* latitudes to tropical latitudes, and there are usually between four and six of them in each hemisphere. The reason for their existence is not clear, but some people believe that they are due to the upper air flow being forced to divert around the great north–south mountain ranges of the Rockies and Andes in the northern and southern hemispheres respectively. Once a wave motion has begun, it is perpetuated around the planet. The waves have considerable variation in amplitude during a year. It is this variation of amplitude, together with their relatively static locations that have a significant effect on the creation of both low pressure and high pressure areas on the surface.

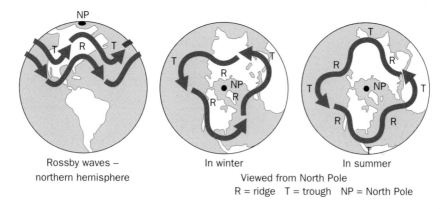

Rossby waves –
northern hemisphere

In winter

In summer

Viewed from North Pole

R = ridge   T = trough   NP = North Pole

Rostow, W. W.: put forward a model of *economic growth* for a country that identifies five stages of development:

1 Traditional society – a subsistence economy based mainly on farming with limited technology or capital to process *raw materials* or to develop industries.
2 Pre-conditions for take-off – an injection of investment which allows both agriculture to be more commercialised, and a single industry to begin to dominate the economy. Technological developments cause a growth in *infrastructure*, including a transport system.
3 Take-off – *manufacturing industries* grow rapidly, with growth being concentrated in one or two parts of the country. Improvements in the transport infrastructure continue, together with some progress in social conditions. Employment in agriculture declines.
4 The drive to maturity – a period of self-sustaining growth. Economic growth spreads to all parts of the country; more industries are developed, and urbanisation takes place on a rapid scale.
5 The age of mass-consumption – a period of rapid expansion of service industries, with a consequent decline in manufacturing.

The key element of this model is the need for an injection of capital, usually from an external source, for example an economically more developed country. This has taken place in many African and Asian countries, with little effect, other than to cause the build-up of huge national debts. The model is now regarded as being too simplistic, as well as being too heavily based on the experiences of Western countries which grew economically in the earlier part of the twentieth century.

rotational movement: a slippage or slide along a curved plane. This type of movement takes place in:
- the **slumping** of **debris** down a slope, producing a tear or scar at the back of the slope, and a lobe of material at the foot of the slope
- the formation of a **cirque** (corrie), where the glacier is thought to pivot as it moves out of the hollow.

running mean: a method by which the overall trend of a set of values in a data set can be determined. It involves the calculation of the **mean** value for, say, the first three consecutive numbers in the set. This is then followed by the calculation of the mean of the second, third and fourth numbers. This is in turn followed by the calculation of the mean of the third, fourth and fifth numbers etc. This example is a three-item running mean, but it may be of any other number such as five. One of the main outcomes of the use of a running mean is that the impact of exceptional values is reduced. However, another is that the final items in the data set will not have a running mean of their own.

3-year running mean

1994	1995	1996	1997	1998	1999	2000	2001	2002	Moving total	Running mean	Middle year
6.1	2.0	2.6	3.3	1.5	5.7	29.2	3.6	0.4			
6.1	2.0	2.6							10.7	3.6	1995
	2.0	2.6	3.3						7.9	2.6	1996
		2.6	3.3	1.5					7.4	2.5	1997
			3.3	1.5	5.7				10.5	3.5	1998
				1.5	5.7	29.2			36.4	21.1	1999
					5.7	29.2	3.6		38.5	12.8	2000
						29.2	3.6	0.4	33.2	11.1	2001

runoff is all of the water which enters a river and flows out of a **drainage basin**. **Through-flow**, **groundwater** flow and **overland flow** all contribute to runoff. It can be quantified by measuring the **discharge** of that river.

rural depopulation: the movement of people out from rural areas and into urban areas. This movement can be attributed to rural **push** and urban **pull factors**. They include:
- rural push factors – small fragmented plots of land which are too small to support a family; crop failures; the use of **mechanisation** reducing the need for **labour** on the land; the clearance of the land by large landowners who wish to redevelop it for commercial purposes
- urban pull factors – the perception of a better lifestyle; the prospect of a job and cash wages; the higher quality of health and education services.

rural development: the encouragement and assistance given to **economic growth** in countryside areas in an attempt to halt **rural depopulation**. In England, the Rural

Development Commission (RDC) is responsible for promoting rural industries and community development. Grant funding is given to small-scale industrial schemes, especially those located in priority Rural Development Areas (RDAs). Similar bodies exist in Wales and Scotland:

- 'Mid Wales Development' (MWD) – promotes the economic and social development of Mid Wales, which includes 40% of the country. MWD offers grants, loans, ready-built factories, and periods of freedom from rent to attract new firms. Village halls and community centres receive grants for improvements, as do theatres and sports centres. MWD assists in the overseas marketing of Welsh hotels, and seeks to improve television reception in the remote valleys
- 'Highlands and Islands Enterprise' – with similar aims in Scotland.

rural landscape: a mental and/or visual picture of countryside scenery. It is almost impossible to define as the landscape is constantly changing, and varies considerably from place to place. The image of a typical English rural landscape is one of rectangular fields, hedgerows, coppices of trees interspersed with farmsteads. However, this is now rare. Meadows have been drained, hedgerows and coppices have been removed, large animal feeding sheds have been erected, and extensive areas of coniferous woodland have been planted. Quiet idyllic villages have become commuter settlements, blurring the distinction between rural and urban.

rural settlement includes isolated farmhouses, hamlets, villages and market towns. However, some of these are difficult to distinguish from urban settlements as townspeople increasingly move out into the countryside.

rural–urban fringe: see *urban fringe*.

A–Z Online

Log on to A–Z Online to search the database of terms, print revision lists and much more. Go to **www.philipallan.co.uk/a-zonline** to get started.

Sahel: the region of Central West Africa which lies between approximately 10°N and 20°N. It includes the countries of Senegal, Mauritania, Mali, Burkina Faso, Niger and Chad. The Sahel is mostly desert margin and has suffered from a number of prolonged *droughts* in recent years.

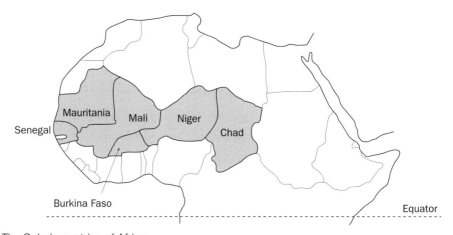

The Sahel countries of Africa

salinisation occurs when *potential evapotranspiration* is greater than *precipitation* and when the *water table* is near to the ground surface. It is therefore a feature of areas with an *arid* or semi-arid climate. As moisture is evaporated from the surface, water containing salts is drawn upwards by *capillary action*. Further *evaporation* causes the *deposition* of the salts on the ground surface. Salinisation has become a major problem in some irrigated areas. Where the irrigation water is unable to drain away, waterlogging has taken place in the soil. The water table has risen up through the soil, bringing salts with it. The roots of plants which cannot tolerate saline conditions become affected and the plants die. The solution to this problem is to provide adequate field drainage together with a constant flow of irrigation water to flush the salts out of the soil.

SALR: see *saturated adiabatic lapse rate*.

saltation: the process by which particles are lifted bodily upwards and forwards before returning back to the surface from which they started. This occurs in:

- desert areas where the wind picks up fine sands and moves them in the direction of the wind. The individual grains bounce along leapfrogging over each other, but rarely reach heights of more than 2 m above the ground
- rivers where the sheer force of the moving water bounces sands, gravels and larger particles along the bed of the river.

In both cases, impact with the returning surface may initiate further particle movements.

salt lakes: are inland enclosed bodies of water that have a high level of salt content in parts per million (ppm). They occur in **arid** and semi-arid climates where the **water table** is very close to the surface. High **evaporation** rates cause substances dissolved in the water to become concentrated. With low **precipitation**, the main supply of water to the lake is from **groundwater** and from tributary streams. Groundwater discharge in these areas is also highly saline as a result of the high **evapotranspiration** rates, which concentrate salts in the ground. The salt content of tributaries depends on the composition of the rocks in the local catchment. The primary mineral constituent of salt lakes is common salt (NaCl).

salt marsh: an area of vegetated tidal mudflats located within an **estuary** or on the landward side of a **sandspit**. Fine sediment accumulates in the sheltered water, which is then colonised by a sequence of salt-tolerant plants as the level of the marsh is raised by further **deposition**. Most salt marshes tend to have a well-developed series of creeks through which the tide rises and ebbs. The sequence of vegetation colonisation which takes place on a salt marsh is an example of the plant **succession** known as a **halosere**.

sample size: the number of respondents or objects in a research survey. The sample size is important as it needs to be large enough to make the data statistically valid. A sample of 20 people, for example, is so small that a different 20 could easily have quite separate views or behave in a completely different manner. When deciding on an appropriate sample size, the following should be borne in mind:
- the higher the sample size, the more expensive the research in terms of cost and time
- the lower the sample size, the greater the chance that **random** factors will make the result inaccurate.

sampling: to make statistically valid inferences, when it is impossible to measure the whole **population**, by selecting a group which will be representative of that population. Such a group is known as the sample. There are three main methods of sampling: **random**, **systematic** and **stratified**.

sandspit: an embankment of sand and shingle (pebbles) which is attached to the land at one end. The unattached end is often curved inwards towards the land, and may have several recurved ridges known as laterals. The highest parts of a sandspit often consist of sand **dunes**. Here sand has blown from the beach, and has been stabilised by vegetation such as **marram grass**. Along the seaward side, ridges of shingle are frequently found, thrown up by the sea during storms. On the landward side, a **salt marsh** may be found where fine silts and mud have been deposited by the mixing of fresh and salt water (**flocculation**). Examples of sandspits include: Blakeney Point in Norfolk and Spurn Head in Humberside.

S

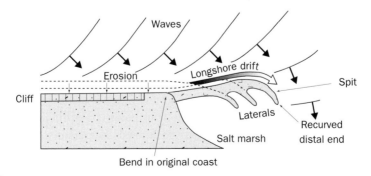

Sandspit

sandur: (plural sandar) the Icelandic term for the *outwash* plain which is to be found in front of the ice margin and consists of *fluvioglacial* sands and gravels crossed by *meltwater* streams.

saprophyte: a group of *organisms*, including some plants, which obtain *nutrients* from dead or dying organisms. Most fungi are saprophytes.

satellite photographs: images that can be transmitted from satellites, particularly useful in the fields of *weather forecasting* and *land use* analysis. This is one example of *remote sensing*. Photographs can be produced using ordinary light but most satellites have sensors that can detect different bands of energy emitted from the Earth. The *infra-red* part of the electromagnetic spectrum is one of the measurements that can be taken and transmitted as a photograph.

saturated adiabatic lapse rate (SALR) refers to the rate at which the temperature of a saturated body of air falls as the air is forced to rise. The temperature fall is caused by heat loss due to the expansion of the body of air. It is lower than the *dry adiabatic lapse rate* because as *condensation* takes place, *latent heat* is released which compensates for some of the heat loss. However, the SALR varies because the warmer the air the more moisture it can contain and so the greater the amount of latent heat released following condensation. It may be as low as 4°C per 1000 m or as high as 9°C per 1000 m. (See also *adiabatic*.)

savanna (tropical grasslands) is the name given to a climatic and/or vegetation type which can be found in parts of tropical sub-Saharan Africa, the Brazilian plateau and northern Australia. The climatic type is also known as the tropical wet and dry, or tropical summer rain.

The main features of the climate are:
- a hot and wet season with temperatures over 26°C, and heavy convectional rainstorms
- a slightly cooler dry season, with temperatures averaging 21°C, and little or no rain
- the length of the dry season decreases polewards, until the desert type climate prevails.

The dry season corresponds to offshore *trade winds* blowing across the area from a dry interior. The wet season occurs when the *inter-tropical convergence zone* moves polewards following the overhead sun. This causes low pressure to develop, and produces strong *convection* and intense rainfall.

The vegetation type is dominated by grassland with both evergreen and *deciduous woodland*. The grass decreases in height polewards with an increasing dry season, and the cover becomes more sparse near to the desert margins with bare ground occurring in between tufts of spiky grass. The trees vary from being evergreen near to the equatorial margin, to being deciduous further away. They have adapted to the drought conditions (*xerophytic*) by having thick barks, small thorny leaves, and extensive root systems. The acacia also has an umbrella-like shape to shade the root area beneath. The baobab stores water within its large trunk. Some people believe that this vegetation type exists because of the actions of humans who have fired the area consistently over thousands of years. The grasses can recover during the wet season, and the trees, such as the acacia and baobab, are fire-resistant (*pyrophytes*).

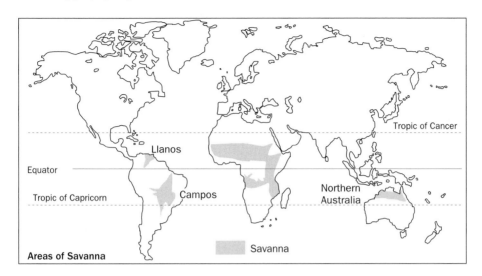

Areas of Savanna

scale: the degree to which a map is smaller than the area that it represents. If the scale of a map is known, a measurement made on the map can be converted to its equivalent on the ground. The scale may be expressed as:

- a representative fraction – for example, the Landranger series of the Ordnance Survey has a representative fraction of 1 : 50,000. One centimetre on the map represents 50,000 cm on the ground or 0.5 km
- a *linear* scale – which on the Landranger series mentioned above would be a 2 cm length on the map for every kilometre on the ground. A linear scale is also usually sub-divided into smaller fractions, such as tenths.

Scale may also be applied to the size of the area of study. It may vary from a local scale, to a national, regional, international or global scale.

scarp: the steep slope of a *cuesta* or *escarpment*. A *spring* is a common feature at the base of a *limestone* or *chalk* scarp.

scatter graph: a graph that shows the relationship between two sets of variables by the distribution of dots. It is usual that the *dependent variable* is placed on the vertical y-axis and the *independent variable* on the horizontal x-axis. Dots are plotted on to the graph using the two sets of data as coordinates. The arrangement of the dots can then be examined to see if there is:

- a positive relationship – as one variable increases than so does the other
- a negative relationship – as one variable increases the other decreases
- no relationship – there is no recognisable pattern to the distribution of dots.

A positive or a negative relationship can be indicated by inserting a *best-fit line*. (See also *regression*.)

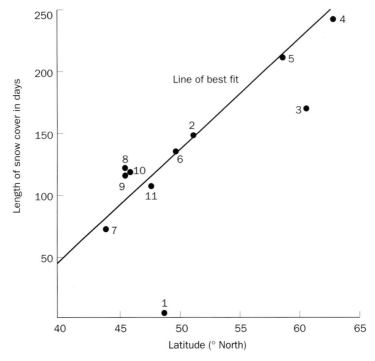

An example of a scatter graph showing the relationship between length of snow cover and latitude for 11 towns in Canada

Schengen agreement: a decision by a number of *EU* countries to remove all border controls such as passport and customs checks on their internal or shared frontiers by 1994. This followed the Schengen Accord, established in 1985, and was actually implemented in March 1995 when seven states, France, Germany, Spain, Portugal, and the Benelux countries abolished border controls. Flights between airports in these countries are classed as 'domestic' enabling business executives and tourists to travel without passports, immigration checks or other border formalities. The countries who are reluctant to join are concerned about the effect on movement of drug smugglers, international criminals, terrorists and illegal immigrants although the participating states accept the need for much tighter 'external' controls. The UK, in particular, was concerned that the agreement would cancel out the advantages of being an island in controlling illegal entry. Italy and Greece have also signed and expect to join and Denmark, Sweden and Finland have expressed interest. Ireland would like to join but has been deterred by the UK's position.

Schumm, S.A.: proposed a method of expressing the particle size of sediment in the perimeter of a river channel as a way of accounting for the nature of the particles making up the bed and banks of the stream. Channels with high percentage values of silt/clay are likely to be in rivers that are relatively narrow and deep and carry a relatively high percentage of their sediment *load* in suspension. Low values of silt/clay are likely to be found in wide and shallow streams carrying most of their sediment as *bedload*.

science park: an *industrial estate* located near to a university or research centre with the intention of encouraging and developing *high-tech industries*. Businessmen and academics can meet to identify and discuss practical applications of new technological research developments. An example is the Cambridge science park, which forms part of the so-called *Cambridge phenomenon*.

sclerophyllous refers to plants which have drought adaptations to reduce the loss of water through *transpiration*. Specifically it applies to leaves which are thick, leathery and have thick cuticles. Such plants are typical of *arid*, semi-arid and scrub environments such as those of *Mediterranean climatic* areas: *maquis*, *garrigue* and *chaparral*.

scree: angular rock fragments which accumulate on a slope below a steep rock face (free face) or summit. The *debris*, the result of physical weathering processes, may be of sufficient extent to produce a scree slope which covers the surface rock. They are common in glaciated mountain areas where *freeze–thaw* acts on steep bare rock outcrops, for example the various large screes above Wastwater in the Lake District.

sea floor spreading is the movement of oceanic crustal plates away from *constructive plate margins* (divergent margins). In the 1960s the mechanism behind the theory of *continental drift* was confirmed by examining the floor of the Atlantic Ocean which revealed a ridge built up by successive eruptions of basaltic volcanic lava, similar in composition to the oceanic *crust*. Evidence showed that the sea floor becomes older with distance from the mid-Atlantic ridge, the point of extrusion along the constructive plate boundary. As new crust is created by the upwelling magma the ocean crust moves away from the boundary. The rocks also reveal evidence of magnetic striping and *palaeomagnetism*. In the Atlantic the rate of spreading is about 1 cm per year. In the last 20 million years the Atlantic has increased in width by over 400 km.

sea level change occurs when the level of the sea in relation to the land varies. Either the land or the sea may change its level. Sea level changes can be classified as being either *eustatic* or *isostatic*:

- eustatic sea level changes apply to the whole world. During times of major glaciation, large volumes of water were stored on the land as ice. As a result sea level worldwide fell by 100–150 m. However, since the great *ice sheets* began to melt, sea level on a world scale has risen as the water has returned to the sea
- isostatic sea level changes concern only certain parts of the world. During the major glaciations, the weight of the ice depressed the Earth's *crust* immediately beneath the ice. This caused a local relative rise in sea level.

Since the ice sheets have melted, the Earth's crust, which had been beneath the ice sheets, has begun to rise. This has caused a local relative fall in sea level. These two types of change often appear to work against each other, although it would appear that isostatic changes are slower to occur then eustatic. Consequently, parts of northwest Scotland and the Gulf of Bothnia in Scandinavia are still experiencing slight falls in sea level some 10,000 years after the last ice sheets disappeared.

secchi disc is a black-and-white coloured disc used in studies of water *pollution*, which is lowered into water until it is no longer visible (this depth is recorded) and raised again until it reappears (this depth is recorded). The average of the two is the secchi depth, which varies with the *algal* build-up and reflects the *trophic* status of the water.

S

secondary data information which is derived from published documentary sources and has been processed, such as processed *census* data, research papers, textbooks etc.

secondary forest: the vegetation which develops following the clearance of the original vegetation and recolonisation through a *secondary succession*. The vegetation may not have reattained the climax stage and secondary forest represents a *seral stage* in the gradual succession back to the climax community. This is found within *tropical rainforest* areas where *shifting cultivators* have cleared the original vegetation.

secondary sector: the sector of the economy which is concerned with the processing of any primary *raw materials* and the resultant manufacture of products. Although this sector supplies a large proportion of the UK's exports it employs less than a quarter of the workforce, a proportion which continues to decrease as sectoral change occurs.

secondary succession: a sequence of plant succession which develops on land that has previously been vegetated. It is characteristic of areas which have been cultivated/grazed and later abandoned, or where forest areas have been cleared by felling or as a result of devastation by fire. Because these sub-seres develop on sites which have already been biologically modified, in that they have some soil and organic matter, the succession is more rapid than in a *prisere*. The pioneer stage may be very short or absent, and the succession tends to be 'telescoped' in time, being particularly rapid on formerly cultivated land where there may be improved soil conditions through drainage, ploughing or the use of organic/inorganic fertilisers. In Britain, birch tends to colonise cleared woodland and after perhaps only 15–20 years can produce large amounts of wind-borne seeds. As it cannot compete with taller shade-casting trees, however, it is gradually replaced by oak or pine as dominants.

Second World: the communist (socialist) countries of Eastern Europe and the former *USSR*. The term was introduced after the Second World War as the Cold War between the capitalist *First World* and the Eastern Bloc developed. It has been of less significance since 1990 with the sub-division of parts of the USSR and its satellite states.

sectoral change: the variation in the relative proportion of the working population in each sector of the economy. As a country begins to develop economically there is a change in the balance of employment in each sector. Initially the *primary sector* dominates with a subsistent economy, but as processing of *raw materials* develops there is an increase in the *secondary sector* and this continues as industry becomes more varied through *cumulative causation* and *linkages*. As the secondary sector increases the primary sector falls. The tertiary sector also starts to rise as opportunities emerge for financial and consumer services. As secondary industry becomes more efficient in its use of *mechanisation* or robotisation, then workers are released and the tertiary sector dominates as both primary and secondary sectors decline. In many economically advanced countries, *deindustrialisation* is leading to further decline in secondary employment and a marked change towards the tertiary and *quaternary sectors*.

sediment: any material ranging in size from fine clays to boulders which has been transported and deposited by an agent of *erosion* such as water, wind, and ice.

sedimentary rock: a rock made up of *sediments* deposited in distinct layers separated by *bedding planes*. They consist of material which is the *debris* produced by the breakdown of other rocks (*igneous rocks*, *metamorphic rocks*) and older sedimentaries. These sediments accumulate in environments such as deep oceans, on the *continental shelf*, and in inland lakes and large river valleys. The overlying weight of sediment causes compaction, cementation and hardening, a process referred to as lithification. Rocks can be subdivided according to their dominant constituents:

- argillaceous – comprise fine material such as clays
- arenaceous – consist of cemented sands
- rudaceous – comprise coarse material e.g. conglomerates
- calcareous – consist of calcium carbonate derived from the remains of shells and corals as in **chalk** and **limestone**.

sedimentation: the **deposition** of **sediment**; the term is usually applied to the accumulation of silts and sands in a water course where the deposition may raise the bed of the channel and pose a flood risk or where sediments infill reservoirs.

sediment cells are sections of the coastline and its nearshore zone within which the movement of coarse sediment (primarily sand and shingle) is largely self-contained. The **longshore drift** of sand and shingle in the nearshore zone has been found to occur in separate sediment cells. There are 11 such cells around the coastline of England and Wales, although smaller sub-cells have been identified within these. Transfer of sediment can take place between these cells. Sections of coast within a cell have similar movements of sediment; as a result, coastal defence systems at one point could affect sediment movement at another. The identification of these cells has an impact on the strategies used for **coastal management**.

sediment sources: there are many different sources for the sediment that is transported and deposited along the coastline. Local **headlands** and cliffs provide material when subjected to wave attack. Deposits of glacial till consisting of clay, sand, gravel and cobbles, both at the coast and inland, are eroded and deposited in the sea. In some areas, volcanic ash and lava are eroded and deposited in coastal areas to create black sand beaches. These geological sources may be referred to as clastic **sediments**. Other sources include the shell fragments and skeletal remains of marine **organisms**; these are biogenic sources.

sediment yield: the total amount of **sediment** being transported by a river over a certain period of time. It is used as a measure of the rate of **erosion** operating in a **drainage basin**. It is calculated by measuring the sediment **load** (for practical purposes this is normally restricted to the **suspended load**) and dividing this by the area of the drainage basin. This is usually expressed as volume per unit area per year, i.e. $m^3 \, km^{-2} \, year^{-1}$ or tonnes $km^{-2} \, year^{-1}$.

It expresses the rate at which material is being removed, or eroded, from the basin. Studies of **erosion** rates in different environments produce conflicting results because it is difficult to allow for the effects of human disturbances, such as cultivation and **deforestation**, which assist erosion. Generally in the humid tropics, in undisturbed areas, the dense vegetation helps to maintain low rates between 15 and 20 $m^3/km^2/yr$, although in disturbed environments very high rates may be found. In the desert regions, rates may be as low as 1–2 $m^3/km^2/yr$. In the temperate and semi-**arid** regions the rates are particularly high, ranging from 15–100 $m^3/km^2/yr$.

segregation is where certain groups of people live apart either because they are forced to do so or for economic and social reasons. Segregation of people can be based on race, wealth or age. (See also **social segregation**.)

seif: a type of desert sand **dune** that is aligned with the **prevailing wind** direction. It is usually asymmetrical in cross-section, and secondary winds, i.e. less frequent and less powerful winds, may help to shape this elongated dune. Some researchers suggest that seif dunes may be the result of migration and coalescence of crescent-shaped **barchan** dunes which become distorted by changing wind directions.

seismic: a term that means 'of an **earthquake**', as in seismic waves, seismic focus and seismology.

self-employment is where workers operate as their own bosses, either working freelance or with the permanent task of running their own businesses. Self-employment has various tax advantages over regular employment especially in claiming expenses that can be offset against income tax bills. The results of the 1991 UK **census** indicated that 13% of the male workforce was self-employed.

self-help scheme is a type of rehabilitation programme implemented in **squatter settlements**, for example in the favellas of São Paulo. Local authorities provide the residents with the materials to improve their existing shelters and may grant the residents rights of ownership. Residents are also encouraged to set up local community schemes to improve medical facilities and education, while local authorities provide basic **infrastructure** in the form of electricity, water and sewerage disposal.

self-sufficiency: where a region/country does not depend on outside help in the supply of a particular commodity. A country, for example, that produced enough grain on its own farms to feed its own people, did not import any, and perhaps even exported some grain, would be said to be self-sufficient in that commodity.

separatism: the attempt by groups within a country to achieve greater **autonomy**, and ideally total independence, from a central government from which they feel alienated. These areas often have distinct languages and culture and are peripheral geographically to the **core** within the nation state. Separatism is common throughout the world, such movements varying from underground, violent and illegal organisations (ETA organisation within the Basque provinces of France and Spain), to peaceful political parties (Scottish National Party/Plaid Cymru, Wales). Examples of such movements include:
- UK – Scottish and Welsh nationalism
- Spain – the Basque provinces and Catalonian nationalism
- France – Breton nationalism
- Canada – Quebec separatist movement
- Russia – Chechen independence (which led to the Russian invasion of 1994–95).

sere is a particular kind of plant **succession**. There are several major types recognised:
- **lithosere** – developed on bare rock
- **psammosere** – developed on a sandy surface
- **halosere** – salt water environment
- **hydrosere** – fresh water environment.

Within the succession each recognisable stage is known as a seral stage.

service sector: see **tertiary sector**.

sesquioxide: chemical compounds, which are common in many soils, resulting from the **weathering** process. They consist of the oxides of the two primary minerals, iron and aluminium.

set-aside: a scheme introduced by the **European Union** in 1992 by which farmers who agreed to take **arable** land out of production would receive compensation. The purpose of the scheme was to reduce over production of arable crops within the EU. The operation of the scheme is controlled by EU legislation. Each year the EU decides on the amount of land that has to be set aside. In the first year of the scheme, farmers had to set-aside a minimum of 15% of their cropped farmland. By 2000 the figure had dropped to 10%. In 2007, following significant rises in grain prices across Europe, the EU decided that for harvest 2008 the set-aside rate would be zero. In 2007 there were approximately 440,000 ha of land in set-aside in UK.

Some environmental schemes are also allowed to count as part of the farmer's set-aside, so converting it to woodland counts towards the requirement. Many farmers have selected set-aside land to provide benefits for wildlife allowing **ecosystems** to develop that are sheltered from the farming areas, for example, strips alongside woodland, blocks adjoining watercourses and larger blocks between crops.

General objections to the scheme have been:
- the rates of compensation are not consistent
- the scheme involved too much bureaucracy with the employment of larger numbers of European civil servants
- many farmers only put their poorest land into set-aside, so the impact on production was less than anticipated.

settlement function refers to the main activities in a place. Function relates to the economic and social development of a centre. Major functions include residential, commercial, agricultural, industrial, *mining*, religious, administrative, tourism/recreational, defensive, market centre, transport centre.

settlement morphology refers to the pattern or shape of settlements. *Morphology* can also be linked with structure where detailed attention is given to the building type, age, construction and function within *land use* zones in the settlement. The pattern or shape made by the *functional zones* within a settlement is known as the functional morphology.

settlement site: see *site*.

settlement situation: see *situation*.

shanty town: see *squatter settlements*.

shape index: a statistical measure for analysing the shape of any geographical area such as a parish, a built-up area or a regional unit. The formula used is:

$$FORMULA: \quad \text{shape index (S)} = \frac{1.27 \times A}{L^2}$$

where A is the area of the unit in square kilometres and L is the long axis of the area drawn as a straight line connecting the two most distance points on the perimeter. The multiplier 1.27 is used so that a circle would produce an index of 1.0 with values ranging down to zero. Maximum compaction occurs when S = 1, i.e. the shape is circular. This is mathematically the most efficient way of enclosing a given area. As an area becomes more elongated the index falls. The single figure, numerical index allows direct comparison of areas of different size and shape.

share cropping: an agricultural system where a farmer has to give part of his crop to the landowner, rather than a fixed rent. It is not unusual for this to be at least 50 to 60%, which gives the farmer little incentive for improvement. It was a widespread practice in the cotton belt of the USA following the civil war and the abolition of slavery.

shield area: see *cratons*.

shifting cultivation is the system of farming that developed in areas of *tropical rainforest* where plots are created from the forest, cultivated and then abandoned. Plots would not then be re-used for many years. It is also known as *slash and burn* and by various local names such as milpa in Latin America, ladang in southeast Asia and chitimene in central Africa (on the wooded **savanna** rather than rainforest). As a farming system, shifting cultivation is very energy-efficient and operates in close harmony with the environment. As the tribes do not work the soil for long periods, **humus** will build up sufficiently for reuse, making this a very sustainable form of **forest management**.

Much of this way of life has been destroyed with the destruction of the rainforest. In the Amazon Basin, for example, land is being cleared for cattle ranches, timber, **hydroelectric power** schemes, highways, reservoirs and other commercial exploitation, particularly the attempts to introduce sedentary cultivators. As a consequence the shifting cultivators are forced deeper into the untouched forest areas or they are forced to live on designated reservations.

shingle is the name given to material of gravel or pebble size that accumulates and forms beaches and offshore **bars**.

shopping centre: a planned development which contains **retail** outlets along with other services such as banking, restaurants and cinemas. Most early centres (1960s and early 1970s) were small and located within town centre redevelopment schemes. Other larger centres developed, such as the Arndale in the centre of Manchester, but only in the mid/late 1980s did the out-of-town location centre begin to appear. This followed the pattern in North America, where huge shopping 'malls' had begun to appear well away from the traditional **central business district** shopping area. The MetroCentre at Gateshead was the first in Britain followed by Meadowhall (Sheffield), Merryhill (Dudley) and Lakeside (Essex). Many city centres have countered this type of development with centres within the traditional CBD, such as Eldon Square (Newcastle) and the Potteries Centre (Stoke-on-Trent). The government in the UK has begun to rethink its policy on this process of **decentralisation** of the retailing function and has announced its intention to discourage further **out-of-town locations** where they 'are likely to make an unacceptable impact on the vitality of traditional town centres'.

shopping patterns: the traditional shopping pattern in the UK was based on a **hierarchy** with the **central business district** or 'high street' at the top, through to district centres and suburban centres, to smaller neighbourhood shopping areas and finally the local corner shop. Two types of shop existed: those selling convenience or low-order goods (newspapers, tobacco, bread) and those selling comparison or **high-order** goods (shoes, clothes, household appliances). The location of these two types of shop was determined by the frequency of visit made by the customer, the accessibility of the shop and the cost of renting the site. This pattern has increasingly broken down in the last thirty years with the emergence of **superstores**, **hypermarkets** and planned **shopping centres**.

shrinking world/shrinkage of space is the consequence of the revolution in transport and communications technologies which reduce the frictional effect of distance on movement and therefore permit space to shrink. The world therefore seems to be smaller in the sense that people, goods and ideas are able to be transferred around the world much faster.

Some of the technological changes which have contributed to this are:
- in transport, the development of giant oil tankers, containerisation and the introduction of roll on/roll off methods have made ocean transfer more efficient and faster. Direct, long-haul flights have brought distant places much closer in term of travel time
- in communications, satellites and high-speed networks have enabled the rapid transmission of live television images as well as personal and business data.

sial: the upper, continental part of the *crust* of the Earth which is composed predominantly of silica and aluminium (from which the term is derived) with a density of 2.6.

significance testing: checking the statistical validity of a result, that is the probability of chance having had an influence on it. This is usually calculated in relation to an objective of 95% certainty that chance has had no effect (5% level). Significance can also be tested to 1% and 0.1% levels.

sill: an igneous intrusion of relatively hard rock which can produce upstanding *scarp* faces or *waterfalls* where the hard layer is crossed by a river. The sill is a sheet of magma which has been injected along a *bedding plane* in sedimentary strata. Examples of features related to sills are the Great Whin Sill in Northern England which is followed in part by Hadrian's Wall and the waterfall at High Force in Teesdale, which is produced by the same intrusion.

sima: the rocks in the Earth's *crust* which consist mainly of silica (Si) and magnesium (Mg). The density of the sima rocks which largely form the ocean floors is between 2.9 and 3.3 and this layer underlies the *sial*.

Simon, J.: an American economist who asserted that increasing world population was compatible with rising living standards, a view that was in direct contrast to those of Thomas *Malthus* and neo-Malthusianism. Simon argued that population increase benefits the world by enlarging the productive workforce and boosting productivity. He thought that shortages of vital resources pose no threat to mankind, and that, far from being a sign of impending disaster, scarcities often led to the development of better resources. Mankind's response to population growth had been, he held, the driving force of civilisation through history. He predicted that while the world would become more populated, it would be more economically stable and less vulnerable to the disruption of resources. People would be richer and their lives less precarious as a result. (See also *Boserup*.)

single European market: the agreement between the *European Union* countries that from 1 January 1993, the trading differences between member countries were to be eliminated so that businesses could treat the whole of the community as their home market. This was intended to be achieved by the abolition of three trading restrictions: *non-tariff barriers*, physical customs controls and different technical standards and taxation levels.

sinuosity: the degree to which a river deviates from a straight line course. The sinuosity ratio is the ratio of the channel length to the straight line distance between the same two points along the centre line of the valley. River courses are commonly sinuous or meandering. (See also *meander*.)

site: the characteristics of the actual point at which a settlement is located and that would have been of major importance in the initial establishment of that settlement and its subsequent growth.

site and services is a process of redevelopment that is implemented in *squatter settlements*. The government or *local authority* provide a site, which may be as basic as a small concrete hut, and essential *infrastructural* services such as water and sewerage facilities. The migrant is given rights of ownership and is then expected to complete the work at his/her own expense. Groups of migrants often undertake this work as a cooperative.

site of special scientific interest (SSSI): sites in the UK designated by the Nature Conservancy Council that have special features which need to be protected. Sites fall into two groups:
- biological SSSIs – the main *habitats* are coastlands, standing and running waters, *peatlands*, lowland grasses, heath and scrub, upland grassland and heath, woodlands
- geological and physiographical SSSIs – they have been selected to represent all aspects of the geology and physiography of the UK.

S

A typical example would be Canford Heath in Dorset, which is one of the largest **heath-lands** in the UK outside of a **nature reserve**, although even here the area is under threat still from the expansion of the suburbs of Poole and Bournemouth.

situation refers to the location of a settlement relative to its surroundings (other settlements, rivers, communication lines and surrounding relief).

skewed result: a **dispersion measure** in which the number of findings is not balanced around the **mean** figure. The findings may be skewed towards values below the mean (**positive skew**) or above the mean (**negative skew**).

slash and burn: see **shifting cultivation**.

sleet is a mixture of rain and snowflakes formed when the temperature higher in the atmosphere is below freezing allowing **snow** to form, this partly melting as it falls through the warmer air beneath.

slope development: this is the result of the interaction of several factors which determine the form of the slope. The major factors are rock structure, **lithology**, climate, soil, vegetation, and the influence of man. All of these factors operate through time, making slope evolution a very complex process. The study of slopes has led to different theories as to how slopes develop through time, the main ones being:
- slope decline – slopes going from the steep to the more gentle
- slope replacement – slope goes backwards but basal slope becomes gentler
- parallel (slope) retreat – the slope retreats but maintains the same steepness.

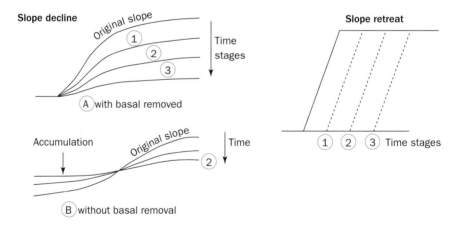

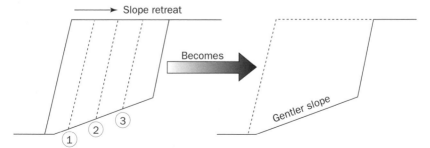

Slope development

slope element: the facets of a slope that go together to give the slope its general character.

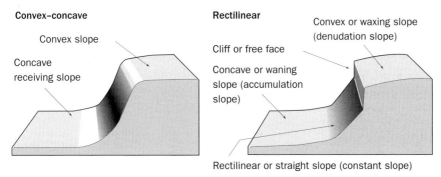

Two slope element models

slumping: a form of *mass movement* on slopes that involves both sliding and flow. Weathered materials on the slope accumulate and on reaching a critical mass, move downwards in a general curvilinear plane producing a rotational movement. Some lubrication by water is evident before the movement can take place unlike a *landslide* which in general tends to be a dry movement.

smog: a mixture of smoke and *fog* produced through the emission by factories of smoke and chemical pollutants added to the output of domestic coalfires. This was a very common occurrence in British cities in the nineteenth and early twentieth centuries, many of the smogs being so thick that the term 'pea soup' was applied to them. In December 1952, a smog in London lasted for several days and was responsible, so it was claimed, for over 4000 deaths. The government then brought in legislation in the *Clean Air Act of 1956*.

More recently, there has been the increase of *photochemical smog*, for which the emissions of car exhausts are seen to be the major culprit. The city of Los Angeles has experienced this phenomenon for a number of years, but photochemical smog occurs over most industrialised and motorised cities when *anticyclonic* conditions prevail and trap the pollutants. Athens is often quoted as the worst city in Europe, but conditions can be equally as bad in British cities, particularly London. This has led to warnings accompanying the *weather forecast* on radio and television.

snow: ice crystals aggregating to form a larger unit, a snowflake, The original ice crystals form under the same conditions as rain, except that as the temperatures are below zero, the vapour forms straight away into a solid. The largest falls of snow occur when air temperatures are just below zero. When it is very cold, the air holds very little water vapour, thus the old saying that 'it is too cold to snow' has some basis of truth.

social chapter is that part of the social charter which was agreed by all twelve member states of the *European Union*, apart from the UK, under the Treaty of Maastrict (1991).

social charter is a set of statutory and non-statutory measures designed to harmonise social legislation within the *European Union*, alongside economic measures. It covers twelve

S

areas: freedom of movement, social protection, vocational training, health and safety, elderly persons, protection of children and adolescents, disabled persons, sexual equality, living and working conditions, employment and remuneration, collective bargaining and the right to strike, information, consultation and participation.

social elite: the social class which has the best living conditions in a population. It may be identified by ascertaining the occupation of the head of the household and/or the level of income of that person. The highest grade band of occupation of head of household is top professional, for example lawyers or directors of large companies. Such people usually have lavish lifestyles and purchasing habits.

social mobility is the movement of individuals from one level in the social *hierarchy* to another. This is often associated with the idea of 'status' or 'class' and the way in which people are perceived by society. People may be upwardly mobile or downwardly mobile; movement reflects factors such as educational attainment, occupation and marriage.

social polarisation is an extreme form of *social segregation* in which different social groups are becoming more socially and geographically distinct. Groups are becoming more polarised in that the level of *inequality* between groups is widening particularly in response to the increasing level of *deprivation* among the poor.

social provision refers to the basic needs of housing, education, sanitation and healthcare that a society ought to provide for its inhabitants. It may also refer to other general needs such as community facilities (e.g. libraries, swimming pools and shopping facilities) and environmental services (e.g. refuse collection and pollution control). The nature of, and demand for, social provision may vary in different areas, and at different times. Demographic change in an area will necessitate changes in the social provision that will be needed. For example, a longer *life expectancy* will result in increased provision of residential care facilities, and greater demands being placed on the healthcare facilities, both locally at day care centres, and more widely in terms of geriatric nursing.

social segregation is the clustering together of people with similar characteristics into separate residential areas in an area. The characteristics may include ethnic and cultural origin, age group, income group, educational background or any other aspect of social class. It is often illustrated by the nature of the housing in an area. Examples could include private housing in comparison with public council housing and semi-detached and detached housing in comparison with flats and bedsit properties. In recent years, social segregation has become more complex with the process of *gentrification*. Affluent social groups have moved into areas associated with lower income groups and have upgraded the quality of the housing stock. Sometimes this has occurred according to a fashion, whereas in other cases it has happened as part of a *regeneration* policy, as in the *London Docklands*.

social welfare refers to the basic conditions in a society and its *well-being*. It reflects the quality of provision of facilities such as housing, education, healthcare, food supply and employment, factors influencing the *quality of life*. It can more specifically be regarded as the systems established by local and national government, as well as by *non-governmental organisations*/voluntary organisations, to support disadvantaged members of society.

soil acidity: the measure of the concentration of hydrogen ions within a soil. It is measured on the *pH* scale. Most British soils are slightly acidic with pH values of 5.5 to 6.5, a value of 7 being neutral. Soils in upland areas tend to be more acidic as the heavier rainfall leaches out the alkaline substances. If soils are highly acidic then iron and aluminium compounds become mobilised which can poison plants and *organisms*.

soil conservation: the maintenance of the ability of a soil to provide an optimum growth of plants. Soil conservation techniques exist to protect a soil from *erosion*, and to maintain or enhance soil *fertility*. They include:
- the addition of animal and/or chemical *fertiliser* to a soil
- the rotation of crops to ensure that nutrient-demanding crops, for example cereals, are alternated each year with soil-restoring crops, for example ley grass or peas
- *afforestation* on steep slopes to prevent soil from being washed away
- terracing on steep slopes and contour ploughing on more gentle slopes. By these methods water is given more time to soak into the ground
- the construction of lines of stones across water courses in dry areas so that the limited supplies of water can be trapped
- reduced sizes of herds in semi-arid areas to reduce *overgrazing*.

soil erosion: the washing away or blowing away of topsoil such that the *fertility* of the remaining soil is greatly reduced. Soil erosion is most rapid in those areas where there is the misuse of the land by people. Some activities which cause soil erosion are:
- the removal of vegetation by either chopping down trees or *overgrazing* by animals. In both cases the soil is exposed to the wind and the rain
- the overcultivation of the soil by growing the same crop in the same field year after year (*monoculture*), thus weakening its structure
- the compaction of the soil by the use of heavy machinery. This reduces the rate of *infiltration* into the soil; water flows across its surface and therefore erodes it
- the ploughing of land at right-angles to the direction of the slope. This encourages *rilling* to take place.

soil fertility: see *fertility (soil)*.

soil moisture graph: illustrates the relationship between *precipitation* and *potential evapotranspiration* for a place over the period of a year. The figure below shows a typical soil moisture graph for a location in West Africa which experiences a tropical wet/dry climate. Precipitation is greater than potential evapotranspiration between June and September, whereas potential evapotranspiration is greater than precipitation between October and May.
- When precipitation is greater than potential evapotranspiration, at first there is some refilling of water into the pores within the dry soil. This is soil moisture recharge. When

S

the soil is saturated, excess water will have difficulty infiltrating into the ground, and may flow over the surface. This is soil moisture surplus.

- When potential evapotranspiration is greater than precipitation, water is at first evaporated from the ground surface and transpired from plants. Water may also be brought up to the surface through **capillary action** and then evaporated. This is soil moisture utilisation. However, eventually the soil will dry out completely, creating a soil moisture deficit.

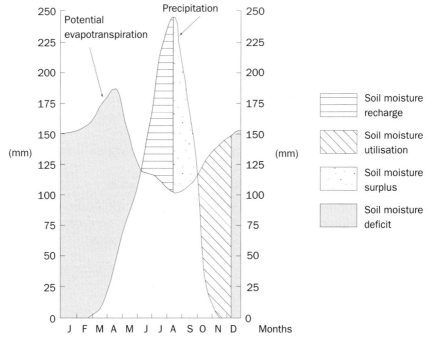

Soil moisture graph for an area with a tropical wet/dry climate in the northern hemisphere

soil profile: shows the variations which occur in the characteristics of a soil vertically down from the ground surface to the underlying **parent material**. A well-developed soil will tend to have a series of horizontal layers within the profile, called horizons, which can be distinguished by their colour, texture or chemical composition.

The main soil horizons which occur in most soil profiles are designated by letters:
- O – the uppermost organic horizon comprising of decomposing organic matter
- A – the topsoil, containing both mineral matter and organic matter
- E – the eluviated horizon, which is having nutrients and other substances washed out of it by downward percolating water
- B – the subsoil, generally illuviated, having nutrients and other substances washed in by downward percolating water.

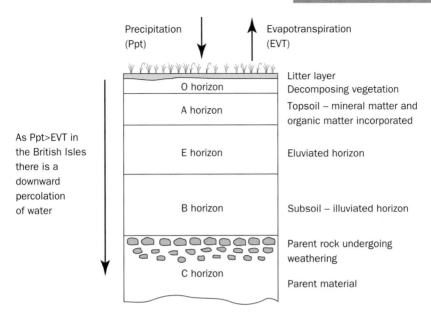

The soil profile of lowland British Isles

soil structure: the manner in which individual particles of soil aggregate together. These aggregates are called *peds*, and they are stuck together by organic matter, and the secretions and mucilages from soil fauna. It is the shape and alignment of the peds which determine the size and number of pore spaces through which water, air and roots can penetrate. They also, therefore, influence the agricultural value of the soil.

There are several types of soil structure, which are not always easy to see in the field. The main ones are:

- crumb peds – small clumps of soil similar to breadcrumbs. They are the most productive, as they are well aerated and well drained
- blocky peds – large brick-like shapes, often enriched with clay
- platy peds – a layered plate-like pattern to the soil, through which water has difficulty draining. It may be created by the use of heavy machinery on a soil which compacts it.

soil survey techniques are used for the collection of data concerning the varying soil types in an area. The following suggestions may be applicable:

- the use of maps – topographic, soil and geological – in order to identify the possible locations of varying soil types
- the use of appropriate equipment – spade, trowel, auger, metre rule, and a camera
- obtaining permission to go on to a landowner's land
- the digging of a pit, or the drilling of an auger, or the use of exposed sections in cuttings or *landslips*
- the description of the soil in terms of its profile and constituent horizons; the varying textures, structures, colours, acidity levels, organic content, stoniness, *land use*, vegetation type etc.
- the returning of the location to its original state – filling in a pit, replacing turf, and expressing thanks to the landowner.

soil texture is the composition of a soil in terms of the varying proportions of different sized mineral particles. Soils consist of particles which are either clay (less than 0.002 mm diameter), silt (between 0.002 and 0.02 mm diameter) or sand (over 0.02 mm diameter). Soils which have a high proportion of sand particles have a sandy texture, and are free-draining and easy to work. Soils with a high proportion of clay particles have a clay texture, and retain water easily. Such soils are heavy, often waterlogged and difficult to work. An ideal soil texture is a loam, which has an even distribution of the three different sized particles.

solar constant: the amount of *solar energy (insolation)* received per unit area, per unit time on a surface at right-angles to the sun's beam at the edge of the Earth's atmosphere. Despite its name it does vary slightly according to sunspot activity, but this is unlikely to influence daily or yearly weather, although it may influence long-term global *climatic change*.

solar energy: see *insolation* and *energy budget (the Earth)*.

solifluction: the movement of soil downslope in *periglacial* areas caused by the summer melting of the surface layer (or *active layer*). During the summer the soil above the *permafrost* melts but the water is unable to drain through the soil as it is still frozen underneath. On slopes as low as 2°, this saturated layer can start to move as a *mudflow*. The end product is often a lobe of material at the base of the slope which may combine with other lobes to give a terrace-like effect. Fossil solifluction deposits are widespread in Britain, providing evidence of colder periods in the past. (See also *head*.)

solution: the removal of dissolved minerals and weathered products by rainfall and percolating *groundwater*.

solution load: the part of the stream's *load* which is dissolved in the water. Almost all rocks are soluble to some extent, and when streams cross over areas which are susceptible to chemical solution, minerals are dissolved. This is most marked on *calcareous soils* such as *limestone* where carbonic acid dissolves calcium carbonate.

South refers to the less developed countries of the tropics and sub-tropics as identified by the *Brandt Report* (1980) and as such is often regarded as being synonymous with the *Third World*. It includes the areas of South and Central America, Africa and most of south and southeast Asia. The South faces particular problems in relation to food supply, the need to reduce poverty and the need to establish production systems through investment in development of resources.

spatial displacement: the movement of an activity from one point in the landscape to another. For example, in relation to crime if one area develops strategies to combat crime such as installing alarms, starting a *neighbourhood watch* scheme or obtains more regular police patrols, then crime may well be displaced to another part of the urban area which has not yet developed these responses.

spatial margins to profitability: the boundaries of the area in which a firm can operate and make a profit. This theory takes into account both revenue and costs. Revenue depends on demand, which in turn will be influenced by population density and level of *disposable income*. Costs include production costs, made up of wages, rents, capital inputs and transport costs. The spatial margins are defined by the intersections of the space-cost curves and the space revenue curves. In the example shown, the revenue is constant across the landscape while costs vary. The optimum location is the point of maximum profit, and in this case is also the least cost point. Ma and Mb indicate where profit gives way to loss on this landscape and they are the margins to profitability. If revenue also varies across the landscape then the point of maximum profit will not coincide with the point of least costs.

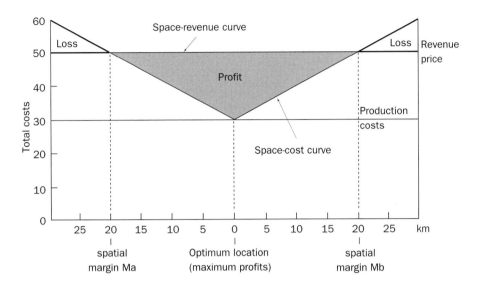

Spearman Rank correlation: a process which produces a numerical value to summarise the relationship between two sets of data. It is based on the ranks of the individual values of the two variables rather than on the values themselves. The process can be illustrated by using the example of the *correlation* between distance from the sea and annual temperature range for 16 weather stations across North America (see next page):

• the individual values of the two sets of data are arranged in rank order. The highest value of each variable is given the rank 1 and successive lower values are given ranks 2, 3, 4 and so on

• where two or more of the values are of equal rank, each of the values has been given the average of the ranks which would otherwise have been allocated

• the difference between the two sets of ranks (d) is then calculated for each pair of variables for each station, and this value (d) is then squared (d^2).

• the value of d^2 is then placed into the formula given, and the correlation coefficient is calculated

• the answer for such a calculation will always be between 0 and ±1. A positive value indicates a direct relationship between the two variables, and a negative value indicates an inverse relationship. The nearer to +1 or −1 the stronger the relationship

• the value of the correlation coefficient, however, can only be ascertained by testing its significance. (See also *significance testing*.)

Station	Temperature range (°C)	Rank	Distance from sea (km)	Rank	d	d²
1	12.5	16.0	10	15.5	0.5	0.25
2	15.0	15.0	110	13.0	2.0	4.00
3	23.0	12.5	400	9.0	3.5	12.25
4	24.0	10.5	520	7.0	3.5	12.25
5	24.0	10.5	630	6.0	4.5	20.25
6	27.0	9.0	980	4.0	5.0	25.00
7	35.0	3.0	1290	1.0	2.0	4.00
8	38.0	1.0	850	5.0	4.0	16.00
9	32.0	4.5	1230	2.0	2.5	6.25
10	28.0	8.0	1140	3.0	5.0	25.00
11	36.0	2.0	250	11.5	9.5	90.25
12	32.0	4.5	490	8.0	3.5	12.25
13	31.0	6.0	350	10.0	4.0	16.00
14	29.0	7.0	250	11.5	4.5	20.25
15	23.0	12.5	100	14.0	1.5	2.25
16	22.0	14.0	10	15.5	1.5	2.25
					Σd^2	268.50

Spearman Rank correlation coefficient Rs $= 1 - \dfrac{6 \times \Sigma d^2}{n^3 - n}$

where n is the number of pairs of items in the sample

$$Rs = 1 - \frac{6 \times 268.5}{16^3 - 16}$$
$$= 1 - \frac{1611}{4080}$$
$$= 1 - 0.39$$
$$= + 0.61$$

An example of a Spearman Rank correlation

specialisation: the division of a work process into separate job functions so that individuals or groups can develop expertise by specialising. Thus people become skilful and productive in particular occupations, and the wider community will benefit from this. At an individual scale, this is being called into question by the Japanese approach of multi-skilling. On a national scale it is still the case that countries specialise in certain agricultural or industrial goods.

However, as the **transfer of technology** takes place on an increasing scale, this too will begin to weaken.

species: this term relates to biological populations in which the individual members can inter-breed with each other, but not with other species. Different trees, plants and animals form species.

sphere of influence: the area around a settlement which comes under its economic, social and political influence. This sphere of influence can be of different sizes for each of the different functions provided by the settlement. Similarly, the sphere of influence will tend to be greater for larger settlements than for smaller ones.

It can be identified:
- theoretically, by the break-point theory of **Reilly**
- by **fieldwork** methods such as:
  - **questionnaire surveys** – asking people in an urban centre where they have come from to obtain a service or to shop or work or asking people in rural areas where they go to for similar services
  - establishing the areas served by certain functions such as delivery areas for electrical or furniture stores; secondary school catchment areas; market areas for estate agents, local newspapers and hospitals; or the areas served by local transport providers.

spit: see *sandspit*.

spontaneous settlement: see *squatter settlements*.

spread effect: the transmission of capital, resources and people from the **core** area to the **periphery**. The term which was coined by **Myrdal** was applied by **Friedmann** in his core-periphery model. As a core area continues to attract wealth, population and **economic growth**, it reaches a point where diseconomies of scale may set in; congestion causes an increase in transport costs, the price of land rises which makes expansion costly, pollution costs may increase; faced with these rising costs organisations may decide to **decentralise** to the periphery. This will develop a spread effect.

spring: the emergence of underground water at the ground surface. It may take a variety of forms:
- a general seepage resulting in a patch of damp ground
- a weak localised flow of water forming a small stream
- the emergence of a river of some size, called a resurgence.

A spring occurs where a **permeable** rock, for example **limestone**, overlies an **impermeable** rock, for example clay. Water passing through the permeable rock cannot penetrate downwards any further, and so has to reemerge. A spring may also appear where the **water table** reaches the surface.

spring-line: a series of **springs** located along a valley side, usually at the same height. In the case of a **chalk** or **limestone cuesta** (**escarpment**), a spring-line will occur on both the **scarp** and dip slopes. Settlements frequently occur at these springs giving rise to a series of villages spaced regularly along the foot of the slope – spring-line settlements.

spring tide: an exceptionally high tide followed by an exceptionally low tide which occurs shortly after each of the new and full moons of each month. Consequently, there are two spring tides each lunar month and they are approximately 14 days apart. They are caused by the moon and sun being in alignment and so exerting a greater gravitational pull.

squatter settlements are established by people who have occupied land illegally and are using it to build their homes which are constructed from anything that the squatters can obtain. They are found in and around every major urban centre in the ***Third World*** and it is not unusual to find that as much as half of the city's population live within such a settlement. Different names are used around the world:

- favelas (Brazil)
- barriadas (Latin America)
- bidonville (north Africa)
- bustees (India).

Squatter settlements take many forms and are found on all types of land. In Hong Kong, the lack of land for building has led to the squatters living on junks (boats) in the harbour and other waterways. They are the result of rapid ***urbanisation*** – the large-scale movement of people from rural areas of a country to the cities. Squatters are not 'passive' in their use of the urban environment, but are actively seeking to better themselves, either to make improvements onsite or to move to a new area in order to upgrade their quality of life. Many of the residents of such settlements hold jobs within the city, particularly in the city centre. In Lima (Peru), for example, to emphasise the positive aspects of many of the newer squatter settlements, the term 'pueblos jovenes' (young towns) has been used for them.

SSSI: see *site of special scientific interest*.

stabilisation: the methods by which fragile and unstable coastal areas are protected from ***erosion***. In the case of ***dunes***, the sand may be stabilised by barriers and fences which act as windbreaks. However, the most common method is the use of vegetation such as ***marram grass*** and the planting of trees as shelter belts. Another solution is to remove the cause of the destabilisation, such as burrowing rabbits, or footpath erosion. In tropical areas, the growth of ***mangroves*** provides an effective protection against storm waves. Their dense root network collects sediment and thereby acts as a stabilising mechanism.

stability: the condition whereby a body of air which if forced to rise, for example, over a mountain range, will return to its original position. Throughout the enforced rise, the body of air is always cooler than the air immediately surrounding it. In terms of adiabatic lapse rates, stability occurs when both the ***dry adiabatic*** and the ***saturated adiabatic lapse rates*** are greater than the ***environmental lapse rate***. Stability may give rise to small cumulus-type ***clouds***, which do not produce ***precipitation***.

stack: an isolated pinnacle of rock standing some distance from a cliff coastline. It is a product of the ***coastal processes*** that at first create a cave, then an arch, the roof of which collapses to leave a stack. An example of a stack is the old Man of Hoy in the Orkney Islands.

stadial: a stage within the general cycle of glacial advance and retreat. Glacial and ***interglacial*** phases are made up of two or more sub-stages or phases. Each glacial phase has been interrupted by short interglacial phases, or interstadials, during which the ice fronts recede. Similarly, interglacial periods were interrupted by colder phases, stadials, during which ***ice sheets*** formed or advanced. These represent shorter term oscillations of ice advance and retreat in comparison to the glacial and interglacial stages. (See also ***interstadial***.)

stakeholder is an individual or a group of people who influence or can be influenced by the actions, decisions and operations of an organisation. Originally a legal term (someone who held money on behalf of others) it came into more common usage under New Labour in the late 1990s with the 'stakeholder economy' – an economy where each member of society has an interest in the state's economic progress. It could be applied to those who have an interest in a development or an issue; those who may influence a decision and be influenced by a decision. For example, in relation to the development of a new retail facility, stakeholders could include potential customers, employees, owners and shareholders, the local community (residents and those unemployed), government, both local and national (planners, potential tax yield), construction companies, environmental pressure groups, product suppliers and financial organisations.

standard deviation is a measure of the distribution of data about a *mean* value. It describes the dispersion of data on either side of a mean value. A low standard deviation indicates that the data set is clustered around the mean value, whereas a high standard deviation indicates that the data is widely spread with significantly higher and lower figures than the mean.

standard error: the potential difference between the *mean* value calculated from a sample of a data set, and the mean value of the total population of the data set. By definition all samples must have a degree of inaccuracy built into them. This inaccuracy is quantified by the standard error. It is usually worked out by calculating the *standard deviation* of the data and then dividing that by the square root of the number of the sample. The standard error can then be used to identify the *confidence levels* of the sample mean (see figure overleaf).

standardisation is the production or use of products or components that are so identical as to allow them to be fully interchangeable. The achievement of standardised parts is a necessary condition for efficient *mass production* to take place. It is also a key element in *economies of scale*, as modern firms use the same components in different products to enable longer production runs to take place and save on design costs.

staple food crop is the produce which farmers mainly grow and on which the local population depends for a major part of their *diet*. In southeast Asia, for example, rice is the staple food crop in many areas.

state industries: in *centrally planned economies (command economies)* the state owns and directs almost all of the industrial development. In China, for example, from 1950 until the early 1980s, the government dictated the site of each factory, the levels of employment, types and methods of production, wage levels, markets and prices within those markets. In such states, individual *entrepreneurs* were not allowed. In economies such as that of the UK, the state has, at various times in the twentieth century, owned and run certain industries which it has taken over (*nationalisation*). In recent years, though, the trend has been to return such industries to the *private sector* (*privatisation*).

state intervention: see *interventionist policies*.

statistical population: see *population*.

stemflow is the water which runs down the stems and branches of plants and trees during and after a rainstorm in order that it can reach the ground surface. It therefore takes place after *interception* has occurred.

Ward	Unemployed %	$x - \bar{x}$	$(x - \bar{x})^2$
1	37.2	16.1	259.21
2	36.5	15.4	237.16
3	21.1	0	0
4	33.9	12.8	163.84
5	28.2	7.1	50.41
6	24.6	3.5	12.25
7	12.0	9.1	82.81
8	19.0	2.1	4.41
9	17.4	3.7	13.69
10	22.0	0.9	0.81
11	17.8	3.3	10.89
12	15.0	6.1	37.21
13	7.3	13.8	190.44
14	9.1	12.0	144.00
15	15.6	5.5	30.25
Σx =	316.70	$\Sigma(x-\bar{x})^2$ =	1237.38
$\bar{x}$ =	21.11	$\Sigma\left(\dfrac{(x-\bar{x})^2}{n}\right)$ =	82.49
		$\sigma = \sqrt{\left(\dfrac{\Sigma(x-\bar{x})^2}{n}\right)}$ =	9.08

Mean	= 21.11
Median	= 19.0
Standard deviation (σ)	= 9.08
Standard error (σ)	$= \dfrac{9.08}{\sqrt{15}}$
	= 2.34

Calculation of mean, median, standard deviation and standard error – an exercise based on unemployment levels in different wards of a town

stewardship: the benevolent management of the environment. It involves the careful control of developments on a wide scale, and seeks to maintain the **sustainable development** of resources. It also involves the identification of priorities for the environment, and the evolution of strategies to achieve them. Stewardship is a concept which has to be undertaken on a macro-scale, across international borders and on a global scale, and herein may be its main weakness.

stone circles are associated with **periglacial** areas and consist of boulders arranged in irregular and interconnected polygonal shapes. They are usually less than 10 m across, and fine sands and silts can be found at their centre. Their formation is due to the creation of an **ice lens**

beneath the surface of the ground. The thickening of the ice at the centre of the ice lens causes the ground to heave upwards. Stones resting on the ground surface are rolled sideways by the heaving ground. On slopes, the circles become elongated because the stones tend to move down the slope. Eventually the polygons become so elongated that stone stripes are created.

storm beach: the highest part of a sand or shingle beach. It is usually above the influence of the high tide having been created during periods of strong winds and large waves. Seaward of the storm beach there may be a series of smaller ridges called *berms*.

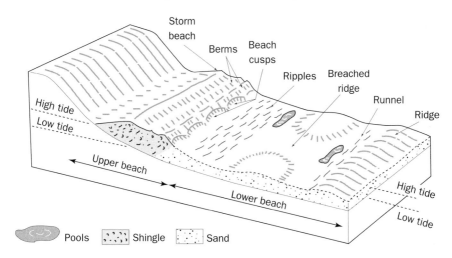

Storm beach

storm surge: a rapid rise in sea level in which water is piled up against a coastline to a level far in excess of the normal conditions under high tides. They occur in those parts of the world where, under certain conditions, water can be funnelled towards a narrow area of sea.

In the case of the storm surge which affected the North Sea coast in 1953, the following events combined to create the crisis:
- an extreme area of low pressure over the North Sea which caused sea level to rise
- strong northerly winds blowing down the North Sea producing waves 6 m in height
- high *spring tides*
- rivers already at flood levels flowing into the North Sea.

Over 2000 people were killed by the surge in eastern England and the Netherlands, and this has prompted a number of schemes to prevent such devastation happening again. Similar surges in the Bay of Bengal have resulted in thousands of deaths in Bangladesh.

storm water disposal: the deliberate removal of flood water from an urban area through a system of under-surface drainage designed to discharge water as quickly and efficiently as possible. One of the consequences of rapid removal of water into neighbouring stream channels is a 'flashy' *hydrograph* with a steep rising limb, a high peak and a shortened *lag time*. Channels fed by urban catchments respond quickly to rainfall input; this can cause *flooding* problems down river of the urban area.

Stouffer, S. A.: proposed *intervening opportunity theory* in 1940.

Strahler, A. N.: devised a method of **stream ordering system** as a means of investigating relationships between geometric properties of a drainage network such as the number and length of streams, drainage density and drainage area.

stratified sampling: samples are selected according to some known background characteristic in the **statistical population**. In studying the distribution of **land use** types in relation to geology, a stratified sample would select points in proportion to the area covered by each type of geology. If 30% of the area was clay, 25% sandstone, 30% **chalk** and 15% **alluvium** then for a total sample of 200 points, 60 would be selected on clay, 50 on sandstone, 60 on chalk and 30 on alluvium. In this way the sample is stratified according to a known factor. In investigating the opinions of the local population on some development, the sample should reflect the various interest groups fairly, i.e. it should account for the sex-age distribution in the population.

stratosphere: a layer in the **atmosphere** which extends from the **tropopause** up to about 50 km above the Earth's surface. In this layer temperatures generally increase gradually with height. The layer is extremely dry with no **clouds** or weather extending above the tropopause although it does contain the **ozone layer** which is concentrated at a height of 25 km (90% of ozone is found below 35 km). The absorption of ultraviolet radiation leads to a warming of the stratosphere with maximum temperatures occurring at the 50 km level.

stratus: a type of **cloud** which is layered and often unbroken. It develops mainly in the lower levels of the **atmosphere** below 2500 m and it forms under stable conditions to produce low grey cloud. This tends to give drizzle but little rain. Higher level stratus is termed alto-stratus.

stream ordering system: a method of classifying the parts of a stream network. A number of systems have been proposed by Horton, **Strahler** and Shreve. Strahler's method is most commonly used. In this the smallest streams, which have no tributaries, are termed first order streams. Where two first order streams join, they produce a second order stream. Where two second order streams join, they produce a third order stream etc. The order only changes when two streams of the same order join, i.e. if a second order stream is joined by a tributary first order stream, it remains a second order stream. Once the order of streams in the network has been designated the basin can be analysed. The number of streams in each order could be counted and a graph drawn to show the relationship between stream order and frequency. This indicates the **bifurcation ratio**. An analysis of this kind allows objective comparisons to be made between **drainage basins** of different size.

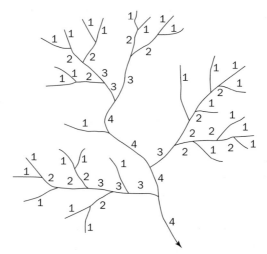

Stream ordering (after Strahler)

structural unemployment occurs when there is a change in demand or technology which causes long-term unemployment. Very often this occurs in particular regions which have been heavily dependent on certain industries such as coal *mining* in South Wales and shipbuilding in northeast England. There is little that governments can do to alleviate such unemployment other than offering retraining for those made redundant.

sub-aerial processes are physical processes – weathering, *erosion*, transport and *deposition* – taking place on the surface of the Earth (literally at the base of the atmosphere). This includes processes such as *freeze–thaw, thermal expansion*, *chemical weathering* as well as forms of surface *debris* movement such as sheet flow, *solifluction*, *mass movement* and *scree* development. These processes operate, for example, on valley sides above the level of river and glacial action and on cliff faces and *headlands*. As such, the term usually excludes processes that are directly linked to the agents of marine, *fluvial* or glacial action.

sub-assembly means putting together a component that will later be part of the final assembly of the finished product. Therefore, on a car *production line*, the assembly of the gearbox would be termed a sub-assembly.

subcontracting: finding a supplier to manufacture part or all of your product. The main circumstances in which this would be used are:
- when a firm is already operating at maximum capacity and therefore cannot meet further demand in any other way
- when there are elements of a product that the firm is ill-equipped to manufacture efficiently
- to cope with seasonal peaks in demand.

subduction occurs when an oceanic *crust* sinks below either an oceanic or a continental plate along a *destructive plate margin* (convergent margin). If subduction is beneath oceanic crust, then *ocean trenches* and *island arcs* are formed; if it is beneath continental crust, then a line of fold mountains, with associated volcanic activity, will form at the edge of the overriding plate. If subduction is so extreme that it closes an ocean so that two continental plates collide, then complex mountain building occurs, as in the Himalayas. As the lithosphere sinks beneath the overriding plate it is colder and more brittle than the *asthenosphere* and this produces stresses. The sudden failure of the lithosphere triggers *earthquakes* which are concentrated along the line of the subducting plate in a zone called the *Benioff Zone*. When the lithosphere has subducted to depths between 80 and 100 km, and temperatures of 1400°C, the rocks begin to soften and melt, and this less dense material migrates towards the Earth's surface as plutons which produce *intrusive* and *extrusive magma*.

subglacial refers to features that lie at the base of, or beneath, a *glacier* or *ice sheet*. Subglacial streams flow in tunnels at the base of the ice and the material which is deposited along their channels is let down during *deglaciation* to produce *eskers*.

sublimation involves a direct change of state from a solid to a gas without passing through the liquid stage.

submergent features are produced along a coastline which has experienced a relative rise of sea level as in the *post-glacial* period. On upland coasts features such as *rias* and *fjords* are formed when former valleys are inundated while in lowland areas broad shallow estuaries are produced. A rise in sea level also causes *aggradation* of the lower river valley which produces an extensive area of easily flooded marsh and mudflats.

subsidence: generally means sinking to a lower level. When applied to land movements it often occurs when underground *mining* has weakened the subsurface structure which results in the surface layer sinking. In meteorological terms, air which descends is subsiding. In *anticyclonic systems* subsidence leads to a warming and drying of the air; stable air tends to subside and bring calmer weather. The subsidence of cold air from valley sides into the valley floor can create a *temperature inversion*.

subsidy: a type of government intervention which takes the form of financial grants given to an industry or group of people to encourage production of a particular commodity, or to generate greater output. It may be directed towards consumers in terms of reducing food prices. Subsidies are often provided in agriculture to raise incomes, e.g. for hill sheep farmers, or through the *Common Agricultural Policy (CAP)* to promote production of crops, e.g. rape seed.

subsistence: a type of agriculture in which the produce is consumed mainly by the farmer and the family who work the land or tend *livestock*. There is little surplus to be sold or traded. Farming may be intensive or extensive in operation and may focus on crop or livestock production. Crop systems include *shifting cultivation*, rice production (southeast Asia) and bush fallowing. Subsistence livestock farming is basically *pastoral* nomadism.

suburbanisation is the outward growth of urban development to engulf surrounding villages and rural areas. As cities increased in population in the nineteenth century and early twentieth century, the physical expansion of the urban area also occurred and neighbouring rural areas were taken over by this outward expansion. This was facilitated by the growth of public transport systems and the inhabitants of these areas were able to travel to work in the urban areas as commuters. Initial growth was *linear* along the railway lines but areas became in-filled as tram, bus and later the private car gave individuals greater freedom of movement. The process of suburbanisation may also be viewed as the suburbanisation of settlements physically separated from the urban area (sometimes called 'extended suburbanisation'). This represents a major trend in the redistribution of population away from city centres; these extended areas have been given a variety of names: dormitory, suburbanised and commuter villages.

succession: a series of changes which take place to a plant community. This may be a *primary succession*, or *prisere*, if the development begins on a surface which has previously not been vegetated, or it is termed a *secondary succession*, or sub-sere, if the original vegetation has been cleared or destroyed naturally.

succulent: a plant which is able to store water in its stem or leaves as an adaptation against drought conditions. Species of cacti have no true leaves, but are formed of thick green stems containing large water storing cells which absorb large amounts of water and swell up. The lack of leaves, and the thick waxy covering to the stem, helps to reduce *transpiration*. The baobab of the *savanna* regions has a huge trunk, up to 9 m in diameter, which has a soft and spongy interior allowing storage of water. The giant euphorbia, which has a stem similar to cacti, is also common in the thorn scrub of East and Central Africa.

sunrise industry: one positioned in a rapidly growing market, usually based on *new technology* and *innovation*.

sunset industry: one believed to be in terminal decline, with obsolete technology and an obsolete product.

superimposed drainage is a system established on a series of younger rocks, lying on a markedly different older series. The drainage will become adjusted to the structural pattern of the younger series. In the course of time, *erosion* will remove the younger layers, leaving the established drainage pattern cutting down into the older rocks, the structure of which bears no relation to the drainage system.

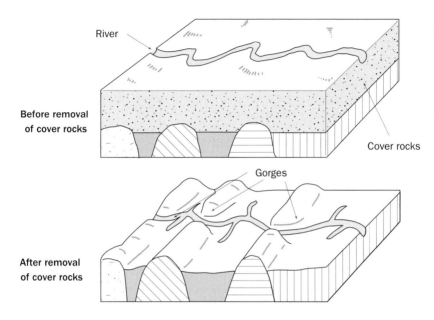

superstore: a large *retail* outlet with substantial car parking space that offers a very wide range of goods within a self-service shop.

supraglacial refers to that which is happening or to be found on the surface of the glacier.

supra-state: the grouping of countries into a much larger organisation. The *European Union* is an example of such a supra-state, which by 2009 consisted of 27 member countries, including the UK.

surf zone: the area on the foreshore of a *beach* between the breakpoint position of the approaching waves and the upbeach (landward) limit of the *swash*. The extent of this area depends on the type of wave, *constructive* or *destructive*, and the angle and nature of the beach material. A steep beach with large sediment quickly absorbs the swash energy which restricts the zone. A shallow angle beach of fine material allows the swash to extend a greater distance up the shore.

suspended load: the material small enough to be carried along in the flow of water in a stream. Some particles, such as silt or clay, can be carried by turbulence within the water; the greater the turbulence the larger the size of the particles which can be carried. The relationship between stream velocity and particle size (calibre) *erosion*, transport and *deposition* has been determined experimentally by Hjulström (see the *Hjulström curve*). The greater the velocity, the greater the size and the amount of material that can be carried in suspension.

sustainable agriculture: a system by which food production can expand but in a way that does not destroy the natural environment or put an excessive strain on resources. Many believe that sustainable strategies in agriculture are the only way forward if all societies are to obtain a more secure food environment.

Sustainable agriculture is based on the following:
- replenishment of soil *nutrients* as much as possible with organic material
- maintenance of the soil's physical structure
- minimal off-farm environmental contamination
- maintenance of *habitats* for pollinators and *biological control* agents
- *conservation* of genetic resources of crop and animal species farmed
- direction of technology to change away from the use of *non-renewable resources* and subsidised energy towards *renewables*
- continual cover of the soil by vegetation
- no increase in soil toxicities
- agriculture to be profitable enough to secure adequate subsistence and income for the farmer's family.

sustainable development: defined as 'development that meets the needs of the present, without compromising the ability of future generations to meet their own needs'. The environment should be seen as an asset, a stock of available wealth but if the present generation spends this wealth without investment for the future then the world will run out of resources. If, however, we use this capital to research and develop new resources for the future, we can build machines that will substitute for the environmental resource. A good example is the construction of solar panels to replace oil and coal. Examples of how this can be put into practice are already being seen:
- paying more for leaded fuel for cars rather than unleaded
- buying cars with catalytic converters
- replacing *fossil fuels* in power generation with *renewables* such as *wind power*, *hydroelectric power*, *solar energy* and *tidal energy*.

sustainable resources: these are resources such as soil, forests and fish stocks. If these are well managed, by replanting or manuring for example, the supplies can be sustained. For each there is a maximum sustainable yield, beyond which extraction rates will exceed renewal rates.

swash: the body of water rushing up a beach after a wave has broken. It may carry material up the beach away from the sea.

Swiss-cheese effect: a distribution across an area which is patchy, the patchiness being described as the 'Swiss-cheese effect', i.e. there are holes in it.

synergy occurs when the whole is greater than the sum of the parts, i.e. when $2 + 2 = 5$. On some *science parks* for example, there can be an intense localised interaction between different firms and *entrepreneurs*, research institutions, local banks and business service organisations. The benefits of these links and relations generates outputs, the sum of which is greater than the sum of the individual parts.

synoptic chart a map that shows the weather for a particular area at one specific time.

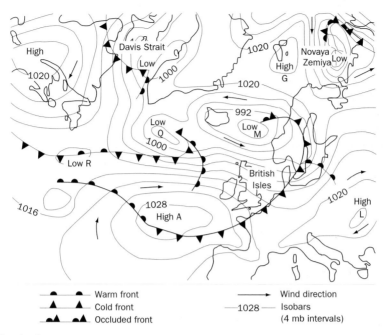

●───● Warm front	──────▶	Wind direction
▲───▲ Cold front	─1028─	Isobars
●▲─●▲ Occluded front		(4 mb intervals)

Synoptic chart

system: any set of interrelated components or objects which are connected together to form a working unit or unified whole. In geography it is usual to recognise two general types of systems:

- closed – there is transfer of energy, but not matter, between the system and its surrounds. The earth is an example of such a system
- open – where systems receive *inputs* and transfer outputs of energy or matter across the boundaries between them. Most natural systems are open ones.

systematic sampling: a method in which the sample is taken in a regular way, i.e. every tenth house.

S

Do you need revision help and advice?

Go to pages 325–53 for a range of revision appendices that include plenty of exam advice and tips.

taiga is a Russian term for the coniferous, or *boreal*, forest which extends across the northern part of the former USSR. This type of forest also covers northern Scandinavia and Canada.

talus is an accumulation of weathered fragments on a slope. This may be a thin layer of *debris* resulting from *weathering* of the underlying rock or it may be a thicker layer resulting from the movement of rock fragments downslope from a rock face. In the latter case it forms a *scree* at the base of a cliff (free) face. The talus may form in cones and is likely to move downhill due to gravity and expansion and contraction; this is termed talus creep.

tariff: a duty or tax imposed by one country on the imports from other countries. This may be used as a device to protect home-based manufacturers from foreign *competition*, particularly when overseas production is cheaper as a result of lower raw material and *labour* costs.

tectonic processes cause movement within the Earth's *crust* which results in uplift or depression, *folding*, *faulting*, warping or plate movement at *constructive* or *destructive plate margins*. The landforms produced by these processes, such as rifts, block mountains, fold systems and *escarpments*, are termed tectonic relief.

teleworking is a form of *homeworking* based in one's place of residence using telecommunications. It is one of a number of flexible working practices that have been introduced since the 1990s and for many companies it gives them greater access to the *labour market*. In the UK, one such example is the Automobile Association breakdown service, which has a number of its staff working from home using telecommunications to keep in touch with stranded motorists and breakdown vehicles. Telebanking is another area where teleworking has expanded.

temperate: a term used to describe conditions which are not extreme, as in a temperate climate, where seasonal temperatures do not display a wide range. The temperate latitudes extend from the tropics to the Arctic and Antarctic circle.

temperate grasslands: extensive areas of natural grassland situated in the dry continental interiors of North America (prairies) and Russia (steppes). They are found in temperate latitudes where the annual rainfall is between 250 and 750 mm. They are also found on continental east coasts between 30° and 40° south of the equator in the Pampas of Argentina and the Canterbury Plains of New Zealand. The Veld of South Africa and the Murray-Darling Basin of Australia are more continental in location. Although grasses grow in a wide range of climates, they only become dominant, as in the temperate grasslands, where trees or shrubs are unable to grow.

The transition from grassland to forest in the cooler northern areas takes place at about the 500 mm annual rainfall isohyet. Further south, in the warm temperate areas the transition occurs around 750 mm isohyet. These grasslands have probably been influenced by fire and human exploitation. In the prairies, the rainfall decreases from east to west and there is a change from tall grass to short grass prairie; there is a corresponding change in soil type with **chernozem** in the drier east and prairie soils in the area of taller grass to the west. Within these soils the spring **leaching** and summer **capillary action** (when higher temperatures produce an **evaporation** demand which exceeds the summer maximum of rainfall) create a layer of calcium carbonate nodules in the soil. This layer develops at shallower depths westward as rainfall decreases. The dominants are grama and buffalo grass; the deep roots may extend down 2 m to the **water table**, binding the soil together. The soil is the major nutrient store as the decay of grasses in the summer, together with high temperatures which favour breakdown, provides a rapid accumulation of **humus** in the soil.

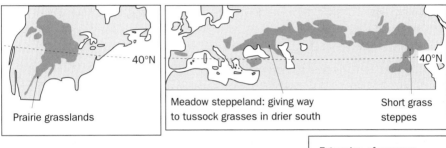

Prairie grasslands

Meadow steppeland: giving way to tussock grasses in drier south

Short grass steppes

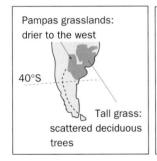

Pampas grasslands: drier to the west

40°S

Tall grass: scattered deciduous trees

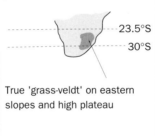

23.5°S

30°S

True 'grass-veldt' on eastern slopes and high plateau

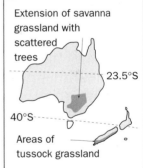

Extension of savanna grassland with scattered trees

23.5°S

40°S

Areas of tussock grassland

temperature inversion: an atmospheric condition in which temperature increases with height rather than the more usual relationship shown by the **environmental lapse rate**.

A number of situations can produce inversions:

- on a clear night when heat is radiated from the Earth's surface, the air near to the surface is cooled by conduction of heat to the cold ground. The lower layer of the atmosphere is therefore cooler than the air aloft
- at night, the colder denser air on the upper slopes of a valley side descends into the valley bottom displacing warmer air aloft. This produces a **katabatic** wind and can develop **frost hollows** on valley floors which can affect sensitive crops
- when a warm **air mass** passes over a colder land surface it will be cooled from below, therefore the normal **temperature profile** will be altered

- when air is subsiding in an **anticyclone**, or at a kata-front in a **depression**, the air is warmed **adiabatically** and is warmer than air at ground level.

Inversions over urban areas can help to trap **particulates** and pollutants in the lower layer of the atmosphere because **stability** increases at the level of the inversion.

temperature profile: the variation of temperature with height above the Earth's surface.

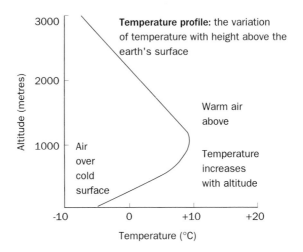

tenure: the means by which property, such as land and buildings, is held by an individual or organisation. In terms of housing, tenure may be by owner-occupancy, or tenancy. Tenancy may take a variety of forms: to a local council, private landlord or **housing association**. It may be furnished or unfurnished.

There are also a variety of forms of land tenure:
- freehold ownership – where the land is owned by an individual
- cash tenancy – where a fixed cash rent is paid by the tenant to a landowner
- **share cropping** – where a tenant gives the owner an agreed share of the produce
- state ownership – where the land is the property of the state and farmers act as tenants of the state. (See also **collective farming**.)

tephra: the ash and pumice that is ejected during a volcanic eruption and then deposited on a landscape. (See also **pyroclastics**.)

terminal moraine: a low crescent-shaped ridge stretching across the width of a valley floor which marks the furthermost point reached by a **glacier**. It is composed of an unsorted and unstratified assemblage of **debris** which has been brought down the valley by a glacier. The snout of the glacier remained stationary for a long period of time, representing a point when **ablation** and **accumulation** were equal. During this time **supraglacial**, **englacial** and **subglacial** materials were deposited at the snout as **meltwater** flowed away. If a glacier retreats intermittently, with several such pauses, then a series of **moraines** across the valley floor are created, each marking the position of the ice front. These moraines are similar in nature to terminal moraines, but are called recessional moraines.

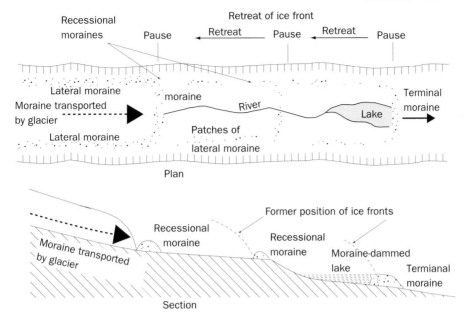

Terminal, recessional and lateral moraines

terrace: a raised level area of ground on a steep hillside. Terracing helps to slow down the rate of surface **runoff** of water, giving it more time to infiltrate. Terraces are common features in areas of intensive rice farming. Here, they act as a means to reduce **soil erosion**, and to make profitable use of slopes which would otherwise be difficult to cultivate. (See also **river terrace**.)

terra rossa: a red-coloured soil developed in areas with a calcium carbonate **parent material**. It occurs in areas which have high levels of seasonal **precipitation** and high temperatures. The parent material is greatly weathered, and silicates are leached out of the soil to leave residual iron-rich deposits. It tends to accumulate in depressions within the **limestone**, and where the vegetation type is **garrigue**. (See also **calcareous soils**.)

terrorism: the use of fear-creating methods to either govern, or coerce a government or society into changing its attitude to an issue. Terrorism often occurs in those areas where the desire for **separatism** is strong. Examples include Northern Ireland prior to 1994, and the Basque region of France and Spain. Terrorist acts frequently take place in large cities in order to gain maximum publicity and disruption.

tertiary sector (service sector) industries provide a service. These include employment in education, health, the police and fire brigade, public transport, retailing, local government, banking and finance, and the armed forces. It is the tertiary sector which employs the most people in an **economically more developed country** such as the UK.

textiles are materials and fabrics that are suitable for weaving. They may be divided into natural fabrics such as wool, cotton and silk, and artificial fabrics such as nylon, acrylic and other synthetic fibres.

Thatcherism: the term given to the range of policies and **attitudes** developed during Margaret Thatcher's period as prime minister (1979–90). The underlying attitudes were based on her dislike of **public sector** activities, extreme faith in the effectiveness of the free market and an authoritarian approach to leadership in general and unionised *labour* in particular. The main policies implied by the term Thatcherism include:

- *deregulation* of *private sector* businesses
- *privatisation* of state and *local authority* controlled organisations
- centralisation of control of the remaining public sector activities
- a switch from direct to indirect taxation
- a desire to promote the *enterprise culture*.

thermal electric power (TEP): electricity produced by means of steam turbines. The heat for steam production comes from the burning of a fuel, such as coal, lignite, oil, natural gas and peat.

thermal expansion occurs in rocks when they are exposed to alternate heating and cooling. The different minerals that make up the composition of rocks have different coefficients of expansion. They therefore respond at different rates to the changes in temperature and this causes stresses in the rock. These stresses can produce minor cracks that can be exploited by *chemical weathering*. This can contribute to the process of *exfoliation*.

thermohaline circulation is the circulation that is caused by differences in the density of sea water. The density of sea water is controlled by its temperature (thermo) and its salinity (haline). The key features of the thermohaline circulation are:

- The Gulf Stream, and its extension, the North Atlantic Drift, bring warm, salty water to the northeast Atlantic and mild conditions to Western Europe. As it moves northwards, this water cools, mixes with cold water coming southwards from the Arctic Ocean and becomes so dense that it sinks to the south and east of Greenland.
- This circulation in the north Atlantic is part of a much larger global system with links to the tropical Atlantic, the south Atlantic, Indian Ocean, Pacific Ocean and the Southern Ocean. Further sinking of dense water occurs near to Antarctica. Below the surface, the denser water from the two main sinking regions spreads out, affecting almost all the world's oceans at depths of 1000 m and below.
- The cold, dense water gradually warms and returns to the surface throughout the world's oceans.
- These movements with the sinking of the cold dense water and the return of warmer water to the surface form a closed loop – thermohaline circulation.

thermokarst is the name given to the landscape that results from *periglacial* conditions where the ground surface is very irregular and hummocky with marshy or lake-filled hollows. The major cause is the melting of masses of ground ice, which results in parts of the surface subsiding. Thermokarst in its natural state is often due to changing climatic conditions, but increasingly one of the major causes has been human development of such areas.

Thiessen polygon: a statistical technique used to convert data measured at specific points into an average value for the whole area. In calculating average rainfall within a *drainage basin* accurate values of rainfall can only be measured where rain gauges are sited. This measurement is effectively a point sample. If several gauges are used across the basin, then polygons can be constructed as a basis for calculating an average:

- the location of the rain gauges within the basin are plotted on a large-scale map
- a straight line is drawn from each gauge to its immediate neighbour

- mid-way between two gauges a line is constructed at right angles to the line joining the gauges
- these lines are extended until they intersect with others to form polygons.

To calculate the average rainfall for the basin:
- calculate the area of each polygon, either by drawing the outline on graph paper, or by using transparent graph paper and counting squares
- sum the area of the polygons to give the total area of the basin.

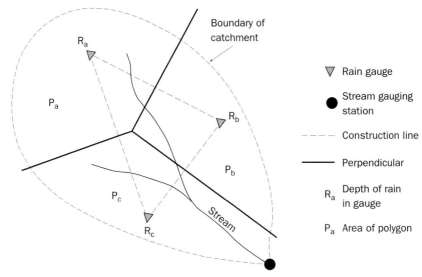

Thiessen polygon

FORMULA: Average rainfall for the basin $= \left[R_a \times \dfrac{P_a}{T} \right] + \left[R_b \times \dfrac{P_b}{T} \right] + \left[R_c \times \dfrac{P_c}{T} \right]$ etc. for all polygons

where R_a, R_b, R_c are the rainfall figures for gauges a, b, c

P_a, P_b, P_c are the areas of each polygon surrounding gauges a, b, c

T is the total area of the *drainage basin*, i.e. $P_a + P_b + P_c$

Third World: the name given to those countries that are also described as *developing*, *economically less developed countries (ELDCs)* or the *South*. It covers a range of countries from the relatively rich *newly industrialised countries (NICs)* to the *least developed countries (LDCs)*.

threatened areas are areas that are sensitive to development and exploitation by human activity. They exist across a variety of scales; from threats of *mining* development in Antarctica and other *wilderness* regions, further extraction of valuable timber and other resources from the *rainforests*, down to issues related to preservation of local coastal landscapes and habitat *conservation* areas such as *wetlands*. It is a generic term that encompasses the full spectrum of ecologically *fragile environments*.

threshold: the minimum number of customers needed to support a good or a service in *central place theory*. It represents the minimum trade area required to provide sufficient sales to cover the costs incurred by the supplier of the service, and therefore shops offering a particular good must be at least far enough apart to satisfy the threshold. This determines, in theory, the spacing between shops offering goods of different *order* within the shopping *hierarchy*. It is also the term applied to the shallow area at the mouth of a *fjord* where the erosive power of the *glacier* was reduced by melting. The threshold may be a rock bar, or it may have some depositional material above solid rock.

throughfall is the water which drips off the leaves of trees during a rainstorm and it occurs when the amount of rain which falls on the *interception* layer of the tree *canopy* has exceeded the capacity of the leaves to hold water.

throughflow is the water which moves downslope through the subsoil. It is particularly effective where further downward *percolation* or *infiltration* is prevented by underlying *impermeable* rock. In the *hydrological cycle* throughflow transfers water from the soil storage zone to the channel at a much slower rate than *overland flow*.

thunderstorm: a violent and heavy form of *precipitation* with associated thunder and lightning. They are produced by convectional uplift under conditions of extreme *instability* which may develop cumulus and cumulonimbus *cloud* up to the height of the *tropopause* where the inversion produces stability. This causes the cloud to spread out to form *anvil clouds*. The updraught through the central area of the towering cumulus system causes rapid cooling and *condensation* which leads to the formation of water droplets, *hail*, ice and super-cooled water which coalesce during collisions in the air. During condensation, *latent heat* is released which further fuels the convectional uplift. As raindrops are split in the updraught, positive electrical charge builds up in the cloud. When the charge is high enough to overcome resistance in the cloud, or in the atmosphere, a discharge occurs to areas of negative charge in the cloud or to earth. This produces lightning; the extreme temperatures generated cause a rapid expansion of the air which develops a shock wave which is heard as thunder.

tidal energy: a *renewable* method of producing energy by using the movements of the *tide*. Schemes could, with reversible blades, harness both the incoming and outgoing tides and therefore maximise the use of the site. Places with the maximum *tidal range* offer the greatest potential. The major drawback is cost, in both economic and environmental terms, and this may explain why only two sites are at present in operation: the Rance Estuary in northwest France and on the Bay of Fundy in eastern Canada. There have been several sites suggested for tidal schemes in the UK, such as the Severn and Mersey estuaries, Solway Firth and Morecambe Bay. The arguments for and against include:

PROS:
- renewable
- very reliable and predictable
- large size; the projected Severn Barrage would provide the same energy as five nuclear power stations
- non-pollutant
- also gain benefits of coastal protection of *estuary*

CONS:
- *flooding* of wetlands bordering estuaries, often the home of many species of birds, particularly those that are migratory
- could have an adverse effect on spawning fish
- construction cost is very high.

tidal range is the difference between the height of the water at high and low tide. This obviously varies, but a mean figure is normally given for each coastal location, the largest ranges being over 10 m, e.g. Bay of Funday (Canada) tidal range of 15 m. Tidal ranges are at their highest at spring tides.

tide: the periodic rise and fall in the level of the water in the oceans and seas, the result of the gravitational attraction of the sun and moon. The moon has the greatest influence, pulling water towards it to create high tide, with a compensatory bulge on the opposite side of the Earth. In the intervening areas, the tide is at its lowest. Tides run in cycles, following the 28-day lunar cycle, therefore every 28 days the gravitational pull is greatest and this gives the highest tides, known as *spring tides*. At the point in the cycle 14 days after this event, tides are at their lowest before they begin to rise again. This lowest point in the cycle is known as the *neap tide*.

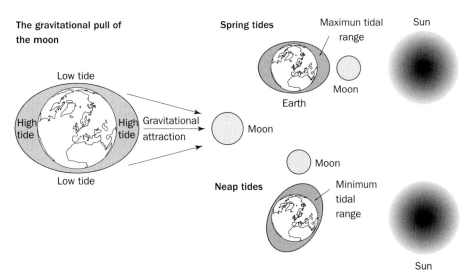

Causes of tides

till: a general term that covers all the materials deposited directly by ice. This will be unsorted and includes clays, sands and rocks. The general term for such glacial deposits was once *boulder clay*, but since many contain neither boulders nor clay, the term till is now in widespread use. Till and *fluvioglacial* material collectively make up glacial *drift*, the term that covers all glacial deposits.

tipping point is the level at which the momentum for change becomes unstoppable. A small increase in temperature, or other environmental change, could trigger a disproportionately larger change in the future. The changes are likely to be irreversible on a human timescale once they pass a certain threshold. One such tipping point is the point at which the disappearance of the summer sea ice in the Arctic becomes inevitable. This would bring about a change in the Atlantic *thermohaline circulation*. The melting Arctic and Greenland ice produces fresh water that would dilute the sea water and slow down the circulation (the North Atlantic Conveyor). If the circulation collapses or only extends to lower latitudes, it could result in very cold temperatures across northern Europe and the destabilisation of the climate of North America.

Epidemics can also reach a tipping point. A sick person infects a few other people, these in turn infect a few others and a point is reached where a very large number of people are infected. A global pandemic is much more likely with inter-continental flights where **infectious diseases** can spread rapidly. This can promote deadly epidemics because people are exposed to 'new' strains of a virus to which they are highly vulnerable because they lack immunity.

Todaro model: an alternative explanation for rural-to-urban **migration** in the **Third World**. Ideas of such movements have in the past focused on the **perceptions** of individuals in that they fall for the deceptive attraction of the 'bright lights'. Michael Todaro argues that people have expectations about the city and these expectations are economically realistic and rational. He argues that people move to the city to find better jobs but are prepared to gamble on the lottery of finding work, as urban employment pays far more than rural work. Therefore, even if the chance of finding work is only 50 : 50, given the difference in wage rates, this is a gamble worth taking.

tombolo: a coastal feature, produced by the **deposition** of sand and shingle, which joins the mainland to an island. It may be formed by the extension of a **spit** due to **longshore drift** or by an offshore bar pushing onshore. Chesil Beach is a spectacular example extending from West Bay near Bridport, on the coast of Dorset, 25 km southeastwards to the Isle of Portland, attaining a height of 14 m above sea level at the southern end.

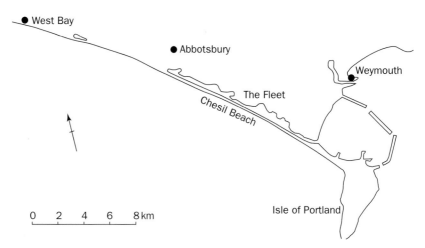

top-down: development initiatives that are instituted by national governments with the intention of reducing regional disparities of wealth. One of the most famous top-down initiatives was the Cassa per il Mezzogiorno in Italy.

topographical map: a map that shows the surface features of an area to scale. This includes both the physical elements, such as relief and drainage, and human features, such as settlements and communications. The Ordnance Survey produces detailed maps such as those for 1 : 25,000 and 1 : 50,000 scales.

topological map: a map that represents a spatial pattern of places and linkages by means of a diagram or graph. Dots or vertices represent places; straight lines or edges represent linkages between places. The original scale, orientation and true distance between places is distorted; the map is drawn to show places that are connected and intersections in the system. The most famous topological map is the one showing the London Underground system. There is no scale and stations are not shown in their true location, but the map is very effective for its purpose – it allows travellers to move efficiently about the system because it accurately portrays the links and interconnections. Topological transformations are produced to allow comparative studies of the connectivity, or degree of interconnection, between networks and the accessibility of individual points in a system.

topset beds are layers of sediment which are laid down at the landward edge of a *delta*. They consist of coarse sands and silts which are deposited horizontally as the velocity of the river reduces on entering the lake or sea.

tor: the most distinctive feature of granite landscapes, e.g. Dartmoor, consisting of an iso-lated exposure of much jointed rock. Their origin has been the subject of much controversy, with two main theories as to their formation:

- blocks of exposed granite were broken up by *frost shattering* during *periglacial* times. The weathered material was moved downhill by *solifluction* to leave the more resistant rock exposed as the tor
- *joints* in granite were widened by chemical action under the surface. Deep weather-ing then occurred during the warmer periods (*interglacials*) when rainwater penetrated the still buried granite. As the joints widened, roughly rectangular blocks were formed. During colder times, solifluction removed the weathered material to leave the granite out-crops. The joint pattern is important, areas of granite where there were few joints being left as the upstanding tor. This is the more widely accepted theory.

Tors also develop on other rock types, e.g. gritstone areas of the Pennines and on the quartzites of the Welsh borderlands.

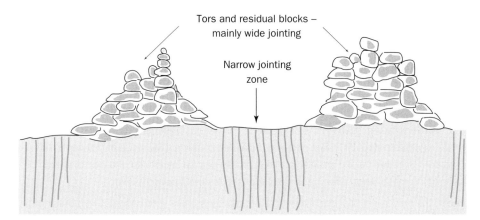

Tors and residual blocks –
mainly wide jointing

Narrow jointing
zone

Tor formation

tornado: an extremely violent and destructive weather system characterised by very power-ful swirling winds which rotate anti-clockwise towards the intense low pressure at the centre with speeds up to 300 km per hour. Their development is linked to extreme *instability* in the atmosphere, *convergence* and vigorous updraughts in the air. They are small in extent; the diameter of the storm centre may be 100 m or less and they may last in an area for only a matter of minutes. The life of the system may only last for a few hours. However, they cause an immense amount of damage along their storm path due to wind speed and rapid fall in pressure. This can cause such rapid pressure differences that buildings explode and *debris* is thrown around in the system. The strong updraughts can uproot trees, lift people, cars and even railway engines. The USA experiences more tornadoes than anywhere else; on average there are 100 deaths per year and damage running into hundreds of millions of dollars. The main area of development is in the Great Plains area of the Mid-West, Gulf Coast and the Mississippi Basin, an area referred to as 'tornado alley'. In this area cool air from the continental interior to the north meets warm, moist air from the Gulf of Mexico; this is particularly effective in spring and early summer, most tornados occurring in May and June, when there is also strong heating of the ground surface which adds to the uplift in the system.

tourism: any *recreational* or leisure time activity that involves an overnight absence from the normal place of residence. The development of tourism can generate employment both directly, in the creation of jobs in the hotel and catering trades, and indirectly, through tourist expenditure on goods and services in the area. Transport, construction and public utilities also benefit through the economic expansion linked to growth. Although the popularity of particular areas may fluctuate making tourism a less reliable base for economic advancement, this sector is rapidly expanding and it receives considerable private and public investment in some areas. The purchasing power of tourists can be a significant contribution to the local economy representing a source of foreign earnings (invisible earnings) and as such it is an export activity. The World Travel and Tourism Council (WTTC) claims that this is the world's largest and fastest-growing industry, and that the current 500 million tourist travellers will increase by 80% in the next 15 years. Although tourism creates economic wealth, it also often brings environmental, social and cultural damage, e.g. polluted beaches, degraded reefs, disappearing indigenous cultures and problems of seasonal unemployment. This has prompted a search for alternative new approaches generally referred to as *ecotourism*.

toxic waste: a source of pollution resulting from the use of metals which produce a poisonous residue or which give off dangerous fumes. Many industrial processes release toxic metals such as lead, mercury, zinc, cadmium or discharge sulphuric acid and cyanide. Some chemical pollutants remain toxic for many years and may become concentrated in the *food chain*. Toxic material may be airborne pollution; mercury may be released into the air along with lead from vehicles which are driven by petrol combustion. These metals are carried by the wind over long distances and are eventually washed out by rainfall to contaminate the soil. Toxic waste dumping, even at official refuse tips and sites, presents a hazard where *leaching* carries waste into streams, rivers and *groundwater* supplies. Discharge of pollutants into sea areas contaminates marine shellfish and fish stocks can be influenced; the concentration of the polluting metal tends to become greater at successive *trophic levels*.

trade: the movement of goods from producers to consumers. The basis of this transfer of goods can be explained by the theory of comparative advantage; countries specialise in activities for which they are best equipped in terms of *resources* and technology. Surpluses can then be traded by a country to provide the income needed to buy in goods which cannot be produced efficiently, or at all, in the home economy. Trade is partly in the form of *raw materials*, foods, beverages, energy supplies and manufactured goods which are termed visible imports or exports. Trade also takes place in services such as finance and *tourism* which are forms of invisible earnings or expenditure. World trade is dominated by *Organisation for Economic Co-operation and Development (OECD)* countries, i.e. the West and Japan, where much of the trade is between the members of this group. These are all economically advanced states and much of the trade is in manufactured goods. There is also a significant movement of fuel oil from the *Organization of Petroleum Exporting Countries (OPEC)* countries, particularly the major exporters in the Middle East, to the advanced economies and to developing world states which are deficient in natural energy resources. Raw materials (ores and industrial crops) and foods and beverages from *economically less developed countries* have traditionally been traded with advanced nations in return for manufactured goods. This is still important, although as countries have begun to industrialise, they have tended to progress from industries which are 'export-oriented', such as basic processing of raw materials, to those which are *'import substitution'* such as the manufacture of textiles, clothing, shoes, ciga-rettes, drinks. This change has led to a decline in the return movement of manufactured items.

trade bloc: a group of countries that share trade agreements between each other, but with *tariff* walls that discourage imports from countries outside the bloc. The *European Union (EU)* is a good example.

trade cycle: see *business cycle*.

trade war: a protectionist battle between governments in which *tariff* barriers against a country's imports leads to retaliation. If a trade war becomes serious enough it can cause a widespread downturn in world trade.

trade winds: the tropical easterly winds which blow in from either side of the equator towards the *inter-tropical convergence zone* from the subtropical high pressure cells. They form the surface component of the *Hadley cells*.

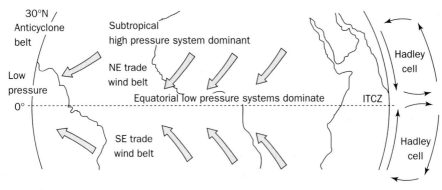

Trade wind belts

transect: an illustrative device for showing features along a line, drawn to show how those features change along the line as distance increases. A typical use would be in urban geog-raphy, where transects can be drawn from the city centre to the outskirts along main roads to show the urban *land use* changes that occur with increasing distance from that centre. Another use would be to show the changing vegetation of a *psammosere* along a sand *dune* area from the beach to the inland ridges.

transfer of technology is the movement of ideas and *innovations* from region to region or from country to country. One of the most significant movements is from *First World* to *Third World* countries but even in First World countries technology is often transferred by *transnationals* from countries such as Japan. Japanese work practices, for example, have been readily adopted by European countries. These include:

- *'just in time' production*
- worker teams who rotate jobs between members, which helps to increase skill levels

- the use of 'envoys' – representatives from a 'buying' industry who are permanently based in the factory of component suppliers so they can pass on directly any required changes to design and specification
- the use of 'milkmen' – representatives from a 'buying' industry who regularly visit the factory of component suppliers to inform them of required changes to design and specification.

transform faults occur where two tectonic plates are moving past one another without any degree of convergence or divergence. These boundaries may be termed *conservative plate margins* or passive margins, e.g. the San Andreas fault system. They are also found on the margins of the spreading *mid-oceanic ridges*; here they lie at right angles to the orientation of the ridge and appear to offset the line of the ridge. They may have been formed as a result of the plate segments spreading at different rates, thus causing faults running from one part of the ridge to a displaced section of the ridge.

transhumance is the seasonal movement of animals by people who live in a fixed settlement to pasture elsewhere. The term is usually applied in mountain regions where animals are moved from their lower winter areas to the summer mountain pastures. It can also be applied to other areas of the world where there are seasonal movements, such as the dry grassland regions on the desert fringe.

transition zone (twilight zone) is the second zone of the *concentric urban model*. In this zone there is to be found the oldest housing in the urban area which is deteriorating into slum property or being invaded by *light industry*. In *developed* countries, the inhabitants of such areas tend to be of the poorer social groups with some immigrants. This zone, together with zone 3 in the concentric model, would constitute the *inner city* in the UK.

translocation is the movement of soil components in any form (*solution*, suspension) or direction (upward or downward). Soil moisture is usually involved.

transmigration programme: the planned movement of people from one area of a country to another. The best example can be seen in Indonesia where, to relieve the population pressure on the main island of Java, people were resettled in the other islands within the country, notably Sumatra, Kalimantan and Irian Jaya. This programme began in the 1930s and by 1990 the total number of migrants had reached 3.5 million. The programme was not without its critics, as the *indigenous* populations of the host islands do not seem to have been treated well, particularly in the case of tribal land rights. There has also been some concern expressed over the destruction of large areas of the *tropical rainforest* of the area in order to give land to the migrants. By 2000 the number of annual migrants had reduced to 22,000 and the State Ministry of Transmigration and Population (SMTP) announced its decision to terminate the programme. The programme had been designed to supply *labour* to timber companies developing tree *plantations* for the pulp and wood-based industries. This was not working effectively because of complications of competing land claims and land ownership and the fact that the average income of the transmigrants was too low to sustain an appropriate living.

transmittable diseases (communicable diseases) can be spread from one person to another. Such diseases can be airborne (spread by droplet infection), faecal-borne (spread by faecal contamination of water, food, utensils etc.) or spread by direct contact from animals (*vectors*) or humans. Once the method of transmission is known, appropriate, and in many cases standard, preventative measures can hopefully be applied, given the right socioeconomic and political conditions.

transnational/transnational corporation (TNC): also known as multinational, this is the term for a company that operates in more than one country. Transnationals in the past were typically associated with mineral exploitation or *plantations*, but since the 1950s these firms were increasingly associated with manufacturing, particularly motor vehicles and electrical goods. Today they cover a wide range of activities including petroleum (BP, Exxon), motor vehicles (Ford, General Motors), electricals/electronics (Philips, Sony, Hitachi), financial services (Barclays, HSBC), food and hotels (Coca Cola, McDonald's, Holiday Inn, the Forte Group). The headquarters of such companies were invariably located within the *First World*, but increasingly there has been the emergence of transnationals that are based within the *newly industrialised countries (NICs)* and *recently industrialised countries (RICs)*. There are a number of positive and negative attitudes towards TNCs:

PROS: • they create employment and therefore raise living standards
 • foreign currency is brought into the country which improves the country's *balance of payments*
 • can lead to a *multiplier effect* and thus an increase in economic activity
 • *transfer of technology:* may help improve the level of expertise of the local workforce

CONS: • the jobs provided may only require low-level skills; production rather than *research and development* and managerial and supervisory positions may be filled by company employees from the home country
 • most or all of the profits may be exported back 'home' to shareholders in advanced economies
 • they are able to exert excessive political muscle; their global operation allows them to get round disadvantageous trade restrictions
 • decisions are made in a foreign country and on a global basis; therefore the company may pull out at any time.

transpiration is the process by which water is lost from a plant through the stomata (very small pores) in its leaves.

transport network: the arrangement or pattern of the lines of transport (road, rail, canal etc.) that link up a number of places.

trench: see *ocean trench*.

trend: a general direction of movement. To describe a trend is to give an overall picture, and not to focus on individual changes.

triangular graph: a graph with three axes, in the form of an equilateral triangle, which allows three separate variables to be plotted. Each axis is scaled from 0 to 100% moving around the triangle in either a clockwise or anti-clockwise direction. The advantages of using this type of graph are:
 • it allows classification of data in that points which have similar characteristics are located in 'clusters' on the graph
 • it is the only graph which enables three variables to be plotted.

However, only data which can be divided into three groups and which add up to 100 in percentage form can be used; absolute values cannot be recorded. It is very useful for example in *soil texture* analysis where % sand, % silt and % clay can be identified or in the structure of employment in terms of *primary*, *secondary* and *tertiary*.

Worked example: the use of a triangular graph

Sample X (within loam) is:
  40% on sand axis
  20% on clay axis
  40% on silt axis

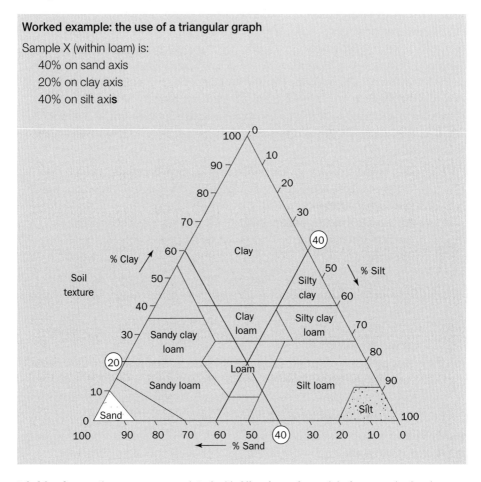

trickle down: the process associated with *Hirschman's* model of economic development by which *economic growth* and wealth are spread from the *core* region of a country to its peripheral regions. The expansion of the core increases the demand for food and resources from the *periphery*. The core may also provide the necessary machinery, fertilisers or *hybrid* seeds to permit the required increases in agricultural productivity. As incomes in the periphery increase, the demand for consumer goods in that area will increase. This may lead to the *decentralisation* of *branch plants* to the periphery from the core. Governments may seek to encourage this decentralisation as the core becomes congested and land prices increase.

trophic levels: the links or stages of the energy transfer model known as a *food chain*. At each trophic level some energy is available as food for the next trophic level. Some of this energy is also lost as excreta, some as the decay of dead *organisms*, and some as respiration.

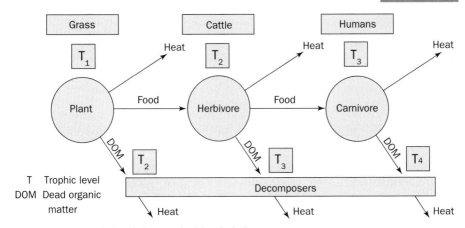

Examples of trophic levels in a typical food chain

tropical: a description of those areas of the world which lie between the Tropic of Cancer and the Tropic of Capricorn. The Tropic of Cancer joins the points on the Earth's surface where the midday sun is directly overhead at the June Solstice (21 June). The Tropic of Capricorn joins the points on the Earth's surface where the midday sun is directly overhead at the December Solstice (21 December).

tropical rainforest: see *rainforest*.

tropopause: the narrow layer within the *atmosphere* which marks the boundary between the *troposphere* and the *stratosphere*. It is located at about 16 km above the equator, and 8 km above the poles. The tropopause is characterised by an inversion where the temperature of the limited amount of atmosphere that exists at these heights begins to increase. This inversion marks the upper limit of the weather variations that exist in the atmosphere.

troposphere: the lowest layer of the *atmosphere* which extends from the ground surface to the *tropopause*. Its main characteristics are that the temperature of the atmosphere within it decreases with height, and that *pressure* decreases with height. Over 50% of the total mass of the atmosphere is located within 6 km of the surface of the Earth.

truck farming: the term used in North America for *market gardening*.

truncated spur: a steep wall on the side of a glaciated valley between two tributary valleys. It marks the point where a glacier has removed a former spur which extended into the pre-glacial river valley. *Glaciers* are less able to negotiate bends in their course and consequently grind back obstacles in their path.

tsetse fly: the African bloodsucking fly that transmits the disease trypanosomiasis (sleeping sickness) which affects both humans and domestic animals, particularly cattle (where the disease is called nagana). The distribution of cattle in Africa as a consequence is very much dependent on the area of infestation of the fly. The most effective controls have been environmental ones, such as the clearing of woodlands, periodic burning of the bush and destruction of wild game. Other attempts through insecticide, aerial spraying and trapping have not had the same success. Environmentalists have become concerned over the amount of vegetation it is proposed to clear in order to open up vast areas to cattle farming. The fear is that once the vegetation is cleared the areas will become susceptible to widespread *soil erosion*.

tsunami: a tidal wave caused by submarine shock waves originating from an **earthquake** or volcanic eruption. The amount of damage it causes is related to the distance from the event which caused it, and the nature of the coastline over which it passes. A gently sloping **continental shelf** allows a tsunami to build to great heights, whereas deep water extending close to a shore minimises wave size. Narrow V-shaped bays also concentrate the wave energy and favour the maximisation of wave heights.

tundra is the name given to a climatic and/or vegetation type which can be found in the most northerly parts of North America and Eurasia (north of 65°N).

The main features of the climate are:
- long and bitterly cold winters with temperatures averaging −20°C
- brief mild summers with temperatures rarely being above 5°C
- a large temperature range of over 20°C
- low amounts of **precipitation**, less than 300 mm, most of which falls as snow
- strong winds blowing the dry powdery snow in blizzards, and creating a high **windchill** factor.

Although the winters are severe and the sea regularly freezes, temperatures do not fall as low as more inland areas due to the moderating effect of the sea. The cold temperatures are due to the short hours of winter daylight in such areas, and although the daylight hours are longer in summer, the angle of inclination of the sun is low. These areas are dominated by high pressure with dry descending air. In summer some **depressions** do penetrate, giving some precipitation.

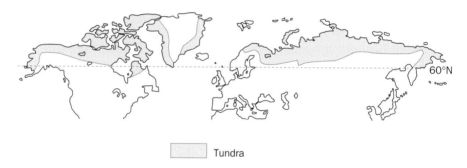

☐ Tundra

The main features of the vegetation type are:
- dwarf species such as cotton grass, mosses and **lichens** growing close to the ground often in a cushion-like form. This is an adaptation to the strong winds which blow
- sheltered places may have dwarf willows and stunted birch trees
- the plants have long dormant periods during the cold dark winters, but grow rapidly in the summer when daylight hours are long. 'Bloom mats' of anemones, arctic poppies and saxifrages burst into life providing a mass of colour
- many plants have small leaves to limit **transpiration**
- the plants are shallow rooted because the soil is permanently frozen at a shallow depth (**permafrost**).

Another feature of tundra areas is the existence of **permafrost**. The surface layer of this may thaw out in the summer producing waterlogged conditions in flat areas.

typhoon: the local name for a tropical storm or **hurricane** in the western Pacific Ocean. Typhoons tend to occur off the eastern coasts of China and Japan.

ubac: a hill or mountain slope which faces north (in the northern hemisphere) and so receives minimum light and warmth. (See also *aspect*.)

ubiquitous refers to *raw materials* that are evenly distributed and therefore have no influence on industrial location decisions. Water is generally considered to be ubiquitous.

UN: see *United Nations*.

underemployment occurs where people nominally have jobs, but jobs that do not keep them fully occupied. This often results in very low wages, and is particularly prevalent in the agricultural sectors of less economically developed countries.

underpopulation occurs when there are too few people in an area to use the resources efficiently. In such circumstances an increase in population will mean a better use of the resources and an increased living standard for all of the population. Canada is often quoted as the best example of an underpopulated country.

UNFPA: see *United Nations Population Fund*.

UNHCR: see *United Nations High Commissioner for Refugees*.

uniformitarianism: a concept, fundamental in geology, that processes which operate at present also operated in the past, and produced the same results. These processes need not have operated at the same rate, nor at the same intensity. The idea was first put forward by James Hutton in 1795.

Union of Soviet Socialist Republics (USSR) was once the world's largest sovereign state. It consisted of 15 republics, of which the Russian Soviet Federated Socialist Republic (RSFSR) was by far the largest. The Soviet Union disintegrated in 1991, leading eventually to the creation of separate countries from the existing republics. The 15 republics were (in order of population size): Russia, Ukraine, Byelorussia, Uzbeckistan, Kazakhstan, Georgia, Azerbaydzhan, Lithuania, Moldavia, Latvia, Kirghizia, Tadzhikstan, Armenia, Turkmenistan and Estonia.

unitary authority is the name given to the local government areas (*local authority*) which now provide the full range of services for their designated areas. They have replaced the former two-tier system of counties and districts in some parts of the UK. For example, the county of Berkshire has been replaced by five unitary authorities (Newbury, Reading, Wokingham, Bracknell Forest, and Windsor and Maidenhead). Other counties to be replaced include Humberside, Avon and Cleveland. On the other hand, unitary Rutland has been re-born out of Leicestershire and York is separate from North Yorkshire. In some parts of Britain there is an unusual mix of authorities. Derbyshire is a two-tier shire county, with the unitary city of Derby in the middle.

United Nations: an international organisation whose stated aims are to facilitate cooperation in international law, international security, economic development, social progress, human rights and achieving world peace. It was founded in 1945. There are currently 192 member states, including nearly every recognised independent country in the world. From its headquarters on international territory in New York, the UN and its specialised agencies decide on substantive and administrative issues in regular meetings held throughout the year. A number of international bodies are part of the UN umbrella, including the *World Health Organization (WHO)*, United Nations Children's Fund (UNICEF) and others, and it has also sought to deal with international poverty through the *Millennium Development Goals (MDGs)*.

United Nations High Commissioner for Refugees (UNHCR): a body of the *United Nations* that has special responsibility for refugees. The UNHCR provides international protection in promoting and safeguarding the rights of refugees with regard to freedom of movement, residence and security against being returned to a country where they may be persecuted as a result of their race, religion, nationality or membership of a political group. The UNHCR can also provide short-term financial assistance to governments which grant asylum to help in the aim of allowing refugees to become self-sufficient members of that community as quickly as possible.

United Nations Population Fund: previously the United Nations Fund for Population Activities (UNFPA), this organisation began operations in 1969 and is now under the administration of the United Nations Development Fund. It supports programmes in four areas: the Arab States and Europe, Asia and the Pacific, Latin America and the Caribbean, and sub-Saharan Africa. It works in more the 140 countries, territories and areas. Around three quarters of the staff work in the field. Some of its work involves the lead in providing supplies and services to protect reproductive health. It also encourages the participation of young people and women to help rebuild their societies who are affected by poor reproductive health which expands out into areas such as prevention of sexually transmitted diseases including HIV/*AIDS*. The organisation works in partnership with other *United Nations* agencies, governments and communities. Working together, the agency raises awareness and assembles the support and resources needed to attain the *Millennium Development Goals (MDGs)*.

upward spiral: a virtuous circle of growth in which one development leads to a cumulative process; an industry attracts population encouraging service provision, which in turn provides greater employment and spending power within the local area. This is a snowball effect of *positive feedback* and self-perpetuating growth; the economy is on an upward spiral as illustrated in the process of *cumulative causation* put forward by *Myrdal*.

urban area: a built-up area which forms part of a city or town.

urban blight: the run-down or physical deterioration of part of an urban area. This is often linked with areas of social and economic deprivation within the *inner city* and may reflect some uncertainty as to how *derelict land* areas are to be redeveloped. In residential districts the growth of non-residential activities such as wholesaling or manufacturing can lead to established population moving out to be replaced by very poor urban groups, the most recent migrants, students and other short-term residents. Often these groups do not have the finance or interest to maintain the external fabric of buildings and blight spreads.

urban climate: a localised weather pattern associated with an **urban area**. Some geographers refer to this as the 'climatic dome' within which the weather is different from that of the surrounding rural areas in terms of temperature, air quality, **relative humidity**, **precipitation**, visibility and wind speed. For a large city, the dome may extend upwards to 250 to 300 m and its influence may well continue for tens of kilometres downwind (also known as the urban plume). Towns and cities differ in their climate from the surrounding rural areas in a number of ways:

- temperatures in the cities tend to be warmer – see **urban heat island**
- atmospheric composition – as there is more pollution in cities, there are higher concentrations of gases such as sulphur dioxide, and **carbon dioxide**. There is also more dust and other **particulates** from industrial processes and the exhausts of motor vehicles
- there is a higher incidence of **cloud**, and therefore lower amounts of sunshine
- the levels of precipitation are greater, with more 'rainy' days. Both **thunderstorms** and the occurrence of **hail** are more likely
- the frequency of **fog** is higher, both in terms of its duration and intensity
- wind speeds are on average lower because of the sheltering effect of buildings. However, within city centres, there is increased turbulence of wind, together with a channelling effect down particular streets (see also **venturi effect**).

urban decay: the decline and dereliction of areas of a town or city. Urban decay is a feature of many **inner city** areas of the UK. Economic investment is low, and many of the traditional industries of these areas have closed due to falling demand, lack of space, or the use of **new technology**. The housing stock is in a poor state of repair. It lacks basic amenities, is overcrowded, and is generally inhabited by people on low incomes who cannot afford improvements. Even where slum clearance has taken place, housing has been replaced by poorly built high-rise flats which are themselves becoming unsightly and unsafe. Many shops lie empty, boarded up or protected from crime by metal grills. Graffiti, fly-posting and vandalism are other visible signs of urban decay. (See also **derelict land**.)

Urban Development Corporations (UDCs) were set up in the 1980s and 1990s to take responsibility for the physical, economic and social regeneration of selected inner-city areas with large amounts of derelict and vacant land. They are an example of what is known as **property-led regeneration**. They were given planning approval powers over and above those of the **local authority**, and were encouraged to spend public money on the purchase of land, the building of **infrastructure** and on marketing to attract private investment. The intention was that private investment would be four to five times greater than the public money initially invested.

The appointed boards of UDCs, mostly made up of people from the local business community, had the power to acquire, reclaim and service land prior to private-sector involvement and to provide financial incentives to attract private investors. In 1981, two UDCs were established: the **London Docklands** Development Corporation (LDDC) and a Merseyside UDC. Eleven others followed in areas such as the Lower Don Valley in Sheffield, Birmingham Heartlands, Trafford Park in Manchester and Cardiff Bay. By 1993, UDCs accounted for nearly 40% of all urban regeneration policy expenditure. Over £12 billion of private-sector investment had been attracted, along with £4 billion from the **public sector**. They had built or refurbished 35,000 housing units and created 190,000 jobs.

c

urban field: a term used to describe the area around an urban centre which is functionally linked to that centre in terms of the provision of goods and services, administration or employment. More recent terms such as *sphere of influence* or market area are now generally used.

urban fringe: the area on the margins of the urban settlement which lies beyond the city boundary but within the urban *sphere of influence*. This area includes a variety of residents and activities. Mixed in with the local rural economy, which is still present but in decline, there are *dormitory settlements* housing middle-income *commuters* who work within the city. The fringe is under pressure from developers which increases land and house prices in the area, and the growth of newer housing areas may lead to a physical segregation, as well as an economic divide, between the *indigenous* population and the new arrivals. Many of these areas are part of the *green belt* and there may be conflicts over the demands for space for further expansion in the urban fringe. With such a concentration of relatively wealthy and mobile population this area provides economic potential for leisure developments and out-of-town shopping centres as well as further residential expansion.

urban function: the use of buildings or *land use* in an urban area. The broad categories include:

- residential, which may be subdivided by age, quality or form of ownership: detached, semi-detached, terraced; council, private, rented; Victorian, inter-war, post-war
- industrial – heavy industry, light manufacturing, *industrial estates*
- commercial – retailing including services, wholesaling, financial
- offices – medical, legal, agencies (estate and travel)
- entertainment – cinemas, hotels, leisure, clubs, public houses
- open land – recreational, parks and gardens, church yard/cemeteries, derelict/waste land.

urban heat island: the zone around and above an *urban area* which has higher temperatures than the surrounding rural area. Cities tend to be warmer than the surrounding rural areas for the following reasons:

- Building materials such as concrete, bricks and tarmac act like bare rock surfaces in that they absorb large quantities of heat during the daytime. Much of this heat is stored then slowly released at night. Some of these urban surfaces also have a high reflective capacity, particularly modern office blocks and department stores with huge windows. Additionally, multi-storey buildings tend to concentrate the heating effect in the surrounding streets by reflecting energy downwards.
- Heat comes from industries, buildings and vehicles, which all burn fuel (*anthropogenic heat*, i.e. caused by human activities). Although they regulate the temperature indoors, air conditioning units release hot air into the atmosphere. Even people generate heat and cities contain large populations in a relatively small space.
- Air pollution from industries and vehicles increases *cloud* cover and creates a 'pollution dome', which allows in the short-wave radiation as well as reflecting it back to the surface.
- In urban areas, water falling on to the surface is disposed of as quickly as possible. This changes the urban water and heat balances – reduced *evapotranspiration* means that more energy is available to heat the atmosphere.

The heat island effect develops best under certain meteorological conditions. The contrast between urban and rural areas is greatest under calm, high-pressure conditions, particularly with a **temperature inversion** in the boundary layer above the city. Heat islands are also better developed in winter when there is a bigger impact from city heating systems. Urban–rural contrasts are much more distinct at night when the impact of **insolation** is absent and surfaces that have absorbed heat by day slowly release it back into the atmosphere. Heat islands are not constant – they vary both seasonally and diurnally.

The heat island effect is also spatially variable. The edge of the island is usually well defined and temperatures change abruptly at the rural–urban boundary. Some climatologists have likened the effect to a 'cliff' in temperatures. From this point temperature rises steadily to a peak in the city centre, where building densities are highest. The rise tends to be gentle, at an average of 2–4°C per km. After the cliff of the city edge, this steady rise has often been referred to as a 'plateau'. Within the plateau, however, there are variations that reflect the distribution of industries, power stations, water areas and open spaces.

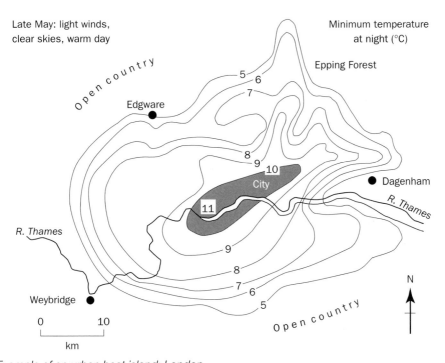

Example of an urban heat island: London

urban heritage tourism involves people visiting an urban area because of its interesting history. This may be due to industrial archaeology such as old factories and docks, canals and railways, as well as historic buildings and former homes of historic figures. These features, once regenerated and marketed, can act as a magnet for the increasing number of urban tourists visiting for day trips and short breaks.

urban hydrograph: a graph that shows river **discharge** against time for an area with an urban catchment. The **interception** layer in an urban area consists of buildings, roads and artificial surfaces which reduce **infiltration** and move water into neighbouring channels very rapidly through drainpipes and storm drains. Less water falls on natural surfaces such as gardens, parks and open spaces, which encourage infiltration and *throughflow*, and consequently a large proportion of the rainfall reaches the channel quickly to produce a 'flashy' *hydrograph.* This has a steep rising limb, a short *lag time*, a high peak discharge and a steep recessional limb. Channels fed by urban catchments are more liable to *flooding* downstream of the urban area.

urban inequality: the spatial variations that exist in the standards of living that urban dwellers enjoy, the range of services to which they have access and the general quality of life that they experience. Inequality reflects *social segregation* within cities and is the result of a wide and varied range of factors. Once established, it becomes one of the major concerns for city managers, as it is likely to cause tension and conflict and can even lead to serious social unrest.

urbanisation is the process whereby the proportion of people living in towns and cities increases. Urbanisation occurs when rural-to-urban *migration* is greater than urban-to-rural, and when *life expectancy* and *natural population increase* are greater in the urban area. Although the process has been going on for centuries, rapid urbanisation really only began in the nineteenth century and in the *First World*. By 1900, England, for example, was esti-mated to be already 80% urbanised. In the early 2000s, the UN has estimated that over 80% of the First World population lives in urban areas compared with around 50% in the *Third World*. Rapid urbanisation has occurred and will continue in the *Third World*, to the extent that it is predicted by 2020 that at least 53% of the population will live in cities. This has led to the development of primate cities (see *primacy*) in such areas and the growth of marked differences between the *core* and *peripheral* regions.

Some of the problems associated with rapid urbanisation include:
- the city is not able to provide enough housing for all of the migrants so they are faced with three alternatives: sleep on the streets; try to rent single rooms; or build their own shelters on land which they do not own
- this has led to the growth of *squatter settlements* or shanty towns, which have prob-lems of overcrowding, sewage disposal, disease, crime etc.
- pressure on services such as refuse collection, health provision, education, police and fire services, power supplies and sewage disposal
- transport systems become overused and the road network is unable to accommodate the increase in vehicular traffic
- the number of jobs available in the city does not match the incoming migration which leads to vast unemployment. In addition, a large number of people work in the *informal sector* and are therefore classified as *underemployed*.

Studies of cities in the less economically developed world, however, have shown that the picture of a bleak *downward spiral* is not always the case. To many migrants, urban life may be very superior. Employment can be found with better wages and many shanty towns are not always areas of deprivation, extreme poverty and disease. Some are very well-organised and not the first destination of recent immigrants to the city. (See also the *Todaro model*.)

urban primacy: see *primacy*.

urban renewal is the replacement of old structures by new ones and the conversion of space and buildings from one use to another. Renewal involves several processes: *redevelopment* and/or *regeneration*, which may involve demolition, and improvement, which seeks to adapt and modernise existing fabric to meet modern needs. Renewal has largely taken place within the *inner cities* of urban areas in both Europe and North America.

urban settlement: the definition of an urban settlement varies from country to country, although the *United Nations*, for their statistical purposes, define it as a centre that has a population of at least 20,000. In Denmark, for example, an urban settlement can have as few as 250 inhabitants, whereas in the USA it has to contain 2500. In some countries the figure is very high. In Japan, for example, an urban area has to contain at least 30,000 people. Geographers have used other criteria in an attempt to distinguish between urban and rural areas. Some of these features include *urban function*, occupations, service provision, *land use* and various social factors.

urban wind: in large cities the wind behaves very differently from the way that it does in surrounding rural areas. The main features of urban winds are:
- urban wind velocities can be up to 30% lower than in rural areas
- periods with no winds at all can be 10–20% higher
- at certain times winds can be extremely strong as high-rise buildings give rise to a funnel effect (see also *venturi effect*). It is not unknown for the winds to be so strong as to blow over pedestrians and cause buildings to sway
- there is often small-scale turbulence and eddying (water blown out of ornamental fountains etc.). This is often the result of the rapid warming of air over concrete and tarmac surfaces leading to the uplift of air, often into strong updraughts, especially alongside tall buildings. There is usually a compensatory downdraught effect on the opposite side of the street.

USSR: see *Union of Soviet Socialist Republics*.

utilities are industries that provide the most basic services such as water and power. Before *privatisation*, it was generally accepted that utilities were state-run, not only because of their importance to the overall economy, but also because they were operated on such a large scale and were therefore natural *monopolies*.

Aiming for a grade A*?
Don't forget to log on to **www.philipallan.co.uk/a-zonline** for advice.

c

valley cross-profile: see *cross-profile*.

valley long profile: see *long profile*.

variable: a measurable characteristic of any person (age, *life expectancy*, *diet*), place (*gross domestic product*, farm size, crop production) or any other thing or quantity.

varve: the annual *deposition* in lakes near to glacial margins, which consists of a layer of silt lying on top of one of sand. The coarser and lighter coloured sand is deposited during spring and summer melt, the darker silts representing the period of decreasing discharge towards the autumn when the finer material will settle. Each combination of light and dark bands thus represents 1 year's deposition. Counting the varves and measuring their respective thickness will give some indication of the glacial history of the region.

vector: a person, animal or plant that is a carrier of a *transmittable disease*, but is not affected by it. It is a potential source of infection for another *organism*. For example, with *malaria* the vector is the anopheles mosquito, which hosts and spreads the malarian parasite into the human population.

vegetation succession: see *succession*.

venture capital is risk capital that often provides a significant investment in a small or medium-sized business. The need for it arises when a rapidly growing firm requires more capital, but the firm is not yet ready to raise money on the stock market. In these circumstances, merchant banks might provide the funds themselves, or arrange for others to do so.

venturi effect is an increase in the velocity of air (or water) due to the constriction of flow. In urban areas, winds blowing in towards the centre are funnelled between buildings producing a venturi effect. Streets in the central areas tend to be narrower than in the suburbs and are flanked by high-rise buildings. They act as 'urban canyons' causing channelling, turbulence and increased velocity. This can be uncomfortable for pedestrians, swirling up litter and making walking along the pavements quite difficult. High wind speeds can also occur in gaps between individual buildings due to funnelling. The scale of the effect is influenced by the size of the gap in relation to the height and geometry of the surrounding buildings. The effect is more pronounced when straight streets are aligned parallel to the wind direction. This effect is also observed in upland areas where there is channelling by the relief, e.g. through a gap between two mountain ranges or where there is terrain-induced acceleration, e.g. over hilltops or low ridges. This is a factor that influences the potential of sites for *wind energy* turbine installation.

vertical integration occurs when two firms that operate in the same industry, but at different stages in the production and supply chain, join together. The integration might come about through a merger or a take-over. Backward vertical integration means buying out a supplier, e.g. a chocolate manufacturer buying a sugar producer. *Forward vertical integration* means buying out a customer, e.g. a chocolate manufacturer buying up a chain of newsagents.

vicious circle: a series of consequences that act to produce a *downward spiral* of events in which the situation worsens. For example, in regions of poor or inefficient agriculture, low yields means that the farmer has no surplus to sell after the family has been fed; no surplus means that there is no income; a lack of income prevents the purchase of better seeds or fertiliser which in turn enforces low yield the following year. The situation gets worse because continuous production without soil replenishment through fertiliser will decrease *fertility* which will reduce yield even faster. Without *aid*, the farmer's only course of action may be to borrow money at high interest, thus causing further problems. This vicious circle is often referred to as the poverty cycle or trap.

virtuous circle: a series of changes that produce an *upward spiral* as one development encourages growth in another. This leads to a concentration of activity in favoured areas such as the *core*. The processes of *cumulative causation* and growth produce *agglomeration* of activity which attracts other industries, population and services. These in turn make the region more attractive for further services and industries.

visible trade is concerned with the import and export of goods. The difference between visible exports and visible imports is called the *balance of trade*. (See also *balance of payments*.)

visibility: the distance that one can see, which depends on the amount of water droplets, smoke and dust present in the atmosphere. Under international codes, if visibility is reduced below one kilometre, then a *fog* is present, as distinct from *mist* or haze.

volcanic plume: this term can be applied to two phenomena:
- the upwelling of *magma* at a *hot spot* within the *mantle* causing the *crust* immediately above it to weaken and to allow the magma to rise to the surface. Such activity takes place in the middle of the Pacific plate, and gives rise to the Hawaiian complex of *volcanoes*
- the cloud of steam, gases and other *tephra* ejected from a volcano during an eruption. The direction of the plume is influenced both by the *prevailing wind* direction, and by the location of the eruption on the volcano.

volcano: the conical mountain created by *extrusive* materials such as lava and ash which emerge or are ejected from a central vent or *crater*. There are a large number of volcanoes, and a variety of types of eruption, both of which make the classification of volcanoes difficult.

One method of classification is by shape and composition:
- a basic shield – a gentle-sided cone composed of layers of basaltic lava flows, for example Mauna Loa, Hawaii
- an acid dome – a steep-sided cone composed of viscous acidic lava which quickly solidifies, for example, Mt Pelée, Martinique
- a composite cone – a volcano built up of alternate layers of lava and ash with steep concave sides. This type of volcano frequently has secondary cones on its slopes. An example is Mt Etna, Sicily

- ash/cinder cone – composed of fragmental material ejected during a series of eruptions, often with relatively steep sides. An example is Paricutin, Mexico
- caldera – a large circular depression within a much larger volcanic mountain. It is believed that it may be produced by a huge explosion in the magma beneath the volcano which removed its summit, or by the subsidence of the upper parts of the cones following such an explosion. An example is Crater Lake, Oregon, USA.

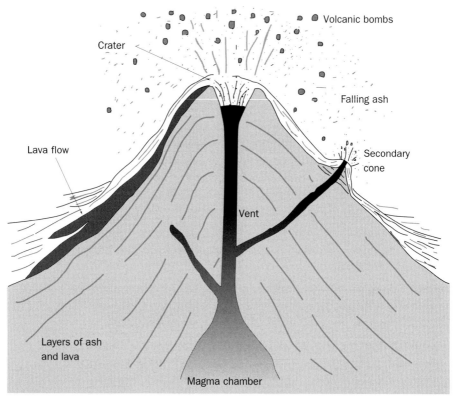

Features of a volcano

voluntary migrants are international migrants who have moved as a result of their own desire and motivation – their own free will. They are sometimes referred to as *economic migrants* because many of them tend to be moving in order to improve their *quality of life*.

vulnerability is a measure of the level of risk or susceptibility to damage in an area resulting from the impact of a *hazard*. A hazard becomes a major problem when it occurs in a vulnerable location. The level of vulnerability is influenced by physical, social, economic and environmental factors or processes. Characteristics of the area that can affect vulnerability include factors such as population density, existence of early warning systems, awareness and education of the public, level of *infrastructure*, evacuation procedures, emergency systems and building design. Assessment of vulnerability is a key element of *risk analysis*. Generally, less developed countries are more vulnerable and at greater risk because they are less well prepared for a hazard event.

wadi: also known as an *arroyo* in the Americas, this is a steep-sided ravine in desert and semi-desert areas, usually streamless, but sometimes containing a torrent (*flash flood*) for a short period after heavy rain. The infrequency of flash floods compared to the great numbers of wadis in some areas, suggests that they were created at a time when storms were more frequent and severe.

Walker cell: an east-west *tropical atmospheric circulation* in the South Pacific first recognised by Gilbert Walker in 1923. This circulation covers a broad area and is associated with strong convectional activity. (See also *el Niño*.)

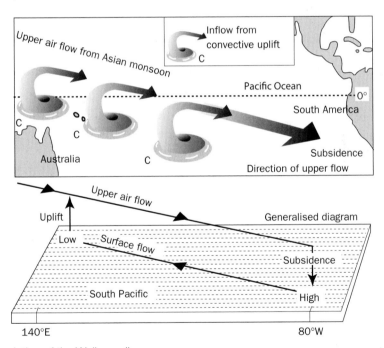

The circulation of the Walker cell

warm front: the leading edge of a mass of warmer air usually seen in a *depression*, where it is the leading edge of the *warm sector*. Such fronts are associated with thickening *cloud* and almost inevitably, some form of *precipitation*. A warm front is not as steep as its cold counterpart, having a gradient of between 1 in 100 and 1 in 250.

warm sector: the area between the two fronts of a *depression* where the highest temperatures are recorded. Over the British Isles, the warm sector of a depression represents tropical maritime air which is pulled in from the southwest and is therefore warm and muggy in summer and mild in winter. **Cloud** tends to be low (**stratus**) or broken, with the **precipitation**, if any, being in the form of light rainfall or drizzle. The low cloud of the warm sector often drifts on to upland areas where it forms hill **fog**. As the **cold front** catches the **warm front**, the size of the warm sector decreases until the warm air is finally pushed aloft as **occlusion** occurs.

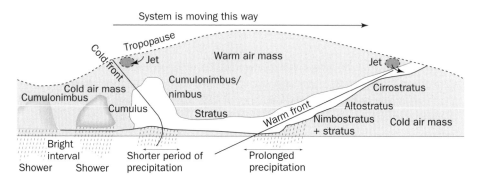

waste disposal: the disposing of the by-products or rubbish (unwanted materials) produced by humans in their use of the Earth's resources and the manufacture of goods. Disposing of waste is an increasingly difficult task; the nature and quantity of modern waste make it so much harder to cope with. Some waste is hazardous and many environmental organisations have increasingly campaigned for greater safeguards in the disposal of such materials.

The places for the dumping of modern waste include the following:

- **landfill sites**, particularly for domestic rubbish and **chemical waste**
- rivers, for low-level chemical waste and the treated water from sewage works
- the sea, although organisations such as **Greenpeace** have been actively campaigning for this to stop
- the **Third World**, for the waste products from the **First World**. Some countries have been willing to take hazardous waste from the First World in return for large payments, although the lower safety standards of such countries has given rise to some serious concern.

Of increasing concern in the UK is the question of the disposal of **nuclear waste**, much of which is dumped in special sites in west Cumbria, close to the nuclear installations at Sellafield. (See also **recycling**.)

water balance: in a **drainage basin** hydrological system, inputs of **precipitation** (P) are balanced by outputs in the form of **evapotranspiration** (E) and **runoff** (Q), together with changes to the amounts of water held in storage within the soil and as **groundwater** (ΔS):

FORMULA: $P = E + Q + \Delta S$

When precipitation exceeds evapotranspiration, this produces a water surplus. Water infiltrates into the soil and groundwater stores (the **water table**); when the pore spaces are saturated excess water contributes to run off. When evapotranspiration is greater than precipitation, evapotranspiration demands are met by water being drawn to the surface of the soil by **capillary action**. Groundwater stores are depleted and runoff tends to be reduced. The annual changes in precipitation and evapotranspiration are shown in a **soil moisture graph**.

water-borne diseases are those that are essentially spread by insects and other very small animals that breed, or are found in water. The major types, all of which are **endemic** to areas of the **Third World**, are:

- **malaria**, spread by the female anopheles mosquito which breeds in stagnant water
- yellow fever, which is also spread by a mosquito, although a different species
- **bilharzia**, which occurs when a parasitic worm living inside freshwater snails, invades the body
- river blindness, caused by the blackfly, which again infects the body with a parasitic worm. The fly lives in fast flowing waters that are highly oxygenated
- guinea worm, which invades the body after water is drunk containing the larvae of the worm.

Typhoid and **cholera** are also caught by drinking contaminated water. Cholera was first traced in London to certain communal water supply stand-pipes, which, once turned off, resulted in a decline in the disease. Typhoid is caused by the intake of water containing a type of salmonella, but can also be spread through the handling by typhoid carriers of food and therefore its subsequent contamination. Good hygiene and the availability of a safe water supply are essential for the control of these diseases.

water budget graph: see **soil moisture graph**.

waterfall: a steep fall of river water where its course is markedly and suddenly interrupted. This may be the result of:

- a resistant rock occurring across the course of a river, and interrupting its progress to a **graded profile**
- the edge of a plateau
- a fault-line scarp
- the overdeepening of a valley through **glacial processes**, producing a **hanging valley**
- the **rejuvenation** of the area, giving the river renewed erosional power which in turn gives rise to **knickpoints** where profiles intersect.

waterlogged describes the state of the soil when all the pore spaces below a certain level are full of water. If this extends up to the surface no water can enter the soil, which results in surface **flooding**. The region below the **water table** is thus waterlogged.

water management is needed to maintain the quality and quantity of the water supply. Countries in the **First World** have the financial resources and the technology to develop large schemes but in the **Third World** there are problems in many areas in developing and maintaining a good water supply. Many water management schemes in such areas have to be multi-purpose in order to justify the cost. Modern schemes also attempt to manage the whole of a **drainage basin**, even if, as is the case with the Tennessee Valley (USA), that basin covers several states. Water management there consists of collection, distribution and the prevention of **flooding**. In the UK, one of the most ambitious water management projects has centred on the Kielder reservoir in northeast England.

water quality: in the 1980s the **United Nations** through various agencies aimed to make clean water and sanitation available to everyone by the end of the decade. Although great strides were made, the biggest obstacles to success have been the limited financial and technological resources in the **Third World** and the rapidly growing populations of those countries, particularly the expansion of squatter settlements within urban areas. Despite attempts to clean up water supplies, over 20 million people still die every year from **water-borne diseases** such as **bilharzia**, guinea worm, **cholera** and diarrhoea. Millions more are affected by the insects that breed in freshwater and cause **malaria** and river blindness.

watershed: the boundary that separates one **drainage basin** from another. In the USA it is known as a divide.

water table: in a rock when all the pore spaces are full, there is a zone created which is said to be saturated (**waterlogged**); the upper boundary of this layer is known as the water table. The water table will move up and down, depending on the supply of water from above and the amount of **evaporation** from the rock and soil.

wave-cut platform: the retreat of a cliff leaves behind a gently sloping area which cuts across the rocks of the coastline regardless of type and structure. As the platform grows, incoming waves break further out to sea and have to travel across a wider area which dissipates their energy and reduces the impact on the cliffline. Some wave-cut platforms are now found at a higher level as **raised beaches**, indicating that there has been a negative change in sea level resulting from a fall in sea level or a rise in the land surface.

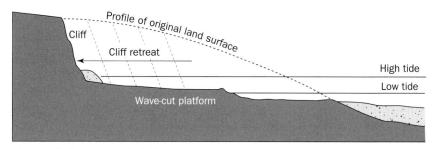

Development of a wave-cut platform

wave formation: waves are caused by the frictional drag exerted by winds as they pass over a water surface. When a wind passes over a water surface it tends to generate waves which move forward in the direction in which the wind is blowing, but there is little or no forward movement of the water itself. The size and energy of the waves produced depend on three factors:

- the velocity of the wind – the stronger the wind, the larger the wave
- the period of time during which the wind has blown – the longer the wind blows, the larger the wave
- the **fetch**.

wave refraction is the tendency for waves to become parallel to the line of a coastline. It is due to the fact that a wave approaching a shore loses energy as the depth of water decreases. This can be illustrated by the example of waves approaching an irregularly shaped coastline, such as **headlands** and bays.

The waves approaching the headland find that the water shallows more quickly and the movement here is slowed down. However, the waves still in the deep water of the bays are unaffected and move more rapidly until they reach shallow water further in. The line of the wave approaching the coastline therefore begins to reflect the shape of the submarine

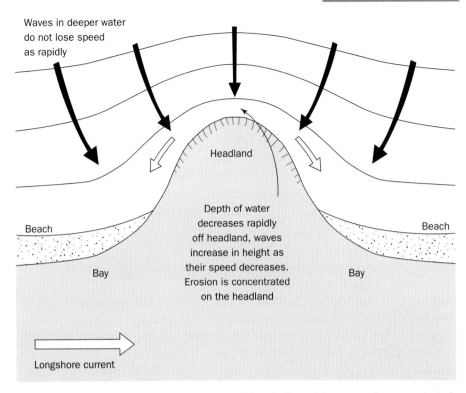

Waves in deeper water do not lose speed as rapidly

Headland

Depth of water decreases rapidly off headland, waves increase in height as their speed decreases. Erosion is concentrated on the headland

Beach

Beach

Bay

Bay

Longshore current

contours. Another consequence is that the erosional effect of the waves is concentrated on the headland. This may also cause a slight local rise in sea level at the headland, resulting in a longshore current from the headland to the bay.

waves are elliptical or circular movements of water near the surface of the sea or any other body of water. They can be described by a range of terminology:

- crest and trough are the highest and lowest points respectively
- wave height is the vertical distance between the crest and the trough
- wavelength is the horizontal distance between two crests
- wave period is the time taken for a wave to travel through one wavelength.

Wave motion decreases rapidly with depth, and there is little movement at a depth greater than half of one wavelength. (See also *constructive waves*, *clapotis* and *destructive waves*.)

weather: the state of the atmosphere at a particular point at a specific time. The weather may be described in terms of temperature, *precipitation*, wind speed, wind direction, *cloud* type, *humidity* and visibility.

weather forecasting: the prediction of what the weather of a particular location is going to be. To do this a great deal of precise information is needed, and it is collected by the Meteorological (Met) Office through observations from satellites, aeroplanes, ships, weather balloons and a network of recording stations. This information is sent to the Met Office at Exeter in Devon. It also receives similar information from weather stations throughout the world via the *World Meteorological Organization (WMO)*. With the help of computers and past records the meteorologists are able to produce charts and forecasts that are constantly updated.

W

weathering: the disintegration and decomposition of rocks in situ by the combined actions of the weather, plants and animals. (See also *chemical weathering* and *physical weathering*.)

Wegener, A. L.: the German meteorologist and geophysicist who formulated the first complete statement of the *continental drift* hypothesis. In 1915 he published his work on the origins of the continents and oceans, suggesting that in the late Palaeozoic all the present-day continents had formed a single land mass, which had subsequently broken apart. The name *Pangaea* was given to this supercontinent. Wegener's theories did not find widespread acceptance, particularly as his suggestions as to the driving forces behind the continents' movements seemed implausible. By 1930 his theories had been rejected by most geologists, and sank into obscurity, only to be resurrected as part of the theories of *plate tectonics* in the 1960s. Wegener died in 1930 on his last expedition to Greenland.

welfare indicators are used to measure the state of *well-being* of an area of a country or for a country as a whole. They give an indication of the *quality of life* of that area or country. They may include measures of:
- health – for example, *infant mortality* rates, numbers of people per doctor, *life expectancy*, nature of *diets*
- incomes – for example, wage levels, unemployment rates, number of households with pensioners, numbers of children in receipt of free school meals
- provision of basic services – for example, households lacking or sharing a bath or toilet
- social indicators – for example, crime rates, levels of vandalism.

well-being is a term that is commonly used with regard to the state of someone's health, but it can be extended to include aspects of *quality of life*, standard of living or *social welfare* within a community or country. Quantitative *welfare indicators* enable measurement and assessment of levels of well-being, but there are also qualitative elements that are difficult to measure such as contentment, happiness and self-esteem.

wells are holes dug into the ground for the purpose of obtaining underground supplies of water. They should be sunk into an *aquifer*, and go some distance below the level of the *water table*. Much of the aid donated to *economically less developed countries* is used for the purpose of drilling tube wells. These have concrete walls and are covered. The water is extracted using a modern pump.

wetlands are areas where the soil is frequently or permanently waterlogged, with the *water table* being at or near to the ground surface. They are *threatened areas* as they have been drained for either *arable* or *pastoral* farming, or for the exploitation of their peat supplies. They provide a distinct habitat and are the home of a variety of rare plant, animal, bird and insect species. The Wetlands and Wildfowl Trust (WWT) has established a number of reserves with designated status as *sites of special scientific interest (SSSI)* around the UK as a means of preserving wetlands and wetland species. The largest urban wetland site in Europe has been developed at Barnes, along the south bank of the River Thames in southwest London. The largest remaining wetland in Britain is the Norfolk Broads, managed by the Broads Authority, but like other wetlands it is under threat. The land is required for farming, and the use of water pumps and irrigation in the area has reduced the water table so much that the land neighbouring the rivers is now shrinking to below the levels of the river channels. Flood-banks need to be reinforced, especially when faced with a *spring tide* or *storm surge*.

wetted perimeter is the total length of a river's bed and banks that are in contact with water when viewed as a cross-section.

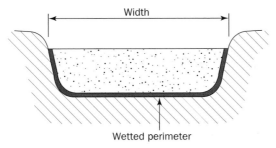

River width v. wetted perimeter

white collar worker: people who work as professional, managerial or clerical staff.

WHO: see *World Health Organization*.

wholesaling: the process of buying large quantities of goods from suppliers and selling on in smaller volumes to retailers or business users. This role is often looked down on because the wholesaler's profit margin appears to make goods more expensive for the consumer. In reality, wholesalers reduce the number of journeys that manufacturers would have to undertake if they wished to service each outlet at which their goods were sold. Wholesalers therefore reduce manufacturers' costs.

wilderness is generally defined as a wild, uncultivated, natural environment that has not been subjected to human activity which has modified, either deliberately or accidentally, the area's *ecology*. It is the 'wild' character of a place that makes it a wilderness and the presence or activities of people does not disqualify an area from being 'wilderness'. Large wilderness parks are currently protected by law and enforcement policies in many countries. People also believe that there is a need for such areas as places where the wildness of nature can be experienced by those seeking 'spiritual refreshment'.

Scientific reasons for establishing wildernesses include:
- the need to maintain the gene pool of wild *organisms* to make sure that genetic diversity is maintained
- keeping animal communities within their natural environments to allow research into their *ecology*
- the need for a pure, natural system as a yardstick against which managed or mismanaged *ecosystems* can be observed and measured.

Some environmentalists argue that by focusing such attention on the more spectacular wilderness landscapes, there is a lack of funding for other *conservation* areas.

wildlife refuge: an area largely free from the impacts of human settlement and development where animals, birds, insects and plants can exist in harmony with their natural environment. They have a high *conservation* value and for this reason people will wish to visit them. The management strategy for such an area must be based on a low *carrying capacity* in that human impact must be kept to a minimum. Examples of wildlife refuges can range from Wicken Fen in Cambridgeshire, an area of wetland with over 5000 species of insects including the rare swallow-tail butterfly, to the nesting sites of the osprey at Loch Garten near Aviemore.

wind chill: the accentuation of cold temperatures when accompanied by high wind speeds. On a cold but calm day, warm human skin heats the air next to it, which in turn passes heat to the next layer and so on until heat is taken away from the body. This is an

W

inefficient heat transfer mechanism as the conductivity of air is low. Therefore, calm air feels relatively warm. A wind, however, pulls heated air away from the surface of the skin, and replaces it with colder air. The faster the air moves the more heat is taken from the body, making the cold air feel much colder than the reading on a thermometer. For example, an air temperature of 0°C with a wind speed of 6 m/sec has a wind chill factor of –10°C.

wind deposition occurs when the moving air no longer has the ability to carry the amount of sand and other particles that it was transporting. A number of *aeolian* landforms are created, for example, *dunes*, *barchans* and *seifs*.

wind erosion is the removal of sand and other particles from the ground surface and the subsequent use of those materials to wear away rocks in other areas by the force of moving air. The main processes involved are *deflation*, *saltation* and *abrasion*. The *aeolian* landforms produced include pedestal rocks, *yardang* and *zeugen*. A pedestal rock is a mushroom-shaped rock that has been eroded more just above its base than at higher levels. This is because the sand grains carried by the wind are rarely lifted more than 1 m above the ground surface and so perform little *erosion* above this height. The shape may be accentuated if there is a less resistant layer of rock lying beneath a resistant layer.

wind formation: winds result from the differences that exist in air pressure between different places. These differences occur because of variations in temperatures between places. When the air temperature of a place increases, the air in that area expands and rises, thus reducing air pressure. Conversely, when temperatures fall the air becomes more dense and air pressure increases. The gradual change of pressure between different areas is called the pressure gradient, and it gives rise to a movement of air from the area of relatively high pressure to the area of relatively low pressure. This movement of air is wind. The strength and direction of the wind is influenced by the *pressure gradient* and by the *coriolis force*.

wind power is one of the few *renewable* sources of energy being developed commercially on a large scale. For it to be efficient, a wind turbine should be located at a point where wind is both regular and strong. Suitable sites include hilltops and coastal areas. It is common to find many turbines grouped together to create a wind farm. The initial cost of setting up a wind farm is expensive, but the power it produces is relatively cheap. Wind farms may also attract small businesses to areas which otherwise would be lacking in job opportunities. However, wind power could never provide large quantities of electricity. There are concerns about the unsightliness of the wind farms that currently exist.

wind transport refers to the movement of sand and other particles through the air. It is greatest where winds are strong, turbulent, blow from a constant direction and blow for a long period of time. It is more likely to take place when the ground is dry, unconsolidated, and sparsely vegetated. The particles on the ground are therefore loose and available to be set in motion.

Wind moves material by three main processes:
- suspension
- *saltation*
- surface *creep* – the rolling forward movement of larger particles resulting from the impact of falling sand grains.

WMO: see *World Meteorological Organization*.

World Bank or the International Bank for Reconstruction and Development, was established in 1947 to provide aid to ***developing*** countries in the form of loans and technical assistance. Originally, the bank supplied loans for capital projects such as those concerned with improving a country's ***infrastructure***, but from 1980 onwards it was allowed to offer help in the case of ***balance of payments*** difficulties, subject to conditions.

world city: a city that has a global significance as an important focal point that is not simply a function of its population size or geographical extent. It is generally accepted that London, New York and Tokyo are world cities, but there is considerable debate about the criteria for designation of a world city. The characteristics that may be used include:

- a strong role in international events and world affairs, e.g. Washington DC, Berlin, Brussels are major capitals of influential nations or unions
- a major international airport that serves as an established hub for several international airlines, e.g. London, New York, Tokyo
- cities that have diverse international cultures and communities, e.g. New York, Los Angeles, Toronto, Chicago, San Francisco, São Paulo
- cities that attract large foreign business sectors, e.g. Hong Kong, Moscow, Shanghai, Singapore
- cities that house international financial institutions, corporate headquarters of law firms/ international companies, and stock exchanges, e.g. New York, London, Paris, Tokyo.

World Health Organization (WHO): a specialised agency of the ***United Nations***, based in Geneva and established in 1948 to further international cooperation for improved health conditions. The work of WHO includes:

- providing a central clearing house for information and research on such features as vaccines, cancer research, nutrition, drug addiction and nuclear radiation hazards
- sponsoring measures for the control of ***epidemic*** and ***endemic*** diseases by promoting mass campaigns involving mass vaccination programmes, instruction on the use of antibiotics and insecticides, assistance in providing pure water supplies and sanitation systems and health education for rural populations. The eradication of smallpox by 1980, for example, was due largely to the efforts of WHO
- encouraging efforts to strengthen and expand public health administrations of member nations.

world heritage site: a UNESCO programme that 'catalogues, names and conserves sites of outstanding cultural or natural importance to the common heritage of humanity'. This can be a monument, building, a whole city, mountain lake, desert or forest. World heritage sites in the UK are very diverse and include Stonehenge, the Giant's Causeway, Blenheim Palace, Durham Castle and Cathedral, the city of Liverpool and the Jurassic coast.

World Meteorological Organization (WMO): a specialised agency of the ***United Nations***, founded in 1951 to promote the establishment of a worldwide meteorological observation system, the standardisation and international exchange of observations and the development of national meteorological services in ***Third World*** countries. Its administrative headquarters are in Geneva.

World Trade Organization (WTO): the organisation that replaced ***General Agreement on Tariffs and Trade (GATT)*** in January 1995.

W

xerophytic describes the vegetation that is adapted to living in dry conditions. Drought-resistant plants have some of the following features:

- thick, corky barks to cut down *transpiration*
- bulbous trunks made of spongy wood which stores water
- widespread branching rooting systems to gather water from the widest possible area
- large proportion of the *biomass* is below the surface
- small leaves to cut down the surface area through which water can be lost
- sunken stomata that only open at night to allow respiration, and to take in *carbon dioxide* for the subsequent production of organic matter during the day
- waxy surfaces on the leaves to help reduce water loss
- reduced surface area, by having spines instead of leaves
- folded leaves with the stomata inside
- deep penetrating roots that will seek water at depth (tap roots).

xerosere is a name applied to a plant *succession* that develops under markedly dry conditions. It may be a succession that develops on a bare rock surface (*lithosere*) or one that forms on sand *dunes* (*psammosere*).

yardangs are typical of desert and semi-desert areas and are extensive ridges of rock separated from each other by grooves or troughs. They are aligned in the direction of the *prevailing wind*. Their height can vary from one metre up to well over 50 m.

yield in agricultural terms is the amount produced, usually expressed in terms of the area involved, e.g. tonnes per hectare.

zero growth is a situation in *demographic* terms where the population reaches replacement rate and there is no growth. It is predicted that in the early years of the twenty-first century this will be the norm for Europe.

This situation can have its drawbacks:
- an ageing population, and the cost of providing for it
- *labour* shortages in many key industries
- the need for those in work to pay for the elderly, particularly in terms of pension provision.

zeugen: in desert and semi-desert areas, these are tabular masses of resistant rock standing out from softer underlying rocks. They are produced by differential *erosion* through the scouring effect of sand-laden winds.

Zipf, G. K.: popularised the *rank–size rule* in books published in 1941 and 1949.

zonal soil: a classification of soils which covers those types that show the maximum effects of climate and vegetation on soil formation. Zonal soils are mature soils in that they have had time to develop distinctive profiles with clear horizons. Examples of zonal soils are *tundra* soils, *podsols*, *brown earths* and *chernozems*.

Geography revision lists

Using A–Z Online

In addition to the revision lists set out below, you can use the A–Z Online website to access revision lists specific to your exam board and the particular exam you are taking. Log on to **www.philipallan.co.uk/a-zonline** and create an account using the unique code provided on the inside front cover of this book. Once you have logged on, you can print out lists of terms together with their definitions, which will help you to focus your revision.

Main concepts required for success in geography AS and A2 examinations

Geography is such a diverse subject which is reflected in the content of the AS and A2 specifications, and therefore in the courses that you, as students, will be undertaking. The following pages set out lists of terms in each of the major areas of the subject (and some important minor areas), which should be revised for examinations. There is also an indication given as to the unit module to which the terms apply, but remember that the examination bodies have placed topics at different levels (either at AS or A2) which will determine the number of terms that are required and, to some extent, the depth to which you are required to show knowledge of them.

We have listed revision terms for 21 geography topics which are to be found in the AQA, Edexcel and OCR AS and A2 specifications. On the website we have also set out revision lists for other examination bodies which may provide the geography exams at your school or college, including:

- Council for the Curriculum, Examinations and Assessment (Northern Ireland) (CCEA)
- Welsh Joint Education Committee (WJEC)
- Cambridge International Examinations (CIE)
- Scottish Higher
- International Baccalaureate (IB)

When approaching your examinations, look up the definition for each word or term, making sure that you understand it and can repeat the definition at a future date. The authors do not claim that the lists are comprehensive, or that every word or term will have been covered in your course, but they do feel that a knowledge of most of the terms on the lists will put you in a very good position for examination success.

Geography revision lists

The areas of the subject that are covered in the following pages are:

1 Glaciation
2 Fluvioglaciation
3 Periglaciation
4 Hydrology/rivers and river valleys
5 Soils
6 Plate tectonics/vulcanicity
7 Meteorology/climatic systems
8 Coastal geomorphology
9 Desert landscapes
10 Ecosystems/vegetation
11 Weathering and mass movement
12 Population and resources
13 Economic activity, development and change
14 Urban settlement and redevelopment
15 Agriculture and food
16 Environmental matters
17 Hazards
18 Geography of health/disease
19 Political geography
20 Energy
21 Tourism

1 Glaciation

AQA	**Unit 1 Option**
Edexcel	**Unit 4 Option**
OCR	**AS Unit F761 Managing physical environments**

Ablation

Abrasion

Accumulation

Arête

Boulder clay

Cirque/corrie

Drumlin

Englacial

Erratic

Extensional flow

Firn

Fjord

Glacial budget

Glacial lake

Glacial movement

Glacial valley

Glacier

Hanging valley

Ice age

Iceberg

Ice cap

Ice core

Ice sheet

Internal deformation

Knock and lochan

Lodgement till

Moraine

Nivation

Nunatak

Pleistocene

Plucking

Pressure melting point

Pyramidal peak

Ribbon lake

Roche moutonnée

Subglacial

Supraglacial

Terminal moraine

Till

Truncated spur

2 Fluvioglaciation

AQA	**Unit 1 Option**
Edexcel	**Unit 4 Option**
OCR	**AS Unit F761 Managing physical environments**

Esker

Fluvioglacial landforms

Fluvioglacial processes

Ice marginal landforms

Kame

Kettlehole

Meltwater

Outwash

Varve

3 Periglaciation

AQA　　　Unit 1 Option
Edexcel　Unit 4 Option
OCR　　　AS Unit F761 Managing physical environments

Active layer

Blockfield

Ice lens

Ice wedge

Patterned ground

Permafrost

Pingo

Scree

Solifluction

Stone circles

Thermokarst

4 Hydrology/rivers and river valleys

AQA　　Unit 1 Core
OCR　　AS Unit F761 Managing physical environments

Abrasion

Alluvial fan

Alluvium

Bankfull

Base flow

Bedload

Braiding

Capacity

Channel efficiency

Channelisation

Channel morphology

Channel processes

Competence

Corrasion

Delta

Discharge

Drainage basin

Estuary

Evapotranspiration

Flooding

Floodplain

Fluvial

Graded profile

Groundwater

Helical/helicoidal flow

Hjulström curve

Hydraulic action

Hydraulic radius

Hydrograph

Hydrological cycle

Infiltration

Interception

Knickpoint

Lag time

Lake

Levée

Load

Long profile

Manning's roughness coefficient

Meander

Overland flow

Oxbow lake

Peak flow

Percolation

Point bar

Pools and riffles

Pothole

Rainshadow

Rapids

Rejuvenation

Rilling

River capture

River cliff

River deposition

River erosion

River regime

River terrace

Runoff

Saltation

Sediment

Stemflow

Suspension

System

Throughfall

Throughflow

Urban hydrograph

Valley cross-profile

Valley long profile

Water balance

Waterfall

Watershed

Water table

Wetted perimeter

5 Soils

AQA   Unit 1 Option on cold climates and desert environments
OCR   AS Unit F761 Cold and hot arid/semi-arid environments

There is no specific unit or option given over entirely to soils. The topic appears in several units, such as those above.

Anaerobic

Azonal soil

Brown earth

Calcareous soils

Calcification

Catena

Cation exchange capacity

Chernozem

Eluviation

Evapotranspiration

Fermentation

Ferralitic soils

Ferruginous soils

Fertility (soil)

Gleying

Humus

Illuviation

Intrazonal

Ionic exchange

Iron pan

Laterite

Leaching

Lessivage

Mor

Mull

Nutrient cycle

Organic matter

Organisms

Parent material

Ped

Pedogenesis

pH

Podsol

Podsolisation

Potential evapotranspiration

Rendzina

Salinisation

Sesquioxide

Soil acidity

Soil conservation

Soil erosion

Soil moisture graph/budget

Soil profile

Soil structure

Soil survey techniques

Soil texture

Terra rossa

Translocation

Water balance

Waterlogged

Zonal soil

6 Plate tectonics/vulcanicity

AQA   **Unit 3 Option**
Edexcel **Unit 4 Option**
OCR   **A2 Unit F763 Global issues (Option A1: Earth hazards)**

Benioff Zone

Conservative plate margin

Constructive plate margin

Continental drift

Core

Crust

Destructive plate margin

Earthquake

Epicentre

Extrusive

Faulting

Folding

Hot spots

Island arc

Lahar

Liquefaction

Mantle

Mid-oceanic ridge

Mohorovicic

Mountain building/orogenesis

Ocean trench

Palaeomagnetism

Plate tectonics

Plume

Pyroclastics

Richter scale

Rift valley

Sea floor spreading

Subduction

Tephra

Transform faults

Tsunami

7 Meteorology/climatic systems

AQA   **Unit 3 Option**
Edexcel **Unit 1 and Unit 2 Option**
OCR   **A2 Unit F763 Global issues (Option A3: Climatic hazards)**

Air mass

Albedo

Anticyclone

Atmosphere

Atmospheric circulation models

Atmospheric particulates

Beaufort scale

Bergeron–Findeisen theory

Clouds

Cold front

Condensation

Convection

Convergence

Cyclone

Depressions

Dew

Dew point

Divergence

El Niño/La Niña

Energy budget

Environmental lapse rate

Evaporation

Evapotranspiration

Fog

Front

Frost

Global warming

Greenhouse effect

Hadley cell

Hail

Humidity

Hurricane

Hygroscopic

Insolation

Inter-tropical convergence zone

Inversion

Jet stream

Katabatic

Microclimate

Mist

Monsoon

Mountain and valley winds

Occlusion

Orographic

Ozone depletion

Ozone layer

Particulate pollutiom

Photochemical smog

PM10s

Pollution reduction policies

Potential evapotranspiration

Precipitation

Precipitation intensity

Precipitation processes

Pressure

Pressure gradient

Prevailing wind

Radiation

Radiation fog

Relative humidity

Smog

Solar energy

Stratosphere

Sublimation

Temperature inversion

Temperature profile

Thermohaline circulation

Thunderstorm

Tipping point

Tornado

Tropopause

Troposphere

Typhoon

Urban climate

Urban heat island

Urban wind

Venturi effect

Weather

Wind chill

8 Coastal geomorphology

AQA	Unit 1 Option
Edexcel	Unit 2 Option
OCR	AS Unit F761 Managing physical environments

Atoll

Backshore

Backwash

Bar

Barrages

Barrier beach

Beach

Beach depletion

Beach nourishment

Berm

Blow holes

Constructive waves

Cuspate foreland

Delta

Destructive waves

Do nothing

Dunes

Emergent coast

Estuary

Eustatic adjustment

Fetch

Groyne

Hard engineering

Headland

Isostatic readjustment

Longshore drift

Mangroves

Refraction

Retreat the line

Salt marsh

Sea level change

Sediment cells

Spit

Stack

Storm beach

Submergent features

Swash

Tombolo

Wave-cut platform

Wave refraction

9 Desert landscapes

AQA   Unit 1 Option

OCR   AS Unit F761 Managing physical environments

Abrasion

Aeolian

Arid

Aridosol

Bajada

Barchan

Deflation

Desertification

Dunes

Endoreic

Ephemeral

Exogenous

Flash flood

Inselberg

Mesa

Oasis

Pediment

Playa

Saltation

Salt lakes

Seif

Thermal expansion

Wadi

Yardang

Zeugen

10 Ecosystems/vegetation

AQA   Unit 3 Option
Edexcel Unit 3 Option
OCR   A2 Unit F763 Global issues (Option A2: Ecosystems and environments under threat)

Biodiversity

Biomass

Biome

Bog (blanket and raised)

Climax vegetation

Desertification

Dunes

Ecology

Ecosystem

Epiphyte

Food chain

Food web

Habitat

Halophyte

Halosere

Heathlands

Hydrosere

Lichen

Lithosere

Mangroves

Maquis

Marram grass

Mediterranean vegetation

Niche

Net primary productivity (NPP)

Nutrient cycles

Peatland

Pioneer

Plagioclimax

Prisere

Psammosere

Pyrophytes

Soil moisture graph

Species

Succession

System

Transpiration

Trophic levels

Tundra

Xerophytic

Xerosere

11 Weathering and mass movement

OCR   A2 Unit F763 Global issues (Option A1: Earth hazards)

Weathering occurs in a number of options within the AQA, Edexcel and OCR specifications, particularly those concerned with cold climates, coastal geomorphology and arid/semi-arid areas.

Avalanche

Bedding plane

Creep

Earthflow

Exfoliation

Freeze–thaw

Hydration

Hydrolysis

Landslide

Landslip

Mechanical weathering

Mudflow

Oxidation

Permeable

Porosity

Pressure release

Scree

Slumping

Solution

Thermal expansion

12 Population and resources

AQA   **Unit 1 Core**
Edexcel   **Unit 1**
OCR   **A2 Unit F763 Global issues (Option B1: Population and resources)**

Baby boom	Malnutrition
Birth control programmes	Malthus
Birth rate	Migration
Boserup	Natural increase
Carrying capacity	Non-renewable (finite)
Club of Rome	Optimum population
Cycle of poverty	Overpopulation
Death rate	Population pyramid
Demographic transition model	Population structure
Dependency ratio	Renewable
Doubling time	Resources
Famine	Simon
Fertility	Social mobility
Gender	Social welfare
Illegal migrants	Sustainable resources
Infant mortality	Underpopulation
Life expectancy	Voluntary migrants
Longevity	Zero growth

13 Economic activity, development and change

AQA   **Unit 3 Option**
Edexcel   **Unit 1 and Unit 3 Option**
OCR   **A2 Unit F763 Global issues (Option B2: Globalisation; Option B3: Development and inequalities)**

(A) ECONOMIC ACTIVITY AND CHANGE

Branch plants	Deindustrialisation
BRICs (RICs)	Enterprise culture
Business cycle	Extractive industry
Business park	Fordism
Clarke–Fisher model	Foreign direct investment (FDI)

Global shift

Globalisation

Heavy industry

High-tech industry

Homeworking

Inertia

Interventionist policies

Inward investment

Kondratieff cycle

Kuznets curve

Labour flexibility

Labour mobility

McDonaldisation

Newly industrialised country (NIC)

Pacific Rim

Parent company

Post-industrial

Primary industry

Quaternary industry

Rebranding

Secondary industry

Self-employment

Stakeholder

Sunrise industry

Sunset industry

Tertiary sector

Transfer of technology

Transnational (TNC)

World Trade Organization (WTO)

(B) ECONOMIC DEVELOPMENT

Aid

Brandt Report

Calorie intake

Core Periphery

Cumulative causation

Cycle of poverty

Debt (Third World)

Development continuum

Development gap

Downward spiral

Economic growth

Export processing zone

Friedmann

Gross domestic product (GDP)

Gross national product (GNP)

Hirschman

Human development index (HDI)

Informal sector

International Monetary Fund (IMF)

Least developed country (LDC)

Malnutrition

Migration

Millennium Development Goals

Myrdal

Neo-colonialism

Organisation for Economic Co-operation and Development (OECD)

Organization of Petroleum Exporting Countries (OPEC)

Poverty

Purchasing power parity (PPP)

Quality of life

Rostow

Self-help schemes

Sustainable agriculture/development

Trade bloc

Underemployment

Well-being

14 Urban settlement and redevelopment

AQA   **Unit 3 Option**
Edexcel   **Unit 2 Option**
OCR   **AS Unit F762 Managing change in human environments**

Some of these terms are also relevant within the modules and units that deal with rural settlement, such as the OCR unit on rural change in AS Unit F762.

Bid-rent theory	Peak land value point (PLVP)
Brownfield site	Pedestrianisation
Central business district (CBD)	Planning process
Commuting	Property-led regeneration
Conservation	Reurbanisation
Counterurbanisation	Rural–urban fringe
Deprivation	Self-help schemes
Enterprise zone	Social segregation
Gentrification	Squatter settlement (shanty town)
Green belt	Suburbanisation
Housing association	Urban decay
Inner city	Urban Development Corporation (UDC)
Megacity	Urban field
Millionaire city	Urbanisation
Multiple deprivation	Urban renewal
New town	World city
Out-of-town location	

15 Agriculture and food

AQA   **Unit 1 Option**
Edexcel   **Unit 4 Option**
OCR   **AS Unit F762 Managing change in human environments**

Agribusiness	Eutrophication
Aid	Factory farming
Appropriate technology	Fair trade
Biodiversity	Famine
Biogeochemical cycles	Farm diversification scheme
Cash cropping	Farmers' markets
Common Agricultural Policy (CAP)	Fertiliser
Defra	Food and Agriculture Organization (FAO)
Environmental stewardship	Food processing

Food retailing

General Agreement on Tariffs and Trade (GATT)

Genetically modified crops

Green revolution

High-yielding varieties (HYVs)

Hybrid

Irrigation

Marketing board

Mechanisation

Monoculture

Nitrates

Organic farming

Overgrazing

Overproduction

Pesticide

Pick your own (PYO)

Quota

Set-aside

Soil erosion

Subsidies

World Trade Organization (WTO)

16 Environmental matters

AQA   **Unit 3 Option**
Edexcel   **Unit 3 Option and Unit 4 Option**
OCR   **A2 Unit F763 Global issues (Option A2: Ecosystems and environments under threat)**

Acid rain

Agro-chemicals

Algal bloom

Biodiversity

Biome

Biosphere reserve

Carbon dioxide

Carbon footprint

Chlorofluorocarbons (CFCs)

Ecological footprint

Ecosystem

Ecotourism

Endemism

Environmental impact assessment

Environmentally sensitive area

Environmental stewardship

Eutrophication

Global warming

Greenhouse gases

Hazard

Nitrogen cycle

Organic farming

Ozone layer

Pesticide

Photochemical smog

Polluter pays

Pollution

Stewardship

Threatened areas

Tipping point

Wilderness

Wildlife refuge

17 Hazards

AQA	**Unit 3 Option**
Edexcel	**Unit 1**
OCR	**A2 Unit F763 Global issues (Option A1: Earth hazards; Option A3: Climatic hazards)**

There are numerous hazards in both the physical and human environments. The following list gives some of the more important general terms connected with hazards, but your revision list will depend entirely on the types of hazards that fall within the course that you have followed. The major hazards that are studied at this level include earthquakes, tsunamis, volcanic activity, landslides, mudflows, avalanches, hurricanes, tornadoes, intense mid-latitude storms, ozone depletion, photochemical smog, global warming, wildfires, drought, river and coastal flooding, AIDS and nuclear waste. You should be certain of the terminologies for each hazard, which can be found in different sections of this revision list.

Disaster

Frequency

Hazard

Magnitude

Multiple hazard regions

Park model

Perception

Recurrence interval

Risk assessment/analysis

Vulnerability

18 Geography of health/disease

AQA	**Unit 1 Option**
Edexcel	**Unit 3 Option**

AIDS

Bilharzia

Cholera

Disease of affluence

Endemic

Epidemic

Epidemiology

Infectious disease

Malaria

Malnutrition

Morbidity

Mortality

Obesity

Pandemic

Poverty

Transmittable disease

Vector

Water-borne disease

World Health Organization (WHO)

19 Political geography

AQA   **Unit 3 Option**
Edexcel   **Unit 3 Option**

European Commission
European Parliament
European Union
Eurozone

Ideology
Millennium Development Goals
United Nations

20 Energy

AQA   **Unit 1 Option**

Edexcel   **Unit 3 Option**
OCR   **AS Unit F762**
Managing change in human environments
Barrage
Biofuel
Biogas
Energy poverty
Energy security
Hydroelectric power
Kyoto Conference 1997

Methane
Non-renewable
Nuclear energy
Nuclear waste
Ocean thermal energy conversion
Organization of Petroleum Exporting Countries (OPEC)
Photovoltaic cells
Renewable
Sustainable

21 Tourism

OCR   **AS Unit F762 Managing change in human environments**

Biosphere reserve
Budget airline
Butler's model
Disposable income
Ecotourism
Green tourism

Human development index (HDI)
Multiplier effect
National Park
Sustainability
Wilderness
World heritage site

Examiners' terms

One of the major pitfalls that face candidates in any examination is their difficulty in interpreting the demands of the questions asked of them. Thorough revision is essential, but candidates also require an awareness of what is expected from them in the examination itself. Too often candidates attempt to answer the question they think is there rather than the one that is actually set. Answering an examination question is challenging enough without the extra self-imposed handicap of having misread the question.

Examiners always try to set questions that are clear in what they ask for, and can be answered by everyone who has followed the course and has prepared adequately for the examination. It is not in the interest of the examiner for a question to be read ambiguously or to be answered by only a few candidates. Several checks are made before the final version is printed to eliminate ambiguity, wordiness or undue difficulty.

Correct interpretation of the **command words** of a question is therefore very important. In an A-level geography examination paper, a variety of command words are used. Some command words demand more of the candidate than others. Some require a simple task to be performed; others require greater intellectual thought and synthesis. There is, therefore, a **hierarchy** of command words. The following is an attempt to describe what is required by some of the major command words used in a geography examination.

1 Identify... Name... State... Give...

These words ask for brief answers to a simple task, such as:
(a) identifying a landform from a photograph
(b) giving a value from a graph
(c) giving a named example of an item.

Candidates are not advised to answer by using a single word. It is always better to write a short sentence incorporating the answer required.

2 Define... Explain the meaning of... What is meant by... Outline...

These words ask for a relatively short answer, usually two or three sentences, where the precise meaning of a term is required. The giving of an example is often helpful. The size of the mark allocation will give an indication of the length of answer required.

3 Describe…

This is one of the most widely used command words. A verbal picture of the distinctive features of the item to be described is required. The account should be factual, **without any attempt to explain**. Usually the examiner will give some clue to what particular aspect of the description is required. Some examples are:

- Describe the characteristics of… – what does the feature look like? For example, in the case of a landform: its shape, its dimensions, its composition, its location in the field.
- Describe the changes in… – often used in relation to a graph or series of graphs. Good use of accurate adverbs is required here, using words such as rapidly, steeply, gently etc. One word to be avoided is 'steadily' as both a steep and gentle gradient can be regarded as being 'steady'.
- Describe the trends in… – again, often used in relation to a graph or series of graphs. In this case, candidates are asked to give more of an overall picture and not to focus on individual changes. However, major exceptions to an overall trend should be referred to by the candidate.
- Describe the differences between… – here **differences only** between two sets of data will be credited. It is better if these are presented in the form of a series of separate sentences, each identifying one difference, rather than a paragraph on one item followed by a paragraph on another item, thus making the examiner complete the task on the candidate's behalf.
- Describe the relationship between… – here, **only** the links between two sets of data will be credited. It is important, therefore, that the candidate establishes the relationship clearly in verbal form and that the link is clearly stated.
- Describe the distribution of… – this is usually used in conjunction with a map or set of maps. A description of the location of high concentrations of a variable is required, together with a similar description of those areas with a lower concentration. Better answers also tend to identify anomalous areas or areas that go against an overall trend in the distributions, for example, an area of high concentration in an area of predominantly low concentration.
- Describe the effect(s) of… – a factual account of the consequences of an event is required. The main point to stress here is that it has occurred either after or as a result of the item referred to in the question.

4 Describe and comment on…

This demands a higher level of response than just 'describe'. Usually, the description is straightforward – the candidate is invited to make a judgement on the description, possibly to offer some explanation, or to infer something that could be responsible for, or develop from, the description referred to. The key point is that the candidate is asked to go beyond simple description and to demonstrate their ability to think like a geographer.

5 Distinguish between…

This requires a definition of the two (or more) terms, with the point of difference between them being clearly identified.

6 Compare….

The candidate has to provide a point-by-point account of the similarities and differences between two sets of information or two areas. Two separate accounts do not comprise a comparison, and candidates will be penalised if they present two such accounts and expect the examiner to do the comparison on their behalf. A good technique would be to use comparative adjectives, for example, 'larger than', 'smaller than', 'more steep than', 'less gentle than' etc.

It is worthy of note that 'compare' refers to similarities **and** differences, whereas the command word 'contrast' just asks for differences.

7 Explain… Suggest reasons why… How might…

These commands ask for a statement as to why something occurs. The command word is testing the candidate's ability to know or understand why or how something happens. In geography examinations, the question-setters have tended to move away from using the command word 'explain' and have moved more towards the phrase 'suggest reasons why'. The reason for this is that there are often varying ways in which geographical change and geographical issues can be explained. In many cases, the 'real' reason is not obvious, nor can it be proven in an easy way.

8 Give an explanatory account of… Give a reasoned account of…

These commands ask for a combination of the demands of a 'describe' question and a 'suggest reasons why' question. The question-setters have recognised that the logical way to present an answer is to provide a description of a feature, together with an explanation for it. This may be in the same sentence or in statements that follow on shortly from each other. One approach to such a command may be to give the reasons first and then move on to the consequent description.

Such questions tend to carry a large number of marks and candidates will be expected to write a relatively long piece of prose. It is important that candidates present a logical account that is both relevant and well-organised.

9 Using only an annotated diagram…

Here the candidate **must** draw a diagram and provide **only** a diagram. Annotations are a form of labelling that provide some additional description and/or explanation of the main features of the diagram. For example, in the case of a hydrograph, the identification of 'a rising limb' would constitute a label, whereas 'a steep rising limb caused by an imperm-eable ground surface' would form an annotation.

10 Analyse…

This requires the candidate to break down the content of a topic into its constituent parts and to give an in-depth account. As stated above, such questions tend to carry a large number of marks and candidates will be expected to write a relatively long piece of prose. It is important that candidates present a logical account that is both relevant and well-organised.

11 Discuss…

This is one of the most common higher level command words. It is used most often in questions that carry a large number of marks and require a lengthy piece of prose to be written. Candidates are expected to build up an argument about an issue, that is, to present more than one side to the issue. They should present arguments for and against, making good use of evidence, with appropriate use of examples, and express an opinion about the merits of each side. In other words, they should construct a verbal debate. In any discussion there are likely to be both positive and negative aspects to it – some people are likely to benefit (the 'winners') and some people are likely not to benefit (the 'losers'). Candidates are invited to weigh up the evidence from both points of view and may be asked to indicate where their sympathies lie.

Sometimes, additional help is provided in the question, such as:

- Discuss the extent to which… – here, a judgement about the validity of the evidence or the outcome of an issue is clearly requested.
- Discuss the varying attitudes to… – here, it is clearly stated that a variety of views over an issue exist and candidates are required to present a debate of them. Often there are a variety of different people involved in an issue. There are those responsible for the decision to go ahead with an idea or policy (the decision-makers) and there are those who will be affected, directly or indirectly. Each of these groups will have a different set of priorities and a different viewpoint on the outcome.

12 Evaluate… Assess…

These command words require an extension to the idea of a discussion as given above. In both cases, an indication of the candidate's viewpoint having considered all of the evidence is required. 'Assess' requires a statement of the overall quality or value of the feature or issue being considered. 'Evaluate' asks the candidate to give an overall statement of value. In both cases, the candidate's own judgement is requested, together with a justification for that judgement.

The use of 'critically' often occurs in such questions, such as 'critically evaluate…'. In this case, the candidate is being asked to look at an issue or problem from the point of view of a critic. There may be weaknesses in the argument; the candidate should not take the evidence at face value. The candidate should question it not only in terms of the evidence itself but also in terms of where it came from and how it was collected. The candidate should also comment on the strengths of the evidence as well as its weaknesses.

13 Justify your answer

The command word 'justify' is proving to be one of the most demanding that candidates face. At its most simplistic, a response to this command must include a strong piece of writing in favour of the chosen option(s) in a decision-making exercise and also an explanation of why the other options were rejected in that exercise.

However, decision-making is not as straightforward as this appears. All of the options in a decision-making scenario have positive and negative aspects. The options that are likely to be rejected will have some good elements to them and, equally, the chosen option will not be perfect in all respects. The key to good decision-making is to balance up the pros and cons of each option and to opt for the most appropriate, based on the evidence available.

A good answer to the command 'justify' should therefore provide the following:

- for each of the options that are rejected – an outline of their positive and negative points, but with an overall statement of why the negatives outweigh the positives

- for the chosen option – an outline of the negative and the positive points, but with an overall statement of why the positives outweigh the negatives.

Developing essay-writing skills

Developing essay-writing skills is one of the most difficult aspects of any examination period. However, you should appreciate that many, if not all, students will find this to be the case. Hence, for you, it is a real opportunity to impress and to score at a level higher than your peers.

There is a set procedure to writing essays, which is detailed below. However, essay writing allows more able students to demonstrate a degree of individuality and flair. As in other assessment situations, it is paramount that the question is fully understood and answered.

In short: **read the question** (RTQ) and **answer the question** (ATQ).

The basic points of an essay plan

All essays should have a beginning (the introduction), a middle (the argument) and an end (the conclusion).

Introduction

This must be present, but it does not have to be too long. One or a few sentence(s) would suffice. In the introduction, you could define some of the terms stated in the question. Alternatively, you could set the scene for the argument to follow. By this we mean that a brief background to the area or the issue concerned could be described. Where the essay requires a particular idea, concept or viewpoint to be developed, an initial state-ment of this could be given here. Finally, if possible, be interesting – grab the reader.

Argument

This must consist of a series of paragraphs that may present different sides of an argu-ment. In principle, each paragraph should develop one point only and each paragraph should follow logically on from the other. Avoid paragraphs that list facts without any meaningful link back to the question. Such answers basically leave the interpretation too much in the hands of the reader or examiner and this is not to be encouraged. Finally, in geographical essays, it is important to make good use of examples, that is, real places (which could be local to you). Make them count!

Conclusion

This should reiterate the introduction but this time be supported by the evidence, argument and facts given in the bulk of the response. A set of summative statements that refer clearly to the question set should be provided. In the case of the command word being 'discuss', the conclusion should point out where the points in the original question are valid and those where there is debate or disagreement.

Should you prepare an essay plan in an examination?

There is debate as to whether an essay plan should be done under these circumstances. The answer is that it is a personal decision. If it helps you to assemble your argument, then the answer is yes, and if you do, it must be brief, taking only 2 or 3 minutes to write. The plan must reflect the above formula – and stick to it. Be logical, but only provide an outline – retain the examples in your head.

Other aspects of essay writing

The following are obvious points about essay writing:

- Always keep an eye on the time.
- Write clearly and concisely. Do not be confused or confusing. Do not have endless sentences, make good use of paragraphing and try not to misuse terms.

How are essay questions marked?

Essay questions in all examinations are marked according to levels and according to fixed criteria. These criteria are based on your:

- knowledge (K) – of concepts, case studies, terminology
- understanding (U) – of processes, theories, ideas, issues
- skills (S) – such as the use of language (grammar, spelling, punctuation (GSP)) and logic, as well as geographical skills such as maps and diagrams.

In many cases, the above criteria are assessed at one of three levels. In general terms, these levels can be summarised as:

Level one

K – generalised material; no or limited case studies or just simple statements; inappropriate or incorrect use of terms; irrelevance

U – weak grasp of concepts and ideas; description only

S – basic; not always logical; clear expression of simple ideas.

Level two

K – some use of case studies, but a lack of depth and detail

U – some understanding of processes by explanation; some evidence of evaluation

S – some illustrative material; logical; GSP is satisfactory, accuracy in the use of English.

Level three

K – good use of case studies – appropriate and detailed; good locational knowledge

U – good understanding of processes and issues; evaluative comment

S – a logical and coherent argument – the essay is said 'to flow'; with almost faultless accuracy in the use of English.

In some cases, level four may exist. This will take place when some overall evaluation of a concept or an issue is required, and is often in response to the higher level of command 'discuss', 'evaluate' or 'assess'. Candidates who have the ability to synthesise – to draw together their argument(s) – will access this level and gain very high marks.

Techniques for learning and revision

The aim of this section is to provide some ways in which you can revise material. It provides a range of methods by which you can increase your ability to absorb, retain and recall factual information. They cannot all work for you, so only choose the techniques that work best. Whichever technique you choose, make sure you use it regularly.

Mind mapping

This technique is useful when trying to learn new information or when trying to pull together material from a range of different areas. For an example of this technique, look at the mind map relating to international migration on page 349. Start off with the term 'international migration' in the centre of a piece of paper and ring it. Then perform a personal brainstorm of the major aspects of international migration (the types, the causes, the effects and the issues) and place these around the edge of the ring like branches, as shown. Then, from each branch add further points that elaborate on the initial term, with further subdivisions if necessary. This logical breaking down of the content helps you to assemble the main points that may influence the process of international migration and could assist in the writing of an examination answer on the topic.

One problem with mind mapping is that the number of initial branches could become too large and then lead on to a very complex diagram. You will need to make the end product manageable for you. Each mind map is personal to you, but it should enable you to remember the key points in a topic and allow you to concentrate on them more quickly.

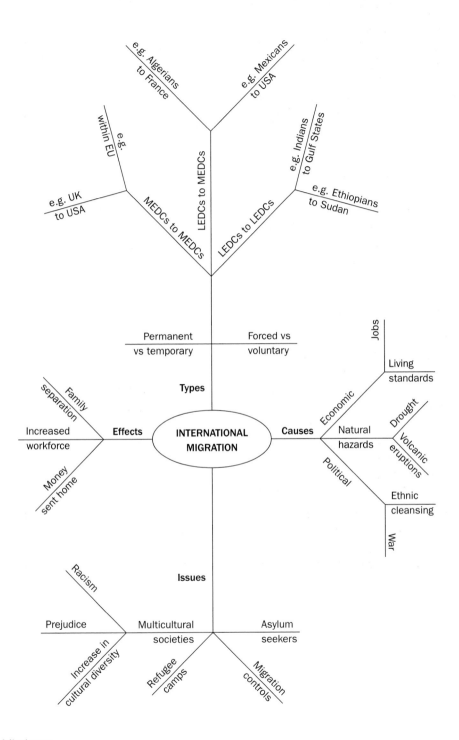

Mind map

Pictorial notes

Some people have the ability to remember facts by where they are located on a page of notes – they can be said to have a 'photographic memory'. Some people exhibit a variation of this by remembering colours and which facts are contained within different coloured areas of their notes.

Another aspect of this form of revision is to convert written notes into a sketch or picture and to use this as a revision tool. An example of this would be to draw a diagram of an anticyclone with subsiding air and for you to add the characteristics of its weather on it. In this way you have a picture of its characteristics, together with descriptive annotations.

Fact cards

A more traditional way to learn material is the fact card approach. Here, you buy a number of small cards from a stationery shop and write your own summary of the key points of a particular topic on the cards. One exercise that you could do is to transfer the contents of one A4 sheet of notes on to one card. The card(s) can then be carried round with you for reading during moments when you would otherwise be doing little, such as travelling to school or sitting in a common room. Try to get hold of a treasury tag, so that you can maintain your cards in a logical order.

Mnemonics and acronyms

These are 'pegs' on which you can hang your memory. You can use either letters or words to construct a mnemonic. One favourite for remembering the rivers that drain into the River Ouse in Yorkshire is **S**heffield **U**nited **n**ever **w**on **a** **c**up, **d**id **t**hey? – SUNWACDT, which is translated as Swale, Ure, Nidd, Wharfe, Aire, Calder, Don, Trent.

Mnemonics where the initial letters make up a word are called acronyms.

As with mind maps, the use of these can be, and possibly should be, personal to you.

Flow diagrams

A flow diagram records inputs and outputs and shows movement or links between places by means of boxes connected with simple directional arrows. You could try to use this approach to summarise the impacts of transnationals (TNCs). The inputs would be the specified changes over time and the outputs would be their effects on the area affected, as either host countries or countries of origin.

Skeleton lists

This is perhaps the most basic of ways to assemble ideas in a logical order before completing the task of writing an answer. Most topics in geography have a list of considerations that can be referred to when looking at factors affecting them.

For example, when considering factors affecting soils, the following may apply:

- soil water
- particle size
- soil air
- weather and climate
- mineral matter
- organic matter
- time
- human activity
- relief.

These lists provide the skeleton around which a discussion of the factors affecting soil could be constructed.

Synoptic assessment

The definition of synoptic assessment in the context of geography is: 'Synoptic assessment involves assessment of candidates' ability to draw on their understanding of the connections between different aspects of the subject represented in the specification and demonstrate their ability to "think like a geographer".'

For all the specifications, synoptic assessment is included in each of the A2 units. Examples of synoptic assessment are:

- decision-making/problem solving/issues evaluation exercises requiring candidates to draw together relevant knowledge, understanding and skills of the specification, to tackle a decision, problem or issue that is new to them
- an essay question covering geographical issues or problems that would require candidates to draw together and apply relevant integrated knowledge, understanding and skills of the specification
- an essay question exploring key geographical concepts through linkages between physical, human and environmental geography
- an assessment on a particular region or area, which is on a scale that allows candidates to draw together and apply relevant knowledge, understanding and skills of processes or concepts of the specification
- answering questions on a fieldwork enquiry or research investigation that has encompassed a variety of themes and issues.

Synoptic assessment by essays

A2 essay questions are designed to be synoptic in nature and offer candidates the opportunity to demonstrate their knowledge and understanding:

- across a range of geographical subject matter
- of connections between the different aspects of geography
- of the importance of both human and physical perspectives on themes and issues.

What is required for a good answer to such questions?

The following list summarises the key elements that are needed:

- There must be consistent relevance to the theme and demands of the question.
- A thorough knowledge and understanding of the subject content should be displayed with accurate use of terminology.

- Case study material should be provided throughout the answer to support the general statements being made. In other words, the answer is evidence-based throughout.
- Clear statements of the relationships between human and physical factors and processes are necessary. These should include an awareness of how human activities and perspectives impact on the physical environment and/or how physical processes impact on the human environment.
- The argument should be logical, organised and balanced.
- The essay should be written effectively with a fluent and cogent writing style. Spelling, punctuation and grammar must be accurate.

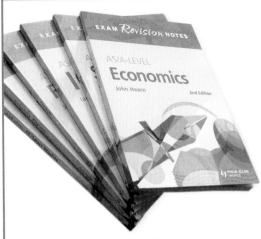